STRUCTURAL
CROSS-SECTIONS

STRUCTURAL CROSS-SECTIONS

Analysis and Design

NAVEED ANWAR
Asian Institute of Technology
Thailand

FAWAD AHMED NAJAM
Asian Institute of Technology
Thailand

AMSTERDAM • BOSTON • HEIDELBERG • LONDON
NEW YORK • OXFORD • PARIS • SAN DIEGO
SAN FRANCISCO • SINGAPORE • SYDNEY • TOKYO

Butterworth-Heinemann is an imprint of Elsevier

British Library Cataloguing-in-Publication Data
A catalogue record for this book is available from the British Library

Library of Congress Cataloging-in-Publication Data
A catalog record for this book is available from the Library of Congress

ISBN: 978-0-12-804443-8

For Information on all Butterworth-Heinemann publications
visit our website at https://www.elsevier.com

Working together
to grow libraries in
developing countries

www.elsevier.com • www.bookaid.org

Publisher: Joe Hayton
Acquisition Editor: Ken McCombs
Editorial Project Manager: Peter Jardim
Production Project Manager: Mohana Natarajan
Designer: Victoria Pearson

Typeset by MPS Limited, Chennai, India

DEDICATION

This book is dedicated to our desire for understanding the physical world around us.

CONTENTS

About the Authors   *xi*

Preface   *xiii*

Acknowledgments   *xvii*

Chapter One   **Structures and Structural Design**   **1**

The Hierarchy of Structures and Their Components   1

Designing the Structures   6

The Building Blocks of Structural Mechanics   24

Cross-Sectional Analysis and Design   35

References   37

Further Reading   37

Chapter Two   **Understanding Cross-Sections**   **39**

Introduction   39

Applications and Classification of Cross-Sections   42

Definition and Representation   47

Cross-Section Properties—An Overview   67

Basic Cross-Section Properties   72

Derived Cross-Section Properties   84

Specific Properties of RC Sections   91

Specific Properties of Steel Sections   95

Numerical Computations of Section Properties   100

Solved Examples   108

Unsolved Examples   128

Symbols and Notation   134

References   135

Further Reading   136

Chapter Three   **Axial–Flexual Response of Cross-Sections**   **137**

Cross-Section Response   137

External Actions and Internal Stresses   140

Axial–Flexural Stress Resultants   153

The General Stress Resultant Equations   157

Extended Formulation of Stress Resultant Equations   169

Computing the Biaxial-Fexural Stress Resultants   175

The Capacity Interaction Surface   189
Biaxial—Flexural Capacity   200
Code-Based Design for Flexure   207
Solved Examples   212
Unsolved Examples   243
Symbols and Notations   248
References   249

Chapter Four   **Response and Design for Shear and Torsion**   **251**
Introduction   251
Basic Elastic Response   253
Response of Reinforced Concrete (RC) Sections   263
Code-Based Shear and Torsion Design of RC Sections   278
Solved Examples   296
Unsolved Problems   324
References   327
Further Reading   328

Chapter Five   **Response and Design of Column Cross-Sections**   **329**
Introduction   329
Complexities Involved in Analysis and Design of Columns   333
Slenderness and Stability Issues in Columns   338
Column Design Process and Procedures   351
Practical Design Considerations for RC Columns   360
Some Special Cases and Considerations   369
Concluding Remarks   372
Solved Examples   372
Unsolved Examples   387
References   390
Further Reading   390

Chapter Six   **Ductility of Cross-Sections**   **391**
What Is Ductility?   391
Action—Deformation Curves   393
Significance of the Moment—Curvature (M—ϕ) Curve   401
Ductility and Confinement of Reinforced Concrete Sections   410
Factors Affecting Moment—Curvature Relationship and Ductility
of RC Sections   423
Concrete-Filled Steel Tubes   430

	Role of Cross-Sectional Response in Nonlinear Analysis of Structures	436
	Solved Examples	444
	Unsolved Problems	475
	References	480
	Further Reading	481

Chapter Seven **Retrofitting of Cross-Sections** **483**
	Overview	483
	Overall Retrofit Process and Strategies for Structures	485
	Retrofitting Techniques for Cross-Sections	494
	Retrofitting of Reinforced Concrete Members	498
	Analysis of Cross-Sections Exposed to Fire	502
	Solved Examples	514
	Unsolved Examples	527
	References	530
	Further Reading	530

Chapter Eight **Software Development and Application for the Analysis of Cross-Sections** **531**
	Introduction	531
	A General Framework for the Development of Structural Engineering Applications	532
	Framework for the Development of Cross-Section Analysis Applications	539
	Introduction to CSiCOL	546
	Mobile Devices and the Cloud for Structural Engineering Applications	553
	Current Research Trends and Future Potential	561
	References	562
	Further Reading	563

Appendix A: Cross-Sectional Properties of Some Common Shapes — *565*
Appendix B: Torsional Constant Factors — *571*
Index — *573*

ABOUT THE AUTHORS

Naveed Anwar is Executive Director/CEO of AIT Solutions (AITS), Director of Asian Center for Engineering Computations and Software (ACECOMS), and an affiliated member of the Structural Engineering Faculty at the Asian Institute of Technology (AIT), Thailand. Dr. Anwar received his BSc degree in Civil Engineering from the University of Engineering & Technology, Lahore, Pakistan, and both masters and PhD degrees in Structural Engineering from Asian Institute of Technology, Bangkok, Thailand. He teaches academic courses and supervises research for masters and PhD students at AIT related to reinforced concrete mechanics, design of tall buildings, and bridge structures.

Dr. Anwar has an experience of over 35 years while completing hundreds of projects related to structural modeling, analysis and design of buildings and bridges, construction project management, structural health assessment, and other related areas. His main area of expertise is the performance-based design and evaluation of new and existing structures, especially high-rise buildings. He has also conducted hundreds of professional trainings, workshops, and seminars, attended by thousands of professionals in more than 15 countries. He is also proficient in the development of computer software for structural engineering applications, including general structural analysis, earthquake resistant design, structural detailing, and is the author of several software and computing tools.

Fawad A. Najam received his BS degree in Civil Engineering from the University of Engineering and Technology (UET, Taxila), Pakistan, and MS degree in Structural Engineering from the National University of Sciences and Technology (NUST), Islamabad, Pakistan. Currently, he is associated with Department of Structural Engineering at the Asian Institute of Technology (AIT) and AIT Solutions (AITS) in Thailand, and has received PhD in Structural Engineering. He is also a member of the Structural Engineering Faculty at the National University of Sciences and Technology (NUST), Islamabad, Pakistan.

His areas of interest include structural dynamics, seismic performance evaluation of high-rise buildings, and structural engineering software development. His research encompasses various practical issues related to seismic design and analysis procedures used in current structural engineering practice. He has also participated actively in various research and academic projects, while working at various appointments including faculty positions, civil/structural engineering consultant, and research associate at various organizations in Pakistan and Thailand.

PREFACE

The cross-sections and their properties are a basic component in almost all aspects of analysis and design of structures. In fact, the primary objective of an efficient design process is to proportion the cross-sections of structural elements to resist the applied load effects and actions. Also, the preliminary dimensions and properties of cross-sections are required to start the modeling and analysis process for the structural system, to prepare initial cost estimates and even to refine the basic space planning and utilization schemes. It is the key role cross-sections play in overall structural behavior, analysis, and design, which motivated authors to write a book focusing on their basic understanding and behavior, with an emphasis on computer applications.

The main objective of this book is to provide a consolidated and consistent information, insight and explanation of various aspects of cross-section behavior and design, irrespective of, but with due consideration to different materials. It discusses theoretical formulations, practical analysis, design computations, and computer applications for determination of cross-sectional response. The main areas of focus and distribution of chapters is as follows:

1. The overall view of cross-section behavior with respect to various aspect of structural mechanics and computing is covered in the Chapter 1, Structures and Structural Design. It discusses the basic concepts and procedures for structural analysis and design, including a discussion of the development of structural design philosophy and progression of design approaches, terminating with a discussion of performance-based design. The chapter also discusses the basic concepts of stiffness, nonlinearity of relationships between deformations and corresponding actions, etc., and lays the foundation for the remaining chapters in the book.

2. The meaning and computation of cross-section properties, from A−Z (area to plastic section modulus) are covered in Chapter 2, Understanding Cross-Sections. It provides the basis for the study and understanding of the structural cross-sections. The anatomy and hierarchy of the cross-section shapes, types, and the materials they are made up, is extensively covered. Several methods for defining the cross-sections, including parametric and general procedures are discussed. The main focus is on computing the geometric properties of various section shapes, using

closed-form solutions as well as more general polygon and mesh-based procedures are discussed. In addition to the formulae to compute the properties, a discussion on the significance of various properties helps to improve the cross-section behavior computer-based methods and several solved examples are included followed by problem for readers to try and explore. The chapter makes use of extensive illustrations and diagrams.

3. The meaning and calculation of stress-resultants using a unified and integrated approach are covered in Chapter 3, Axial-Flexural Response of Cross-Sections. This treatment of cross-sectional behavior is unique and not covered in most of the books on this subject. A set of general equation as well as computation procedure is presented to determine the capacity interaction surface of generalized cross-section made up of any number and configuration of materials. Several solved examples are presented, together with flow charts to demonstrate the code-based procedures for design of beam-column sections.

4. The shear and torsion stress distribution and design calculations for reinforced concrete sections are covered in Chapter 4, Response and Design for Shear and Torsion. Several detailed flow charts are included to demonstrate the procedures used in ACI, BS, and Euro codes for design of cross-section subjected to shear and torsion, followed by solved examples.

5. The design of column cross-sections, slenderness effects, buckling, and relevant behavioral aspects of columns are covered in Chapter 5, Response and Design of Column Cross-Sections. It also discusses the role of boundary conditions on slenderness and provides guidelines about when and how second-order effects can be considered in design process for columns. It provides an overview of column design approaches as prescribed in ACI and BS codes and various practical design considerations and guidelines for proportioning and detailing of RC columns are also included.

6. The ductility considerations, confinement of reinforced concrete sections, and related issues for cross-section analysis and design are covered in Chapter 6, Ductility of Cross-Sections. The definition and development of action-deformation curves especially moment–curvature $(M - \phi)$ curve is discussed extensively. The procedure to generate the moment–curvature curve, using the general formulation of axial–flexural response developed in Chapter 3, is demonstrated. Various factors such as confinement, rebar distribution and axial load effect on the ductility are also shown through examples.

7. The review of various techniques and technologies to retrofit the cross-sections are covered in Chapter 7, Retrofitting of Cross-Sections. The

methods to calculate the capacity of retrofitted cross-sections using the formulations presented in Chapter 2, Chapter 3, and Chapter 4, are presented. Evaluation of sections subjected to high temperatures is also discussed in detail.

8. The discussion about various latest software and applications for computer-aided analysis of cross-sections are covered in Chapter 8, Software Development and Application for the Analysis of Cross-Sections. A brief introduction of CSiCOL (a comprehensive software package for the analysis and design of reinforced and composite column sections) is also included. This chapter also introduces various new ideas and scopes related to development of mobile applications for structural analysis on both Android and iOS platforms. Various applications of cloud computing and component-based software engineering in analysis and design of structures, are also introduced.

Over last two decades, the senior author has had the opportunity to work with and contribute entirely or partially to the development of software tools dealing with the behavior and design of cross-sections and can be used in conjunction with the material presented in this book for practical applications. Some of these include:

1. Structural Design Library, SDL-1, by Technosoft Computer Applications, Lahore, Pakistan, (1993).
2. Structural Designers Library, SDL-2, by Asian Center for Engineering Computations and Software ACECOMS, Asian Institute of Technology, Thailand, (1996).
3. General Engineering Assistant and Reference, GEAR, by Asian Center for Engineering Computations and Software, ACECOMS, Asian Institute of Technology, Thailand, (1997).
4. RISA Section, by RISA Technologies, California, USA (2000).
5. CSI Section Builder, by Computers and Structures, Inc. (CSi), Berkeley, California, USA, (2001).
6. CSiCOL, by Computers and Structures, Inc. (CSi), California, USA, (2008).

The information presented in this book is derived and based on the development of these software, and the senior author gratefully acknowledges the contribution of the organizations who released these software for their support, encouragement, and permission to use some of the material. The readers of this book can be the senior undergraduate students in civil engineering, graduate students in structural engineering, and practicing structural engineers, interested in developing the basic understanding of the behavior of cross-sections. It can also be a reference for those desiring to develop or understand the software tools for cross-section analysis. This book is not intended to cover the conventional literature review, or duplication of information readily available in textbooks or reference books. It will also not cover the derivation and proofs of equations and formulae, instead, adequate references have been included for readers interested in these aspects.

The readers of this book are encouraged to visit the web site www.structuralengineering.info to get additional and updated information related to the material presented in this book. The readers are also welcome to contact the authors directly for their comments and queries regarding the behavior of cross-sections, their modeling, analysis, and design.

Naveed Anwar
nanwar@ait.ac.th
dr.naveedanwar@gmail.com
Fawad A. Najam
fawad.ahmed.najam@ait.ac.th
fawad.najam@gmail.com
November 2016

ACKNOWLEDGMENTS

As with most books covering technical subject areas, this book also would not have been possible without active support and contribution from my family, friends, colleagues, co-author, and the publisher.

The concept and content of this book was developed by the authors, based on my previous book titled "Understanding Cross-Sections." A close collaboration and mutual support amongst both the authors made the present book a reality, which greatly expands on the previous work. The content of this book is developed primarily through two endeavors, one during teaching of the graduate courses in the Structural Engineering programs at the School of Engineering and Technology at the Asian Institute of Technology (AIT) and second while working and associating with Computers and Structures Inc. (CSI) during the development and application of their software. I would therefore like to first express my deep gratitude to Prof. Worsak Kanok-Nukulchai, the President of AIT, and my mentor and advisor for over 30 years in various capacities. His advice, guidance, and support have been an important foundation for many aspects of my professional carrier. Second, I would like to acknowledge the tremendous support, advice, inspiration, and friendship of Ashraf Habibullah, the President and CEO of CSI and for providing me with an opportunity to learn and work with him and other colleagues at CSI on the state-of-art development of structural engineering software, which is extensively referred in this book.

I would also like to thank some of my colleagues and friends for their contribution and advice regarding the development of the various materials in this book including Prof. Pennung Warnitchai, my colleague at AIT, Engr. Jose A. Sy (Boy Sy), President of SY^2, a friend and a creative practicing engineer, and especially Engr. Muhammad Akbar, my close friend and colleague for nearly 40 years.

I would especially like to thank and acknowledge the help and contribution of people who helped in the preparation of the manuscript of this book. In addition to the co-author whose contribution is fundamental to the realization of this work, the tireless work by Ms. Khattiyanee Khancharee for preparing all of the illustrations is greatly appreciated.

So is the help by Mr. Salman Ali Suhail for preparing the solved examples, Ms. Suthathip Thanakorn for typing part of the manuscript, Ms. Maria Shahid for proof editing, Mr. Shehezard Rifthie for helping to prepare the flow charts and some contribution in Chapter 8, and all the team members of AIT Solutions for their support.

And finally and most importantly, I would like to acknowledge the unflinching support, understanding, and love of my wife Farah Naveed, which was not only essential for writing of this book, but also without which all other endeavors in my life would not have been possible. The support of my sons, Shayan, Numan, and daughter-in-law Kanika, is always valuable and important. The upbringing and love accorded by my mother and late father have been the foundation for my life and career, all leading to the writing of this book.

Naveed Anwar

In addition to all those helping hands, highly motivated co-workers, and contributors mentioned above, I would like to extend my gratitude towards the faculty, staff, and officials at Asian Institute of Technology (AIT) Thailand, Higher Education Commission (HEC) Pakistan, and National University of Sciences and Technology (NUST) Pakistan, for their continuous support and facilitations extended over last few years, which helped in the completion of this work. Special thanks is due to Prof. Pennung Warnitchai for his competent guidance, training, and support for over last 5 years.

Thanks is due to Dr. Liaqat Ali, Associate Dean of NUST Institute of Civil Engineering (NICE) Pakistan, for his encouragement and support which always remained with me in the form of self-confidence. I am obliged to Prof. Syed Ali Rizwan, my teacher, mentor, and a colleague for his fatherly kindness, gracious guidance, and frank criticisms. I am grateful to Col. Adnan Ahsan and Dr. Zafar Mahmood for their friendly advices, invaluable suggestions, and discussions.

I would also like to thank my dearest friends and colleagues Dr. Hassan Fazliani, Dr. Adnan Nawaz, Dr. Irshad Qureshi, Dr. Tahir Mehmood, Dr. Saeed Zaman, Dr. Irfan Ahmad Rana, Ms. Meghla Clara Costa, Dr. Faisal Rasool, Dr. Salman Ali Suhail, Mr. Amir Izhar Khan, Dr. Saqib Jamil, Dr. Saad Saleem, and Dr. Bilal Afsar Jadoon for their motivation which kept me going. I am also obliged to Ms. Kashfia Hussain, Ms. Ummi Fadillah Kurniawati, and Ms. Sirisouk Inthavong for their moral support, motivation, cheers, and help whenever I needed it the most.

This book would not have been possible without the support of my sister Fatima, my mother Shabnam, and my father Najam, who continuously support, love, care, and motivate me for always going an extra mile and seek something higher. I am overwhelmed in thanking them and apologize for long working hours and lost weekend gatherings.

Fawad Ahmed Najam

CHAPTER ONE

Structures and Structural Design

A structure is a body or assemblage of bodies in space to form a system capable of supporting loads due to natural and man-made effects

THE HIERARCHY OF STRUCTURES AND THEIR COMPONENTS

Physical Structures

The word structure (pronounced as "strək(t)SHər") is a noun as well as a verb. As a noun, it means "the arrangement of and the relations between parts or elements of something complex," and as a verb it refers to the process to "construct or arrange according to a plan; give a pattern or organization." This term is used extensively in literature, and many disciplines signify these properties and processes.

When applied to the physical and built environment, the term "structure" means an assemblage of physical components and elements, each of which could further be a structure in itself, signifying the complexity of the system. The discipline of "Structural Engineering" refers to the verb part of the definition, dealing with the ways to arrange and size a system of components for construction according to a plan and serving the intended purpose. The primary purpose of any structure is to provide a stable, safe, and durable system that supports the desired function within the physical environment, of which the structure is a part of. The role of the structural engineer, therefore, is to "conceive, analyze, and design" the structure to serve its purpose.

There is hardly any aspect of our built environment or human activity that does not rely on physical structures. Buildings that provide useful places to live and work, bridges that provide us means to move across obstacles, factories that are needed to manufacture almost everything we need, dams to store water to generate power and irrigate lands, transmission towers to distribute electricity; all require structures to function, and structural engineers to design such structures. The role and importance of structural engineering and structural engineers is often underestimated and misunderstood. While a well-conceived and well-designed structure is the backbone of our built environment, a poorly conceived, designed, or constructed structure poses a serious hazard to the safety and wellbeing of people and property. Collapse or failure of structures can claim a large number of lives and result in extensive economic loss. The role of structural engineers is, therefore, critical for overall economic development as well as for improving the community resilience to disasters.

The physical structure, or structure for brevity, can be assembled in infinite ways using very few basic element types or forms, some of which are shown in Fig. 1.1. As is evident, these are derived from or are consistent with basic geometric primitives such as line, curve, plane, surface, and solid. This compatibility can often be used to blend the form, function, and structure.

The assemblage of these components can be of many types and configurations. Some are made entirely of the skeleton-type members, some from surface-type members, and some from solid-type members, but most structures are a combination of more than one member type. Based on the member types, the structures can be broadly categorized as

- cable structures,
- skeletal structural,
- spatial structures,
- solid structures, and
- a combination of the aforementioned categories.

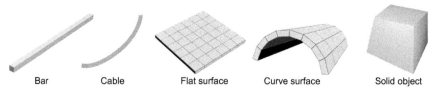

Bar   Cable   Flat surface   Curve surface   Solid object

Figure 1.1 The basic member types and forms that can be used to create structural systems.

Cable Structures: Using Cables as the Main Member Type

These structures primarily transfer forces and internal actions through tension in individual cables or a set of cables. The shapes or geometry of cable profile often govern the behavior. Examples of such structures are:

- Cable nets and fabric structures
- Cable stayed structures
- Cable suspended structures

Skeletal Structures: Using Beam-Type Members

These structures are composed of bar members that mainly resist the loads and forces through a combination of tension, compression, bending, shear, torsion, and warping. Almost all of the discussion in this book is relevant to the cross-section of such skeletal members, often called ties, struts, beams, columns, and girders. Typical application of skeletal structures can be found in the following:

- Truss structures
- Framed structures
- Grid and grillage structures

Spatial Structures: Using the Membrane/Plate/Shell-Type Members

These are structures created by spatial or surface type of elements and they transfer loads through a combination of bending, compression, torsion, and in-plane and out-of-plane shear of the element surface. The cross-sections of such members are generally rectangular. Some of the discussion in this book can be applied to surface members as well. Examples of such structures include:

- General shell structures
- Dome-type structures
- Slab, wall structures
- Silos, chimneys, and stacks
- Box girder bridges

Sometimes, surface members and structures can be created by using a large number of skeletal members, covered by a skin or cladding, combining the fabric, cable, and bars to create surfaces.

Solid Structures: Using the Solid-Type Members

These structures comprise of solid bodies or members in which the forces or loads are transferred through the member bodies. Such members or structures do not have a cross-section in the conventional sense. Some of the structures are:

- Dams, thick arches, thick tunnels
- Pile caps, thick footings, thick slabs, pier heads, large joints, etc.

Mixed Structures: Using One or More of the Basic Element Types

Most often, the real structures are composed of one or more types of basic elements. For example, a typical building is made of columns and beams (skeletal), slabs and walls (shell), footings (solid) structures, etc.

There are several other types of structures that are either a combination of the basic forms or especially developed for a particular application or usage. These include stressed ribbon bridges, fabric structures, skeleton spiral structures, floating offshore structures, pneumatically inflated structures, etc.

Structural Members

The structural members are used as one way to define the boundaries of a certain amount of material having a definite geometry and configuration and serving a particular function in the overall structure. In continuation of the preceding text, the members can be geometrically categorized as: general continuum, regular 3D solids, two-dimensional forms, and one-dimensional members. A continuum may be considered as a three-dimensional volume with any arbitrary shape and size. If a solid block is extracted from a continuum such that all three dimensions (x, y, and z) are of the same order, then it is termed as a "regular 3D solid." However, if any two dimensions of this extraction are much larger than the third, it may be termed as a "surface-type" member. Similarly, if two dimensions are much smaller than the third, it may be called a "line-type" element. These line-type members are the main focus of our attention in this book and will be discussed in detail in subsequent chapters. Fig. 1.2 shows different types of structural elements categorized based on geometric proportions.

Member Cross-Sections

Cross-sections are generally associated with line- or beam-type members, where the length is much longer than other dimensions. In this case, the projection of the member on a plane perpendicular to length is called the cross-section, and the material of the member is "exposed" in the cross-section (Fig. 1.3). Although sections of the surface-type members such as

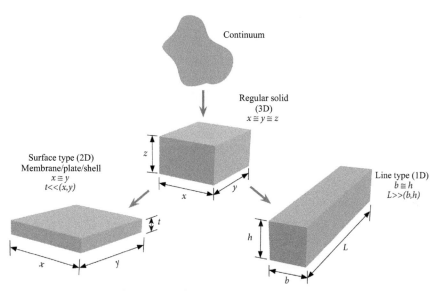

Figure 1.2 Geometrical hierarchy of structural members.

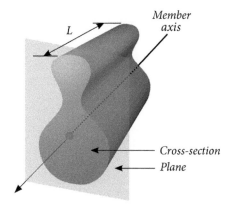

Figure 1.3 Cross-section of a "line-type" member.

slabs and shells can also be considered as cross-sections for the purpose of analysis and design, most of the discussion in this book is targeted to understand the cross-sections of line-type members.

The design of a line-type member often means the design of its cross-section that requires the selection of appropriate dimensions, proportions, and materials used at the cross-sectional level. In fact, the behavior of the member in this case is primarily governed by behavior and properties of its cross-section and material exposed in the cross-section.

The Structural Materials

Material is the basic ingredient that forms the structural members and consequently, the structures. However, by "structural material," we mean the material for which mechanical properties are usually defined for the purpose of structural analysis and design. Therefore, we consider "concrete" as a structural material rather than its constituents such as cement, water, and aggregates. From this perspective, the book discusses the behavior of cross-sections of members made up of the following structural materials and their combinations:

- Concrete
- Reinforcing steel
- Prestressing steel
- Hot-rolled structural steel
- Cold-formed structural steel
- Aluminum and other metals
- Structural timber

There are several properties of the structural materials that can be used and manipulated for the analysis and design of structural members and are discussed in more detail in Chapter 2, Understanding Cross-Sections.

DESIGNING THE STRUCTURES

Since the start of the formal approaches and procedures for carrying out the structural design, there have been many developments in the underlying principles and the implicit and explicit design objectives. Starting with putting limits in the allowable (working) stresses in various materials to achieve indirect safety factors, the design process slowly evolved within last few decades to more explicit consideration of different load and capacity factors. The recognition of the difference between brittle and ductile failure, and the introduction of capacity-based design approaches, led to the more comprehensive performance design using high level of analysis sophistication, and more explicit linkage between demand and performance. The most recent emphasis is on risk-based design, and a more integrated and holistic approach within the framework of consequence-based engineering. This section discusses a brief account of the progression of these design approaches and their impact on the cost, performance, and the final objective of public safety.

The Design Objectives and Philosophy—A Historical Overview

Structural design is the process of proportioning the structure to safely resist the applied forces and load effects in the most resource-effective and friendly manner. The term "friendly" refers to the aspect of design dealing with environmental friendliness, sustainability, ease of construction, and usability that are not explicit part of the strength consideration. Resources refer to the use of material, labor, time, and other consumables that are used to construct and maintain the structure.

Ideally, the role of structural design is straight forward. It is the transformation of the effects of various environmental and man–made actions (including constraints on materials, dimensions and cost etc.) into the appropriate material specifications, structural member sizes, and arrangements (Fig. 1.4).

The basic objective is to produce a structure capable of resisting all applied loads without failure and excessive deformations during its anticipated life. The very first output of any engineering design process is a description of what is to be manufactured or built, what materials are to be used, what construction techniques are to be employed, and an account of all necessary specifications as well as dimensions (which are usually presented in the form of drawings). The second output is a rational justification or explanation of the design proposal developed based on either full-scale tests, experiments on small physical models, or the mathematical solution of detailed analytical models representing the behavior of real structures.

The process of structural design has passed through a long and still continuous phase of improvements, modifications, and breakthroughs in its various research areas. The structural analysis and design philosophies for new and existing buildings have a fascinating history. Perhaps, the first ever achievement in the history of structural design was the "confidence" by virtue of which early builders were able to convince themselves that the resulting structure could, indeed, be built and perform the intended

Figure 1.4 The conceptual role of structural design.

function for the entirety of its intended life. Hence, the job of the very first engineers can be thought of as "to create the confidence to start building" (Addis, 2003). Over the course of the history, various scientists, mathematicians, and natural philosophers presented revolutionary ideas which resulted in improved understanding of structures and built environment. With the developments in different areas of practical sciences, the task of building design was gradually divided among more and more professionals depending upon esthetic considerations, intended functions, materials, optimum utilization of space, lighting, ventilation, and acoustic preferences. The visual appearance, sense of space, and function (or the architecture) became a distinct concern during the 15th and 16th centuries. About a century later, designers first began to think about the load-bearing aspects of structures in terms of self-weight and other sources of expected loading. Thinking separately about the role of individual materials and resulting structures grew during the late 17th and 18th centuries following Galileo's work. The idea that the esthetics should be given proper importance independent of the materials and load-bearing characteristics of the structure prevailed during the late 19th and the early 20th centuries.

Table 1.1 presents a brief timeline of some of the major developments which led to modern techniques and methodologies for analyzing and designing structures.

It is worth noting that historically an understanding of how structures work was never a phenomenon that required detailed knowledge of mathematical procedures and laws of mechanics. A common misconception is that various new structural forms and shapes were first devised by mathematicians (and experts of geometry) and later taken up by builders and engineers. In fact, the opposite is true, with perhaps just one exception, i.e., the hyperbolic paraboloid (whose structural properties were discovered in 1930s). In the last few centuries, artists, sculptors, and builders have displayed a remarkable understanding and skill of converting materials into structures (some of which are still standing today remarking the testimony of their expertise).

The Role of Building Codes

A building code is a properly documented set of rules and guidelines specifying the minimum standards for constructed facilities. The main purpose of building codes is to protect public health and to ensure safety

Table 1.1 Important Historical Developments Related to Structural Analysis and Design

Year (CE)	Development
1452–1519	Earliest contributions from Leonardo da Vinci
1638	Galileo Galilei examined the failure of simple structures and published his book "Two New Sciences"
1660	Robert Hooke presented the Hooke's law which is the basis for elastic structural analysis
1687	Isaac Newton published his document "Principia Mathematica" containing the famous Newton's laws of motion
1750	Leonhard Euler and Daniel Bernoulli developed Euler–Bernoulli beam theory
1700–82	Daniel Bernoulli introduced the principle of virtual work
1707–83	Leonhard Euler developed the theory of buckling of columns
1826	Claude-Louis Navier published a document analyzing the elastic behavior of structures
1873	Carlo Alberto Castigliano presented his theorem for computing displacement as partial derivative of the strain energy
1874	Otto Mohr formalized the idea of a statically indeterminate structures
1922	Timoshenko corrects the Euler–Bernoulli beam equation and presented "Timoshenko's Beam Theory"
1936	Hardy Cross developed the moment distribution method, an important innovation in the analysis and design of continuous frames
1941	Alexander Hrennikoff solved the discretization of plane elasticity problems using the lattice framework
1942	R. Courant presented solution of problems by dividing a domain into finite subregions
1956	J. Turner, R. W. Clough, H. C. Martin, and L. J. Topp introduces the term "finite element method" and published work which is widely recognized as the first comprehensive treatment of the method

and general welfare as they directly govern the construction and occupancy of buildings and other structures. The building code becomes law of a particular jurisdiction when formally enacted by the appropriate governmental or private authority. The complete process of planning, design, construction, and operation of buildings are guided by various building standards, guidelines, codes, and design aids. Improving the building code quality in terms of addressing real-life problems and enforcement would directly help cities to improve their environmental sustainability and disaster resilience.

Historical Development

The earliest known written building code was the Babylonian law of ancient Mesopotamia (also known as the code of King Hammurabi who ruled Babylon from 1792 BC to 1750 BC). It was found in 1901 in what is now Khuzestan, Iran. Consisting of 282 laws, with scaled punishments, "an eye for an eye and a tooth for a tooth," this code is one of the oldest deciphered writings of significant length in the world. It is currently on display in the Louvre Museum in France, with exact replicas at the University of Chicago in USA, Theological University of the Reformed Churches in The Netherlands, Pergamon Museum in Germany and National Museum of Iran in Tehran. It contains detailed accounts of laws pertaining to builders as well as construction conflicts. Similar accounts can also be found in other historical texts including the Bible book of Deuteronomy and works of ancient Greek philosophers.

The modern era for development of building regulations started with "Rebuilding of London Act" which was passed after the "Great Fire of London" in 1666 AD. In 1680 AD, "The Laws of the Indies" were passed by the Spanish Crown to regulate the urban planning for colonies throughout Spain's worldwide imperial possessions. The first systematic national building standard was established with the London Building Act of 1844. Various regulations regarding the thickness of walls, height of rooms, the materials used in repairs, the division of existing buildings, and the design of chimneys, fireplaces, and drains were included. In the United States, the City of Baltimore passed its first building code in 1859. In 1904, a Handbook of the Baltimore City Building Laws was published which served as the building code for 4 years. In 1908 AD, a formal building code was drafted and adopted. Currently, The International Building Code (IBC) has been adopted throughout most of the United States. It is a model building code developed by the International Code Council (ICC) (Rossberg & Leon, 2013).

In the European Union, the European Committee for Standardization developed a set of harmonized technical rules for the structural design of construction works, known as Eurocodes. More recently, various international organizations, research agencies, and educational institutions have developed standards and guidelines pertaining to specialized areas of building design. The most famous among such organizations are American Concrete Institute (ACI), American Society of Civil Engineers (ASCE), Applied Technology Council (ATC), Federal Emergency Management Agency (FEMA), and National Earthquake Hazards Reduction Program (NEHRP).

Disaster Resilience and Environmental Sustainability in Building Codes

Buildings constructed today are likely to govern future cities and consumption patterns for the next 2−3 decades. The way buildings are designed, built, and maintained nowadays will influence the sustainability of cities and the health and safety of its residents for decades to come. Therefore, a lot of problems cities are coping with can be addressed by enforcing and improving the building codes. Disaster resilience, energy efficiency, and prevention of diseases are all issues that are influenced by building codes.

Resilience to earthquakes, for example, is an important issue linked to construction which has a direct impact on human life. Earthquakes of approximately the same intensity may result in very dissimilar amount of loss of life and property in different cities depending upon the standards of earthquake safety being adopted. The same applies to extreme weather events, e.g., cyclones and hurricanes, which are likely to occur more often with climate change; a lot of damage can be prevented by constructing safer buildings. Environmental and energy consumption issues are also among the most important considerations in building design. Designing and constructing buildings utilizing resources efficiently is one of the best ways to achieve sustainability goals in a city. By incorporating disaster resilience and environmental design in the building codes, future buildings can be made more people- and environment-friendly, thus decreasing the carbon footprint of cities and other negative impacts on the environment.

Typical Structural Design Process

A typical structural design process is carried in three interdependent steps, as shown in Fig. 1.5.

a. Conceptual design
b. Modeling and analysis
c. Design and detailing

The process starts with a conceptual design involving the selection of primary shape and form of structure as well as selection of gravity and lateral load–resisting systems. It is probably the most important step in affecting the overall performance and the final outcome. Complete architectural functional plan is developed in this phase. Once the right structural system is defined in terms of configuration of the materials and the types of elements to be used, the next stage is to determine the

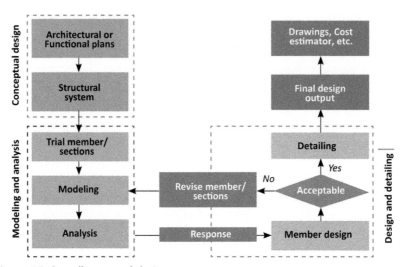

Figure 1.5 Overall structural design process.

expected response of structure under all kinds of loadings. Trial sections are assumed to start the process, and an idealized model is prepared using commercially available computer software. The level of sophistication in the development of computer model is a major consideration starting from fully idealized elastic finite element model to a complicated nonlinear model with specialized inelastic components. The selection of analysis procedure is another important decision to make. Various codes and standards guide the practicing engineers about both modeling and analysis in terms of do's and don'ts. The assumed dimensions and material properties are used to determine the capacities of sections and members and compared against the computed response quantities (moment, shear, etc.), or the response quantities are directly used to determine the required member size, section size, reinforcement, etc. In either case, it is important to verify whether assumed sizes or required dimensions are acceptable. If they are not adequate, the dimensions and models are revised, and the process is repeated until a satisfactory solution is found.

The last phase comprises of detailing and connection design in the light of results obtained from analysis. The design information is converted into a form suitable for actual construction. Drawings and complete plan are prepared for sending to site engineers for proper on-site implementation. It is evident that the materials, the dimensions, and the properties of the member cross-section are essence of the entire design process.

Although most approaches toward the design process can be outlined as discussed earlier, it is worthwhile to note here that the actual member design process can proceed in an alternative (in fact, reverse) direction as well. The conventional approach, starting from the loads and ending at the determination of stresses and strains, and complying with certain limits imposed on these, is shown in Fig. 1.6 (Left). A second approach is to start with the known limits and capacities of the material stresses and strains, to work out the capacities of sections, members, and the structure, and then to determine the load-carrying capacities, as shown in Fig. 1.6 (Right). These calculated load-carrying capacities can then be compared with applied loading directly with the provision of an adequate factor of safety. The first process is typically used in the traditional design of new structures, while the second one is used in the evaluation of existing structures, or for the verification of design (especially in capacity-based design). Each of the step in Fig. 1.6 is in fact a subprocess comprising of several steps. For example, the determination of the "actions" for member design from loads requires definition of several load cases, load combinations, result envelopes, actions sets, etc. as shown in Fig. 1.7.

It is important to note that sometimes, the design envelopes (the maximum and minimum values) cannot be used as design action; rather "action sets" are needed to account for the interaction between action, such as axial—flexure and shear and torsion. The cross-sections are finally

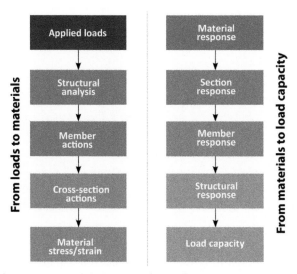

Figure 1.6 The response and design (two faces of a coin).

proportioned and checked for the design actions. The entire process of determining the design actions from loads is typically carried out by a structural analysis software.

Analysis and Design Levels

Fig. 1.8 illustrates various levels of structural design based on the order of rigor used or depending upon the degree of sophistication in computations, starting from rigorous analytical to simplified empirical procedures including shortcut methods using convenient-to-use design aids. The theoretical structural response can often be described through partial or complete differential equations and considerations for equilibrium. These procedures are, however, complex and limited in applications, hence leading to the development of semianalytical, closed-form equations and solutions, developed and simplified for particular applications. However, with the advent of computers and latest computational tools, rigorous numerical procedures were developed using full three-dimensional

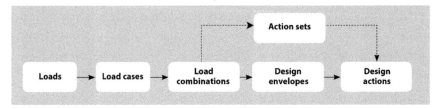

Figure 1.7 From "loads" to "design actions."

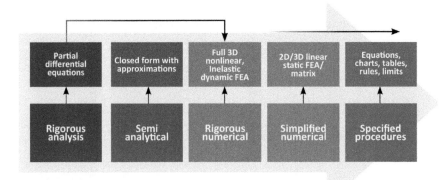

Figure 1.8 The development of various design levels in the reducing order of rigor and complexity.

analysis capabilities. These were implemented in various forms, some simplified for adoption to early computers with limited capabilities, and some for specific applications. The design codes and guidelines have traditionally provided equations, charts, tables, and graphs, derived from analytical as well as physical tests to aid the structural engineers in their routine design work. The structural engineers continue to use these analysis and design levels, depending on the need, complexity of structure, availability of tools, skill of the designer, and the requirements of the projects. It is important to note that the understanding of properties and behavior of cross-sections is essential in all of these levels and approaches.

Traditional Approaches to Structural Design

As mentioned earlier, the primary objective and purpose of structural design is to create a safe and a functional system. This can be ensured by specially focusing on the stability, strength, ductility, and serviceability of the system for different loads and environmental effects, both short term and long term.

At the cross-sectional level, the prime concern is to "balance the external actions with internal stress resultants with adequate margin for safety" and then to check for other limits on

- deflections, deformations, and vibrations;
- durability through checks on crack width, fire protections, permeability, and chemical attacks; and
- ductility and other special considerations.

The margin for safety has been at the root of different design approaches and design methodologies developed since the beginning of the 20th century. Initially, the safety was assured by using rather an arbitrary limit on material stress and cross-section sizes [allowable stress design (ASD) or working stress design (WSD) methods]. Later, partial factor of safety concept for loads and material stresses or action types were developed (Fig. 1.8). A brief introduction of the three traditional approaches of structural design will be discussed.

Working Stress Design

WSD or also known as ASD is the traditional method of structural design not only for reinforced concrete structures, but also for steel and timber. The method primarily assumes that the structural material behaves as a linear elastic manner and that an adequate safety can be ensured by suitably restricting the material stresses induced by the expected "working loads" on the structure. As the specified permissible stresses are kept well below the material ultimate strength, the assumption of linear elastic behavior is considered justifiable. The ratio of the strength of the material to the permissible stress is often referred to as the "factor of safety." There are some obvious issues with this assumption of linear elastic behavior and also the assumption that the stresses under working loads can be kept within the "permissible stresses." A lot of factors may be responsible for inadequacy of these assumptions, e.g., long-term effects of creep and shrinkage, the effects of stress concentrations, and other secondary effects. This design approach usually results in relatively large sections of structural members, thereby (being conservative) providing better serviceability performance under the usual working loads.

Ultimate Strength Design

With the growing realization of the shortcomings of WSD approach in reinforced concrete design and with increased understanding of the behavior of materials (e.g., reinforced concrete) at ultimate loads, the ultimate strength emerged as an improved alternative to WSD. Here, the stress condition at the onset of impending collapse of the structure is analyzed, and the full nonlinear stress—strain curves of concrete and steel (or other materials) is considered. The safety measure is introduced by an appropriate choice of the load factor (defined as the ratio of the ultimate load to the working load and may vary from 1.2 to 2). The ultimate strength method makes it possible to assign different load factors to different types of loads under combined loading conditions. It generally results in more slender sections and often economical designs of beams and columns, particularly when high-strength reinforcing steel and concrete are used. However, the satisfactory strength performance at ultimate loads does not guarantee satisfactory serviceability performance at the normal service loads. The designs sometimes may result in excessive deflections and crack widths under service loads, due to the slender sections resulting from the use of high-strength materials.

Fig. 1.9 shows various design approaches based on the application of factor of safety to either strength or applied load.

Limit State Design Concept

Limit state design concept is an advancement over both WSD and ultimate strength design approaches. This approach, unlike WSD (which is based on calculations at service load conditions only) and ultimate load

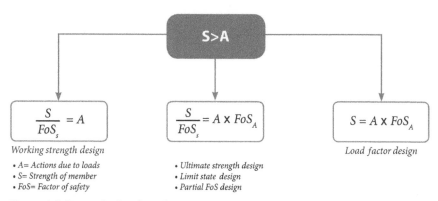

Figure 1.9 Proportioning for safety.

design (which is based on calculations at ultimate load conditions only), aims for a comprehensive and rational solution to the design problem, by ensuring safety at ultimate loads and serviceability at working loads. This philosophy uses more than one safety factor attempting to provide adequate safety at ultimate loads as well as satisfactory serviceability performance at service loads, by considering all possible failure modes. The term "limit state" refers to a state of impending failure, beyond which a structure ceases to perform its intended function satisfactorily, in terms of either safety or serviceability (i.e., it either collapses or becomes unserviceable). So there are two types of limit states: (1) ultimate limit states (which deal with strength, overturning, sliding, buckling, fatigue fracture, etc.) and (2) serviceability limit states (which deal with discomfort to occupancy and/or malfunction, caused by excessive deflection, crack width, vibration leakage, and loss of durability). Table 1.2 presents some of the commonly used limit states in design of steel and reinforced concrete structures.

The basic idea involves the identification of all potential modes of failure (i.e., significant limit states) and determination of acceptable levels of safety against occurrence of each limit state. Factors of safety (Fig. 1.10) are applied at each step starting from characteristic values of both material strength as well as applied loading up to the full member design level. To account for uncertainty in the loading, the expected loads are multiplied by load factors that increase the force demands. For example, the gravity load for demand calculation might be 1.2 times the calculated dead load plus 1.6 times the expected live load. To account for uncertainty in component strength, the estimated strength capacities are multiplied by

Table 1.2 Some Common Limit States

Types of Limit State	Description
Ultimate limit states	• Loss of equilibrium • Rupture • Progressive collapse • Formation of plastic mechanism • Instability • Fatigue
Serviceability limit states	• Excessive deflections • Excessive crack width • Undesirable vibration
Special limit states	Due to abnormal conditions and abnormal loading such as • Damage or collapse in extreme earthquakes • Structural effects of fire, explosion • Corrosion or deterioration

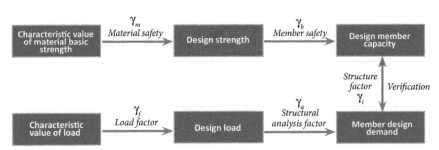

Figure 1.10 The value of safety factor tells us how much confidence we have on our knowledge.

capacity reduction factors (or resistance factors), typically between about 0.75 and 0.9. Components that are especially important to the integrity of a structure may be assigned smaller capacity reduction factors. In some cases, the calculated demand on a component may be multiplied by a demand increase factor. The details can be found in design codes, standards, and guidelines. It should be noted that the traditional approach is to make the structure strong enough to resist the external loads with essentially elastic behavior. It is also important to satisfy serviceability requirements, which usually means providing enough stiffness to control deflections and vibrations. This whole process is essentially "strength-based" or "force-based" where the structural analysis can be elastic and its main purpose is to calculate force demands on the structural components. However, if the force demand in a substantial proportion of

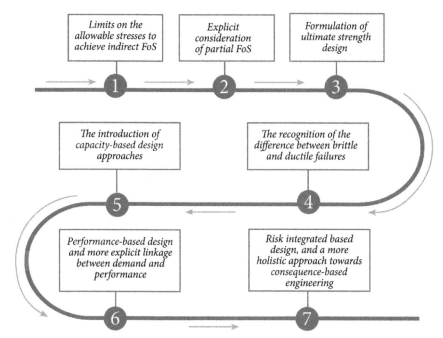

Figure 1.11 Evolution of our understanding of structures.

the components in a structure are close to their force capacities, there could be a significant inelastic deformation of the structure as a whole. Hence, the behavior of a structure could be significantly inelastic under the design loads, and elastic analysis may not necessarily be accurate.

Fig. 1.11 shows a historical overview of important developments in our understanding of structural behavior.

Shortcomings of Traditional Building Codes

With the advent of innovative structural systems, complex geometries, and advanced construction techniques, the requirement from building codes to handle various new aspects is also increasing. Currently the traditional codes govern the design of general, low- to medium-rise, and relatively regular buildings built with traditional construction materials. They are not specifically developed for tall buildings (having total height > 50 m). Moreover, they are prescriptive in nature with no explicit check on intended outcome. They are also not expected to cover new structural systems and shapes. Mostly, the prescribed analysis and design procedures are based on elastic theory, neglecting some of the key aspects of nonlinearity, e.g., realistic demand distribution. The intention to propose simplest and cook-book-type procedures does not provide the opportunity to exploit the potentials of recent computing tools.

Another limitation of traditional building codes (for seismic design) is that the structural performance objectives are considered implicitly. For example, a structure is expected to have minor damage against a low- to medium-intensity earthquake, and substantial damage is allowed for high-intensity ground shaking. There is no explicit verification specified or required in traditional building codes whether these performance objectives are achieved or not.

From Force-Based to Displacement-Based Design

The continuous improvement in understanding structural behavior, especially against earthquakes and strong winds, have resulted in a transition of seismic design philosophy from force-based to displacement-based approach. As mentioned in the preceding section, there is a high probability that a small earthquake will occur during the life of the structure, and a low probability of a large earthquake. For a small earthquake, it seems reasonable to design the structure to remain essentially elastic. However, for a high-intensity earthquake, it is uneconomical to design the structure to remain elastic and a common practice is to allow for substantial inelastic behavior. Hence, for a large earthquake, the elastic strength demand on a structure is likely to exceed its strength capacity. However, the maximum displacement of the structure may still be acceptable, and although some structural components become inelastic, the structure can perform satisfactorily. For those components that become inelastic, the concern for design is deformation, not strength. For satisfactory performance, the deformation demand on an inelastic component must usually be smaller than its ductility limit (Powell, 2010). In a displacement-based design, a practicing engineer starts with displacement as a basic input (not forces) and determines the design forces based on that preselected design displacement. This provides an explicit control over displacements and hence all deformation-related phenomenon. In another latest design approach, "capacity design," the force demands are calculated for components that are required to remain elastic, while the demand-to-capacity (D/C) ratios for components which are allowed to yield are determined in terms of displacements (or deformations). With the advent of all these ideas, the deformation-based approach gained popularity as it provides a clear interpretation of structure's condition and the results are physically more meaningful. This quest for explicitly achieving the design goals soon led the profession to what is now called "performance-based design" and is discussed next.

The Performance-Based Design

As mentioned earlier, in traditional building codes (for seismic design), the structural performance objectives are not considered explicitly. It should be noted that satisfying one design level does not ensure that other design levels will be satisfied as well. Serviceability design only ensures that deflections and vibrations, for service loads, are within limits but provides no information whatsoever about strength. Similarly, strength design ensures that a certain factor of safety against overload is available within a member or a cross-section but says nothing about what will happen if load exceeds the design level. Practicing engineers started to realize the importance of a methodology focusing rigorously on achieving the intended performance instead of fulfilling definite rules to implicitly account for desired functionality (Fig. 1.12).

This realization has led to a relatively recent paradigm shift in current approach toward analysis and design of building structures, termed in latest guidelines and standards as "performance-based design (PBD)." It refers to the methodology in which structural design criteria are expressed in terms of achieving a set of performance objectives. It ensures that the structure as a whole reaches a specified demand level including both service and strength design levels. It is the practice of thinking and working in terms

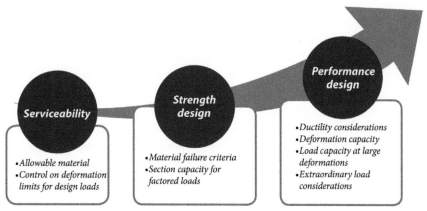

Figure 1.12 From serviceability to performance.

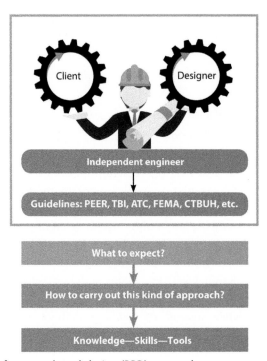

Figure 1.13 Performance-based design (PBD) approach.

of ends rather than means. In PBD, owners and engineers can work together to achieve the best possible balance between construction costs and the structure's ultimate performance (Fig. 1.13). The basic idea is to relate the level of structure's damage to measurable engineering demand parameters. It is similar to associating "numbers, which can be crunched" with "physical extent of damage." For example, the performance objectives set for a building can be related to the level of its damage, which in turn, can be related to its displacements and drift values. Although it is not always possible to quantify the damage as it is greatly influenced by a lot of other factors, mostly displacements and drifts serve as reasonable indicators. That is why sometimes engineers also use the term "displacement-based design" in place of PBD (which ideally should be thought of as a subset of PBD because the performance target can be any response parameter attached to a certain threshold). Since the approach gained popularity among the engineering community around a decade ago, there have been a lot of attempts to develop procedures to correlate damage of various structural systems to response quantities taking into account possible uncertainties and ground motion characteristics. This approach requires the structural designers to go beyond code's cook-book prescriptions and make them able to predict structure's response in the case of future extreme events. This also requires sophisticated structural modeling and simulation using state-of-the-art computer software and sometimes laboratory testing also. While earthquake engineers are sufficiently contributing and exploiting the potentials of this design philosophy, it can also be applied to floods, hurricane, and other natural disasters.

Usually the PBD process starts with a linear elastic analysis for code-based design loadings. The structure is initially designed to remain elastic under a lower level of intended loading termed as design-basis load. Then a nonlinear dynamic analysis is performed for a suit of ground motions, and peak response quantities are extracted. The last stage of the process is interpretation of results, i.e., converting "numbers we have already crunched" into "meaningful outcome for decision-making." ASCE 41-06 provides acceptance criteria in terms of plastic rotations and other demand parameters for each member type, analysis type, and for each performance level (immediate occupancy, life safety, and collapse prevention as shown in Fig. 1.14). Since each of these performance levels are associated with a physical description of damage, obtained results are compared and evaluated based on these criteria to get performance insights.

PBD may not be a guarantee (of structural safety) in itself, but is a successful attempt to answer "What will happen if ...?"-type questions. These novel ideas coupled with advance computational tools have taken structural engineering practice to an advance level of creating optimized, reliable, and cost-effective structures. The question "Is my structure safe?" can now suitably transformed into an optimization problem which can be answered with the help of smart decision-making tools and techniques for consequence evaluation.

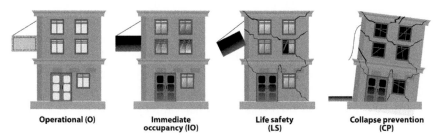

| Operational (O) | Immediate occupancy (IO) | Life safety (LS) | Collapse prevention (CP) |

Figure 1.14 Various performance levels in PBD (based on FEMA 451 B).

THE BUILDING BLOCKS OF STRUCTURAL MECHANICS

The structural modeling and analysis is the key step in any design methodology. This section deals with basic concepts related to structural mechanics and significance of simple ideas in understanding complex structural responses to various types of loadings.

Basic Concepts and Relationships

There are six basic concepts that lie at the foundation of theories governing the behavior of structures, from analysis to design:
- Loads and load effects
- Actions
- Deformations
- Strains
- Stresses
- Stress resultants

Loads are the actual physical excitations that may act on the structure, e.g., gravity, wind pressure, dynamic inertial effects, and retention of liquid. Loads and their effects can lead to actions (which are basically the idealized forces acting on the members), e.g., bending moment, shear force, etc. Actions can lead to deformations, which again are idealized into various components such as rotation, shortening, and shearing angle. Deformations cause strains which are basically normalized deformation at the cross-section material or fiber level. Strains may lead to stresses in material fibers, which generally have a correspondence with the strain through material stress—strain model. The stresses can be summed up in any particular manner to determine the internal stress resultants.

In general, for a structure to be in static or dynamic equilibrium, the internal stress resultants should be in equilibrium with the actions due to loads. An alternative way of looking at the same linkage is that the actions cause stresses in the member cross-sections. These stresses cause strains, which can be summed up to determine deformations. So the relationships between actions, deformations, strains, and stresses can be used in many ways to solve the particular problems at hand. Fig. 1.15 illustrates this whole process starting from loads and ending on stress resultants.

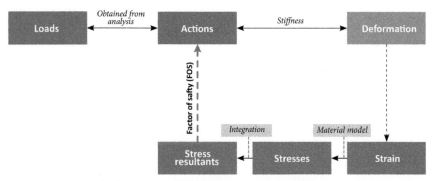

Figure 1.15 Overall relationship of main structural quantities in design process.

A brief description of the relationship between these quantities is given here, without the explicit mathematical formulations that are adequately covered in many texts on structural theory and analysis.

1. Action–deformation relationship: Defining an action–deformation relationship means linking the deformations produced in a member due to applied actions or linking the restraining actions with applied deformations. These relationships involve the entire stiffness of the member and may be either linear or nonlinear. One action can produce more than one deformation, and one deformation may be caused by more than one action.

2. Deformation–strain relationship: Deformation–strain relationship means linking deformations with corresponding strains. Each deformation produces a particular strain pattern or profile on the cross-section. A particular strain may be the result of several deformations. For example, axial strain may be produced due to axial deformation as well as flexural curvature. This relationship is defined primarily by the cross-section's stiffness and may be linear or nonlinear.

3. Stress–strain relationship: Stress–strain relationship means linking strain to corresponding stress. Generally, this relationship is used at a material level, indicating material stiffness and its behavior. For example, Hooke's law describes the stress–strain relationship for a linear elastic material but in general, this relationship is nonlinear for most materials.

4. Stress resultant–action relationship: This last relationship is the expression of equilibrium and completes the cycle of all relationships. In fact, this relationship is the basis for strength design of structural members, which states that "the internal stress resultants should be in equilibrium with external actions with adequate margin for safety."

The Concept of Stiffness

The Structural Equilibrium and Role of Stiffness

Let us consider a structure subjected to an arbitrary force (F). This force will produce an arbitrary deformation U. If we compare the structure to a simple elastic spring (Fig. 1.16), then a simple linear relationship between the force and deformation exists. This linear dependence is the "stiffness" which is the resistance to its deformation.

In a real structure, this resistance or stiffness comes from four sources, as shown in Fig. 1.17:

1. The overall resistance of the structures to overall loads, called the global structure stiffness, is derived from the sum of stiffness of its members, their connectivity, and the boundary or the restraining conditions.
2. The resistance of each member to local actions called the member stiffness is derived from the cross-sectional stiffness and the geometry of the member.
3. The resistance of the cross-section to overall strains is derived from the cross-sectional geometry and the stiffness of the materials from which it is made.
4. The resistance of the material to strain derived from the stiffness of the material particles.

The simplest form of Hook's law ($F = Ku$) can be generalized to include several deformations, several actions, and several stiffness relationships and represented in a matrix form. It then becomes the basis of the "stiffness matrix method" of structural analysis and more generally, the "finite element method."

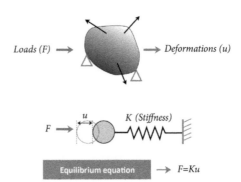

Figure 1.16 Conceptual state of equilibrium and role of stiffness.

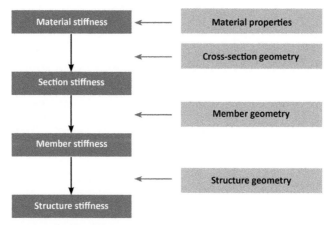

Figure 1.17 The overall stiffness of the structure is derived from the geometry and connectivity of the members and their stiffness. The member stiffness is derived from the cross-sectional stiffness and member geometry. The cross-sectional stiffness is derived from the material stiffness and the cross-sectional geometry.

The Nonlinearity of Response and Stiffness

The equilibrium equation $F = Ku$ is based on the assumption that the relationship between the force and deformation is linear and infinite. This means that a very large force can produce a corresponding very large displacement, and an infinite force can produce an infinite displacement. The equation also suggests that if the force is decreased, the deformation will be reduced and zero force will return the structure to the original undeformed state and that a negative force will produce exactly same negative displacement as in the positive direction. However, in reality, almost none of these assumptions or behaviors is true. The relationship between force and displacement for a real structure can be highly nonlinear and inelastic with no single value of stiffness describing its behavior (Fig. 1.18). In such cases, the stiffness varies at different states of deformation throughout the loading history. The complete equilibrium condition and corresponding equation should reflect not only the nonlinear and inelastic behavior, but also the effect of force being applied fast enough to produce deformation with velocity and acceleration so that the total equilibrium should include effect of inertia and damping.

Fig. 1.19 shows all possible states of equilibrium for a structure, whereas Fig. 1.20 shows the nonlinearity inherent in various stages of stiffness contribution. The detailed discussion of this nonlinear dynamic equilibrium is beyond the scope of this book, but it emphasizes the

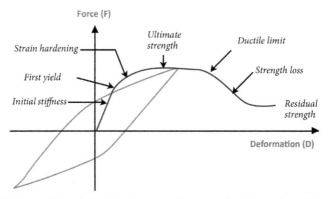

Figure 1.18 A typical relationship between force and deformation. This may also hold for relationship between stress and strain.

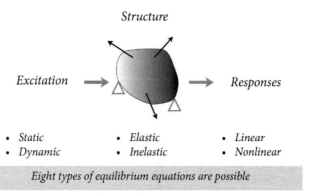

Figure 1.19 The equilibrium conditions for a typical structural system.

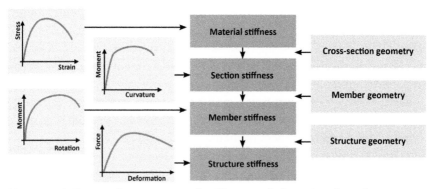

Figure 1.20 The nonlinear nature of stiffness and the role of nonlinear cross-sectional response.

importance of stiffness and cross-sectional analysis in the overall under-standing of behavior. The nonlinear nature of material and section beha-viors (especially for axial−flexural case) will be discussed in Chapter 2, Understanding Cross-Sections, and Chapter 3, Axial−Flexural Response of Cross-Sections.

The Concept of Degree of Freedom (DOF)

Although it is possible to conceive situations where deformation can occur without external actions (such as thermal variation), in general, an external action (a generalized force, moment, or torque) is needed to pro-duce deformation in a structural member. If the actions can be general-ized in terms of their components, we can say that, in general, those action components produce corresponding deformation components. If we assume that the materials are behaving linearly and elastically, we can end up with a simple spring representation for each deformation compo-nent. That is, the action (F) and deformation (U) in a particular sense are proportional and related to each other by the corresponding stiffness.

Member Cross-Sections and the DOFs

As mentioned earlier, a beam member can be broadly defined as a struc-tural component having one dimension significantly larger than the other two. The longer dimension becomes the member axis, and the dimen-sions in the plane perpendicular to the longitudinal axis define the cross-section. The cross-section exposes the materials used in the member. For a member placed in a general three-dimensional space, each point in a member can move in an infinite number of ways. However, if the cross-section is assumed to be rigid in its own plane, all these movements can be completely defined in terms of seven idealized directions, referred to as the DOFs at each section on the centerline of the member, with respect to three orthogonal axes. Using the right-hand rule, if we orient the axis system as shown in Fig. 1.21, these DOFs become

- movement along the member axis, u_z
- movement along the x-axis, u_x
- movement along the y-axis, u_y
- rotation about the longitudinal axis, r_z
- rotation about the x-axis, r_x
- rotation about the y-axis, r_y
- out-of-plane movement (distortion) of the cross-section's points along the longitudinal axis, w_z

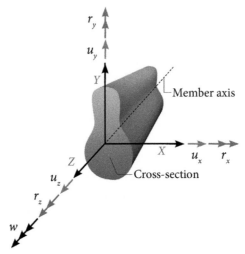

Figure 1.21 Degrees of freedom (DOF). Each section on a beam member can have seven DOFs with respect to its local axis.

It is important to note that in the above definitions, the cross-section is assumed to be rigid in its own plane. This means that the dimensions and shape of the cross-section before and after deformation remain the same. This assumption is mostly true for solid sections. For thin-walled sections and for large box girders, the section may distort in its own plane and some additional considerations may be needed to evaluate its behavior. It is also generally assumed that the member centerline passes through the geometric (or the plastic) centroid of the cross-section. This assumption is generally true for members having length more than at least five times the average size of the cross-section. These aspects will be further discussed in Chapter 2, Understanding Cross-Sections.

DOFs, Deformations, Strains, and Stresses

Each DOF at the cross-sectional centroid is associated with a corresponding deformation in the member. Each deformation in the member produces a corresponding strain profile in the cross-section. Each strain profile generally produces a corresponding stress profile in the cross-section material(s). These relationships are shown below.

u_z → axial deformation → axial strain → axial stress

u_x → shear deformation → shear strain → shear stress

u_y → shear deformation → shear strain → shear stress

r_z → torsion → shear strain → shear stress (may also produce axial stresses and strains)

r_y → curvature → axial strain → axial stress (may also produce shear strains)

r_x → curvature → axial strain → axial stress (may also produce shear stresses and strains)

w_z → warping → axial strain → axial stress (may also produce shear strains)

As illustrated above, sometimes one DOF may be related to more than one stress and strain component. In some cases, strain may not produce any stress, such as the unrestrained thermal expansion or free shrinkage produces elongation and corresponding strains, but does not result in any stresses in the cross-section's material.

Each deformation may produce more than one strain and stress component and each stress component may be produced by more than one deformation.

Stress Resultants and DOFs

Material stresses in the cross-section (or stress components along the reference axes) can be summed up to obtain the total resultants. These stress resultants (as shown in Fig. 1.22), when determined with respect to the member axes and the corresponding DOFs at the cross-sectional centroid, can provide useful information related to the "capacity" of the cross-section.

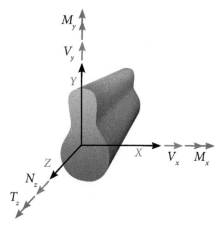

Figure 1.22 Stress resultants and Degrees of Freedom (DOF).

We will discuss this aspect in detail in Chapters 3, Axial−Flexural Response of Cross-Sections and Chapter 4, Response and Design for Shear and Torsion.

Linear, Elastic Stiffness Relationships

Consider a single beam-type member with six DOFs at each end (ignoring warping). Two different types of action−deformation relationships can be derived for this example. In the first case, the total deformations in a particular DOF can be computed from all actions that contribute toward this deformation, assuming that all nonparticipating DOFs are locked. The second type of relationships can be used to compute the restraining actions needed to prevent deformations in all contributing DOFs, while all other DOFs are assumed to be locked. These two approaches are the basis of the flexibility and stiffness matrix (or force and displacement methods) for structural analysis, respectively. Some references discussing the matrix analysis and finite element analysis are given at the end of this chapter.

Deformations for Applied Actions: Flexibility Relationships

For linear elastic elements, it is possible to develop first type of relationship (relating the applied actions with corresponding degree-of-freedom deformations) using principles of mechanics of materials. All we need to know is the stiffness quantity relating each action−deformation pair. For a simple beam element with applied actions on its right end (and assuming the left end fully restrained), the three possible actions are shown in Fig. 1.23. These are the axial load (P), shear force (V), and bending moment (M), corresponding to three assumed DOFs on its right end. Table 1.3 shows the deformations in this beam member for few cases of applied actions.

It is important to note that in all flexibility relationships, actions are related to deformations by various cross-sectional properties (e.g., A, I, and J), material properties (e.g., G and E), and length of member (L).

Figure 1.23 Three possible "actions" corresponding to deformations on three DOFs on right end.

Table 1.3 Some Linear Elastic Flexibility Relationships for a Simple Beam Element

Case	Illustration	Flexibility Relationships
Axial deformation Δ under lateral force P	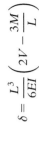	$\Delta = \dfrac{PL}{AE}$
Vertical deformation δ and rotation α under vertical force V only		$\delta = \dfrac{VL^3}{3EI}$ $\alpha = \dfrac{VL^2}{2EI}$
Vertical displacement δ and rotation α under moment (M) only		$\delta = \dfrac{ML^2}{2EI}$ $\alpha = \dfrac{ML}{EI}$
Vertical displacement δ and rotation α under combined shear force (V) and moment (M)		$\delta = \dfrac{L^3}{6EI}\left(2V - \dfrac{3M}{L}\right)$ $\alpha = \dfrac{L^2}{2EI}\left(\dfrac{2M}{L} - V\right)$
The rotation along the member (θ) due to torque (T) applied at the ends, excluding the effect of warping	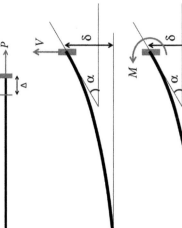	$\theta = \dfrac{TL}{GJ}$

Restraining Actions for Assumed Deformations

For the derivation and development of the stiffness matrix and finite element methods for structural analysis, it is often convenient to develop second type of relationship (involving assumed deformations and the restraining actions needed to "prevent" that deformation). These are actually inverse of the relationships that are used to compute deformations for applied actions. For the same example beam element, the restraining actions against assumed deformations for few common cases are shown in Table 1.4.

The Member Stiffness and Cross-Sectional Properties

It can be seen from the above relationships from member stiffness that the cross-sectional properties play an important role. In fact, if we assume the member to be of unit length, then the above relationships represent the cross-sectional stiffness. For example, for a unit length member, the cross-sectional stiffness (force required to produce unit deformation) for axial action will become

$$P = EA$$

However, if the cross-section is made up of more than one material, for example, a reinforced concrete section or a composite section, then a single value of E cannot be used. In this case, the equivalent axial stiffness of a unit length member must be computed in terms of a summation of stiffness of each part of the cross-section, and the equation becomes:

$$P = \sum_{i=1}^{n} E_i A_i$$

Table 1.4 Some Restraining Actions Related by Corresponding Deformations Through Linear Elastic Stiffness

Case	Actions for Assumed Deformations
Axial force P due to axial deformation Δ	$P = \dfrac{EA}{L}\Delta$
Shear force V at the restraining end for deflection δ and rotation α at the other end	$V = \dfrac{12EI}{L^3}v + \dfrac{6EI}{L^2}\alpha$
Moment M at the restraining end for deflection δ and rotation α at the other end	$M = \dfrac{6EI}{L^2}v + \dfrac{4EI}{L}\alpha$
Restraining torque T due to axial rotation θ	$T = \dfrac{GJ}{L}\theta$

The cross-sectional stiffness defined in terms of equivalent cross-sectional properties can then be used in the overall member's stiffness relationships for various DOFs. To compute the cross-sectional stiffness, we need the material stiffness often represented by constants like E and G for each material and the geometric configuration. Chapter 2, Understanding Cross-Sections, deals with the definition of cross-sections and computation of properties.

CROSS-SECTIONAL ANALYSIS AND DESIGN
The Significance of Cross-Sections in Design Process

The cross-section of beams and columns affects the planning, structural, and construction aspects of an engineering project. In planning context, it affects the space utilization, visibility, lateral clearance, water flow, wind resistance, esthetics, etc. Structurally, it affects axial load capacity, moment capacity, rebar layout, structural stiffness, structural performance, joint design, connections, and foundation design. It also affects rebar cage fabrication, formwork cost and its reuse, construction techniques, and efficiency. If one particular shape may be suitable from structural considerations, it may be unsuitable from planning or construction considerations. Understanding the behavior of various types of cross-sectional shapes and configuration is important for the overall design process. Moreover, as shown in the proceeding section, the cross-sectional properties (combination of the geometric and the material properties) are the key parameters to compute the member stiffness. A detailed discussion of cross-sectional properties, their significance, and computation will be covered in Chapter 2, Understanding Cross-Sections.

Cross-Sectional Analysis

The term cross-sectional analysis broadly refers to the determination of response of the cross-section, either to the imposed external loads, or the imposed deformation, or the self-generated deformations. The response may follow the pattern shown in Fig. 1.20, or may remain linear—elastic, depending on the level of load or deformation or the history of loading or deformation. For simplicity, the response of cross-section is often separated into

- axial—flexural response
- shear—torsion response

It is, however, important to note that in most cases, a full interaction between all six (or even seven) DOFs can occur due to the principal stress and strains, linking the axial and shear deformations.

Axial–Flexural Response

This response pertains to the combined effect of axial load and moments, or a cross-sectional axial deformation and rotation. As both axial load and moment produce axial strain and stress, and axial strain can lead to axial and moment resultants, there is a strong interaction between the two actions and must be considered together. This combined response is discussed in detail in Chapter 3, Axial–Flexural Response of Cross-Sections.

Shear and Torsion Response

This response is related to the combined effect of shear force and torsion or torque, both of which produce shear stresses and strains. Again, due to strong interaction between the two, the response for shear and torsion is often considered together and is presented in Chapter 4, Response and Design for Shear and Torsion.

Ductility of Cross-Sections

The ductility of cross-section, in terms of axial–flexural response, is a specialized behavior that requires consideration of both nonlinear and inelastic aspects and is often treated separately. The ductility is generally characterized by axial load–deformation and moment–curvature responses (although the ductility of shear deformation also exists in many materials). The details of this characteristic of cross-sections will be discussed in Chapter 5, Response and Design of Column Cross-Sections.

Cross-Sectional Design

The design of cross-section, as opposed to the analysis, refers to the appropriate sizing of the cross-section, its geometry, dimensions, and material specification. For steel sections, the design is often a simple task, where the appropriate section from a list of available sections is matched with the required moment or shear demand. For reinforced concrete and prestressed concrete, this may be an iterative process or may require separate determination of concrete section and reinforcement quantity and distribution. The choice of different concrete strengths and several possible combinations of section shape and rebar layout, especially for columns, makes the design a complicated or at least tedious task which needs appropriate software tools to be used. Various aspects of cross-sectional design for flexural, shear, and torsion demands are discussed in the corresponding chapters.

REFERENCES

Addis, B. (2003). Inventing a history for structural engineering design. In S. Huerta (Ed.), *Proceedings of the first international congress on construction history*, Madrid, 20−24 January 2003. Madrid: I. Juan de Herrera, SEdHC, ETSAM, A. E. Benvenuto, COAM, F. Dragados.

Powell, G. H. (2010). *Modeling for structural analysis: Behavior and basics*. Berkeley, CA: Computers and Structures.

Rossberg, J., & Leon, R.T. (2013). *Evolution of codes in the USA*. Retrieved from < http://www.nehrp.gov/ > .

FURTHER READING

Boresi, A. P., Schmidt, R. J., & Sidebottom, O. M. (1993). *Advanced mechanics of materials* (Vol. 6). New York, NY: Wiley.

Cook, R. D., Malkus, D. S., & Plesha, M. E. (2001). *Concepts and applications of finite element analysis*. New York, NY: John Wiley & Sons. International Edition, ISBN-13: 978-0-471-35605-9, ISBN: 0-471-35605-0.

Ghali, A., Neville, A., & Brown, T. G. (2003). *Structural analysis: A unified classical and matrix approach*. New York: Spon Press, Taylor & Francis Group.

Hibbeler, R. C. (2001). *Engineering mechanics*. Upper Saddle River, NJ: Pearson Education. ISBN: 0130910198, 9780130910196.

Hutton, D. V. (2005). *Fundamentals of finite element analysis*. New Delhi: McGraw-Hill Education (India) Pvt. Limited. ISBN: 0070601224, 9780070601222.

Kaveh, A. (2004). *Structural mechanics: Graph and matrix methods (No. 1)*. Hertfordshire: Research Studies Press Ltd.

McGuire, W., Gallagher, R. H., & Ziemian, R. D. (2000). *Matrix structural analysis* (2nd ed.,). Lewisburg, PA: Faculty Books. Book 7, Bucknell University. ISBN: 9781507585139.

NEHRP (2006). Recommended provisions design examples, Topic 7. (Concepts of Seismic-Resistant Design, FEMA 451B). Building Seismic Safety Council for the Federal Emergency Management Agency.

Oden, J. T., & Ripperger, E. A. (1981). *Mechanics of elastic structures*. Hemisphere.

Popov, E. P. (1998). *Engineering mechanics of solids* (2nd ed.,). Lebanon, IN: Prentice Hall. ISBN: 0137261594,9780137261598.

Przemieniecki, J. S. (1985). *Theory of matrix structural analysis*. Mineola, New York: Courier Corporation, Dover Publications Inc. ISBN: 0486649482, 9780486649481.

Wang, C. K. (1973). *Computer methods in advanced structural analysis*. New York, NY: Intext Educational Publishers. ISBN: 0700224297, 9780700224296.

CHAPTER TWO

Understanding Cross-Sections

INTRODUCTION

Definition of Cross-Sections

A brief introduction of cross-sections of line-type members, such as beams, columns, and trusses, was given in Chapter 1, Structures and Structural Design, together with their significance in determining structural response and design.

A cross-section is defined as the projection of a member on a plane normal to its longitudinal axis, as shown in Fig. 2.1. In conventional elastic theory of structural mechanics, this plane itself is assumed to remain undeformed and remain normal to the deformed axis of the members, even after any axial—flexural deformation. This assumption, however, is not valid in all cases and will be discussed further in various chapters of this book.

For the purpose of determining the behavior and response, several cross-sections along the length of the member can be considered, each treated independent of others and based only on properties and actions stated at that location.

Significance of Cross-Sections

The cross-section of beams and columns affects the planning, structural design, and construction aspects of the work. In planning context, it affects the space utilization, visibility, lateral clearance, water flow, wind resistance, esthetics, etc. Structurally, it affects axial load capacity, moment capacity, rebar layout, structural stiffness, structural performance, joint design, connections, and foundation design. It also affects rebar cage fabrication, formwork cost and its reuse, construction techniques, and efficiency. If one particular shape may be suitable from structural considerations, it may be unsuitable from planning or construction considerations.

Structural Cross Sections
DOI: http://dx.doi.org/10.1016/B978-0-12-804443-8.00002-6

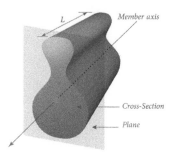

Figure 2.1 Cross-section of a "line-type" member.

As discussed in Chapter 1, Structures and Structural Design, there can be two ways we can perceive and proceed with structural analysis and design. The first one is starting from anticipated loads and performing analysis to determine structural response. The second way is to start from allowable material response and calculate back the value of loads the structure can carry, provided that the response remains within assumed allowable limits (often referred to as the capacity). In both these approaches, the stiffness determination is a necessary requirement. The very first thing to deal with is the formulation of global stiffness of structure, as one cannot relate actions and response at any level without estimating the corresponding stiffness in-between.

The global structural stiffness has many components and levels. At the very basic level, material stiffness (characterized as the Young's modulus of elasticity E in the simplest terms) combines with the corresponding geometric parameter (accounting for the stiffness provided by the cross-sectional shape) to result in cross–sectional stiffness. For example, the axial stiffness of a cross-section with area A, which is composed of a material with elastic modulus E is EA, can be interpreted as the resistance of this cross-section to axial strains. This cross-sectional stiffness combines with complete member geometry and dimensions to result in member stiffness (the resistance of each member to local actions). Long and slender members have smaller stiffness compared to short and thick ones. Similarly, member stiffness also accounts for the type of member and the desired action to resist applied loading. An arch will have different overall member stiffness compared to a cable (even with same cross-sectional stiffness). In an overall structure, a number of members may be arranged in a variety of ways governing the complete structural configuration. Stiffness of individual members combined in a particular manner (depending upon structural configuration) to result in global structural stiffness which inherits the dimensions/properties of all members, cross-sections, and materials. So this global structure stiffness may be referred to as the overall resistance of the structures to overall loads, which is derived from the sum of stiffness of its members, their connectivity, and the boundary or the restraining conditions.

For almost all structures, the definition of cross-sections, the determination of their properties, and stiffness forms the basis for the global stiffness and response. The significance of the understanding of cross-section behavior is, therefore, essential for developing the understanding of the structural behavior and to carryout analysis and design process.

Developing an Integrated Understanding

Due to the significance of the member cross-sections in the overall structural design process, as discussed above, as well as in the actual construction, it is important to develop a thorough understanding of the cross-sections. The structural cross-sections and their behavior may be viewed differently by the structural designers, the theorist, or researchers. The structural designers and engineers are typically interested in the sizing of the section shape, dimensions, and the material specifications for a particular structural member, whereas the theorist and the researchers are involved in developing either theoretical formulations, doing physical testing or numerical simulations, or even developing computational techniques and tools.

However, it is important that an integrated approach be used in the development of the understating of the cross-section definition, their representation, and for the determination of their response from various perspectives, so that various applications and design needs are addressed in the theoretical development, as well as various theoretical developments are appropriately applied to the practical design applications.

In this chapter, a deeper and integrated understanding of the cross-sections is presented, which discusses various applications of the cross-sections, as well as how cross-section geometry and materials are represented, both for specific cases and in generalized form. The chapter focuses on understanding the various types of cross-section properties, from definition, significance, and computational point of view. Definition of cross-sections using polygons and use of polygon meshes to compute properties is also introduced.

Chapter 3, Axial-Flexual Response of Cross-Sections, will focus on the understating and determination of axial—flexural response useful for the design of beams and columns of reinforced concrete, composite, and prestressed cross-sections.

APPLICATIONS AND CLASSIFICATION OF CROSS-SECTIONS

Overview

The definition of cross-section can be viewed from two perspectives. First, for the purpose of analysis or design as a structural designer/engineer and second, from the theorist's point of view for the purpose of developing computational models and procedure for determining the behavior and response. For example, from the engineer or a designer's point of view, a simple rectangular cross-section and a hexagonal or a Tee shape may mean different types of sections. However, from the theorist's point of view, they all may be represented and defined by using a general polygon. Similarly, from the engineer's point of view, the concrete and steel are completely different materials but from the theorist's point of view, both materials may be represented by some simple properties and general stress—strain curves, or a general material model.

First, we will look at the application and classifications of the cross-sections from the structural designers' point of view, followed by generalized representations.

Classifications Based on Types of Structural Members

Cross-sections are typically associated with the line-type member, such as a girder, beam, joist, column, shear walls, and truss. The classification of cross-sections is necessary as the procedure and rules for the design are traditionally handled differently for different types of sections. For example, beam sections are typically designed for moment about the major axis, and shear along minor axis, and sometimes for effects of torsion, while the column sections are often designed for axial load, biaxial bending, and shear about along the two axes. The truss member sections on the other hand are designed for axial compression and tension, and sometimes for bending. A typical structure, some of its members, and their sections are shown in Fig. 2.2.

This classification helps the designer to focus on certain types of section properties and response and ignore others. This focus makes the designer's task easy when applied to typical cases. Most of these classifications were developed for aiding simplified hand calculations, or using design aids and prescribed rules in the design codes.

However, many of these classifications may fail to handle special cases, such as an inclined beam (or tilted columns) or for handling the design of general frame elements based on the output form a three-dimensional structural modeling and analysis software where all types of line members are treated in the same manner, and typically deal with six (or seven) degrees of freedom (DOFs).

Classifications Based on Geometry

The cross-section classification based on the shape is often used in the derivation of the equations for design as well as for the determination of properties. It is customary to discuss the design of RC rectangular and flanged or circular cross-section members using separate formulations, design charts, and equations. Similarly, the design procedures for steel columns vary considerably for symmetrical, unsymmetrical, tubular, and open sections.

The shape classification can be based on basic geometry (see Figs. 2.3 and 2.4), such as

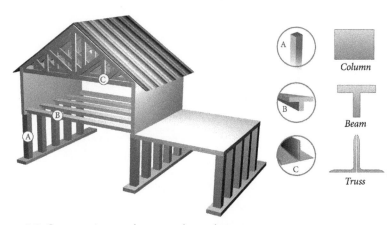

Figure 2.2 Cross-sections and structural members.

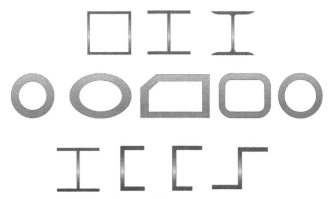

Figure 2.3 Various geometrical shapes of cross-sections.

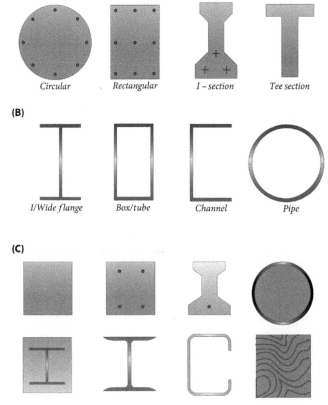

Figure 2.4 Cross-sections made of different materials. (A) Concrete sections, (B) Steel sections, (C) Other materials/composite materials

- rectangular shapes
- flanged shapes, angular corners
- flanged shapes, rounded corners
- circular, elliptical, regular polygonal shapes
- general polygonal forms
- boxed and tubular sections
  The section shapes can also be classified in terms of their symmetry:
- Doubly symmetrical, which are symmetrical about both orthogonal axes
- Axisymmetric, which are symmetrical about one of the axis
- Unsymmetrical, which are not symmetrical about any axis
- Asymmetrical, which are antisymmetrical about the axis

Classifications Based on "Compressed Zone"

Another type of characterization of cross-section is based on the dimensions and stiffness of the parts of the section that are in compression. This classification is generally applicable to steel and other metal sections and includes

- slender sections, where buckling failure of the section parts occurs before material yielding;
- compact sections, where material yielding occurs before buckling failure; and
- plastic section, where only material yielding and failure can occur, and buckling is prevented.

Slender Sections

If no part of the section can reach the prescribed strain (usually the yield strain) before reaching the local buckling stress, or if the buckling strain for all parts of the section is smaller than say, yield strain, then the section is considered slender. In such cases, the usual assumption of linear strain distribution and plane sections is invalid and cannot be used as the basis for determining stresses or stress resultants. This also means that the geometric properties of the entire section cannot be used for determination of the cross-section elastic response. Most of the cold-formed sections, using metal plates or sheets, fall into this category.

Compact Sections

Sections in which all parts or points on section can reach the yield strain before reaching the buckling strain are often termed as "compact." However, with increased deformations, some components or points will undergo buckling and will not be able to form a "plastic" strain flow. Most of the hot-rolled steel sections fall into this category.

Plastic Sections

If all the points and components in the cross-section can not only reach yield strain without buckling, but also continue to remain unbuckled and in plastic state of strain flow until all points or components in the section have at least reached the yield strain, then such sections are termed as plastic sections. Some of the hot-rolled sections, especially the heavy H or W sections, or sections made up of thick plates can behave plastically.

Classifications Based on Material Composition

This is the most extensively used classification that affects not only the application of the cross-section, but also divides the theoretical treatment, design procedures, and usage of the members in various structure types. The material classification is summarized in Fig. 2.4. As can be seen, the sections can fall into the following main categories of materials (see Fig. 2.5):
- Unreinforced, reinforced concrete
- Prestressed concrete
- Concrete filled tubes
- Fiber-reinforced concrete
- Ferrocement
- Steel—concrete composite
- Hot-rolled steel, cold-formed steel
- Structural timber

Classifications Based on Method of Construction

The cross-sections can also be classified based on the way they are constructed or fabricated and put together. This may include the following:
- Cast-in-place using forms
- Precast using forms
- Extrusion of metals or other materials
- Hot-rolling of metals
- Cold-forming of metal sheets
- Built-up using shapes and plates, through welding, riveting, or bolting
- Encased into other shapes
- Jacketed on top of other shapes
- Extrusion of concrete members
- Using spraying or splashing of materials

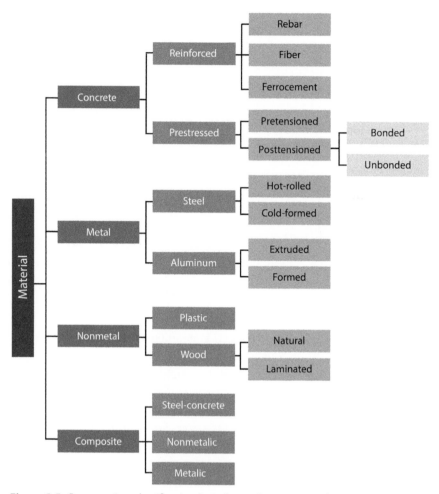

Figure 2.5 Cross-section classification based on primary material composition.

Fig. 2.6 shows some of the typical shapes formed through various methods of construction.

DEFINITION AND REPRESENTATION
Overview

The cross-section for any structural member are defined by two separate parameters, i.e.,

- The geometry of the shape (or sets of shapes) and
- The materials they are made of.

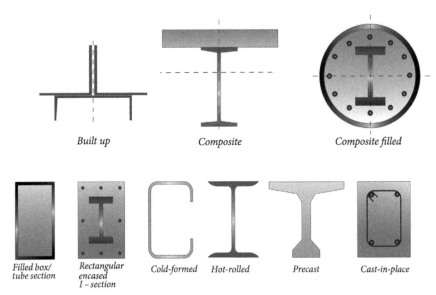

Built up   *Composite*   *Composite filled*

Filled box/   *Rectangular*   *Cold-formed*   *Hot-rolled*   *Precast*   *Cast-in-place*
tube section *encased*
                I – section

Figure 2.6 Some examples of composite and encased sections.

For very simple cross-sections, the geometry and material can be defined at the same time. For example, a steel section can be specified as an I-section W12 × 18 of Grade A-36 or a similar terminology. This definition, together with the information about standard shapes and material specifications, is mostly sufficient for the purpose of defining the properties needed for analysis and design. However, for reinforced concrete and general composite cross-sections, much more information is needed. Each cross-section is unique in some respect such as in terms of shape, concrete strength and properties, distribution and properties of reinforcement, the amount of confinement, the age of concrete, the long-term effects of creep, and shrinkage. Now, we will discuss some of the ways the cross-sections can be defined.

Defining Geometry

From the structural engineer's point of view, the cross-section geometry can be defined in several ways based on the practical applications, such as

- standard sections,
- parametrically defined simple sections,
- built-up and composite sections, and
- sections made up to complex arbitrary shapes.

The standard Cross-Sections

These are sections that have been standardized by the manufacturers, design codes, the construction industry, or by project specifications. The most common example of standard sections is the rolled steel shapes. Almost every country has a local list of standard sections. The section database from AISC, BS, JIS, DIN, and similar standards are quite intensive in providing the detailed dimensions and properties of various steel sections. Another type of standard sections are the precast concrete sections used in prestressed girders, piles, and slabs. The use of these standard sections for the design of members is relatively easier because the dimensions, properties, both geometric and material, and sometimes even the capacities are precomputed and are well documented. The following is a list of typical standard sections, with geometry and material specifications, some of which are shown in Fig. 2.7:

- Hot-rolled steel sections (I, W, C, L, O, etc.)
- Cold-formed steel or metal sections (C, Z, etc.)
- Reinforcing bars (6–40 mm in dia)
- Precast girders (I, T, U, etc.)
- Hollow core slabs and similar components
- Precast concrete piles of various shapes
- Metal deck sections, corrugated roofing sheets

The Parametrically Defined Simple Sections

These sections, often made up of single shapes, are defined by a few dimensions and parameters. For example, circular shape defined by a diameter, or a rectangular shape defined by a width and a height. The engineers or structural designers are used to dealing with these shapes in a more or less standard manner, similar to the standardized shapes and sections. The main difference is that the general parametric shapes have no limit on the value of these parameters, whereas the dimensions of standardized sections are fixed by the manufacturer and availability of sections. Common parametric shapes include Tee, I, L, pipe, etc., in addition to the rectangular and circular shapes. The use of these parametric shapes in section design is quite common. Several design aids and computer programs have been developed that use these parametric shapes. For example, the interaction curves in several handbooks are generally developed for normalized properties and dimensions of rectangular and circular sections. Fig. 2.8 shows some of the parametrically defined sections.

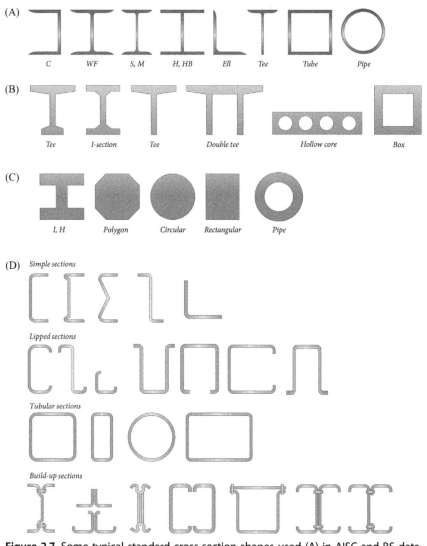

Figure 2.7 Some typical standard cross-section shapes used (A) in AISC and BS databases, (B) in precast, prestressed girders, and slabs, (C) in precast concrete piles, and (D) cold-formed steel sections.

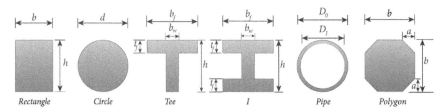

Figure 2.8 Some typical parametrically defined cross-section shapes.

The Built-Up and Composite Sections

Often the simple shapes are combined together to create built-up and composite sections, to increase the strength or stiffness or both. The difference between the built-up sections and composite sections is that the former are made up of shapes having the same material properties, whereas the later are made up of shapes having different material properties. The term "built-up" is often used for steel sections made up of either using several standard sections together or by using standard sections and steel plates or even made up entirely from plates. The built-up and composite shape definition is incomplete unless the material properties of each part are also defined. Figs. 2.9 and 2.10 show some examples of the built-up and composite sections respectively.

Sometimes concrete cross-sections are also made from two different concrete types. In this case, the cross-section is treated as a composite again. The use of different concretes may be due to strength considerations or confinement considerations. A common example is precast concrete girders, used to support cast-in-place concrete slabs, often of lower strength and of different age. An existing section may be strengthened or retrofitted with a higher strength concrete, or a confined concrete core

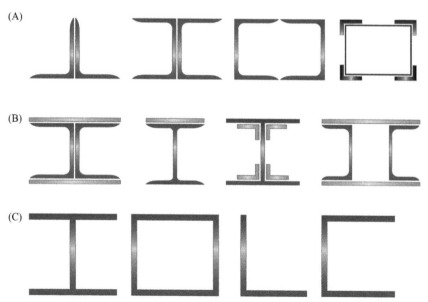

Figure 2.9 Some typical built-up shapes and sections (A) made from standard shapes, (B) made from standard shapes and plates, and (C) made from plates only, by welding.

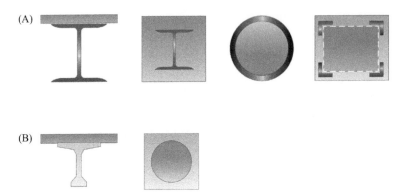

Figure 2.10 Some typical composite sections. (A) Concrete–steel composite; (B) concrete–concrete composite.

may be encased by an unconfined cover zone. Not many computing or design aids and tools are available at this time to analyze or design general composite cross-sections. It is relatively easier to handle sections where shapes of different materials are completely contained in another shape than partially overlapping shapes of different material properties. The definition and analysis of such shapes is further complicated if holes are also located inside the composite sections.

Complex and Arbitrary Shapes

In many projects, there is a need to use members with cross-sections that are of unusual or complex configuration and composition. Bridge piers, and even sometimes building columns, especially, shear walls, may have shapes and sections that do not fall in any of the standard or parametrically defined shapes. For the purpose of analysis and design, such sections can either be modeled using the general polygons, or curved segments or a combination of basic parametric shapes, merged together to form complex shapes. Some of these are shown in Fig. 2.11. Handling of these complex shapes in a generalized manner will be specifically discussed in this chapter.

Defining Materials

Why Material Behavior is Important?

Defining the materials in the cross-section is just as important as the definition of the cross-section shape. As discussed in Chapter 1, Structures and Structural Design, the term "structural material" means the material for which mechanical properties are usually defined for the purpose of

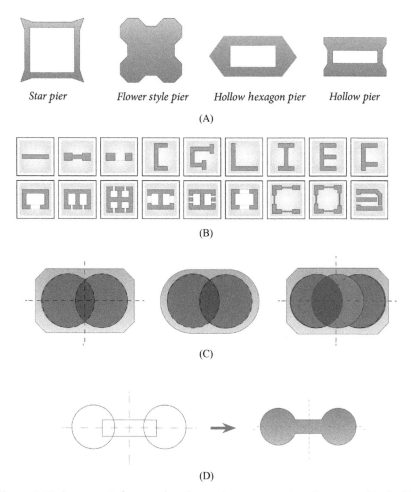

Star pier   Flower style pier   Hollow hexagon pier   Hollow pier

(A)

(B)

(C)

(D)

Figure 2.11 Some typical unusual and complex cross-section shapes used in buildings and bridges. (A) bridge piers (CSI Section Builder), (B) shear wall stacks (ETABS 2015), (C) CALTRANS bridge piers (CSI Section Builder), and (D) merging of shape (CSI Section Builder).

structural analysis and design. The constituents of a construction material, whose individual properties are not required, cannot be referred to as structural materials. For example, cement alone is not a structural material but concrete is an important and complicated structural material. Steel is an example of a simple and homogeneous structural material. Currently, material modeling is considered one of the most difficult parts in the field of structural analysis, modeling, and design. The most important material property, which we need for cross-section analysis is its stiffness which should not be confused with strength. Material lies at the base of understanding complete structural behavior. As described in Chapter 1, Structures and Structural Design, the structure can be considered as a big spring, having a stiffness coming from its component members. These members primarily get their stiffness from cross-sections which ultimately are made up of materials. Similarly, most of the failure or limiting criteria are based on the stresses and strains computed for the materials in the section and hence form the basis for design.

Basic Properties of Materials

For structural analysis, a material is defined in terms of its properties. Several types of properties are used to model the materials, ranging from simple stiffness and strength limits to full anisotropic, visco-elastic, or visco-plastic property matrices. The properties that are relevant for the basic analysis member design can be summarized as:

- modulus of elasticity, E
- Poisson's ratio, ν
- limiting strain, ε_{max}
- characteristic strengths such as f_c', f_y, f_u, f_{cu}
- stress—strain relations
- coefficients of thermal expansion
- mass density
- creep and shrinkage coefficients
- relaxation properties
- ductility ratios
- shear modulus, G (derived from E and ν)

Directional Behavior-Based Classification of Materials

Three types of material classifications are often considered to simplify the definition and computational complexity for a particular structure type and response needs. These are shown in Table 2.1.

Isotropic materials are those which have same behavior in all directions and properties in one direction are uncoupled with the properties in other direction. Most materials are assumed to be strength isotropic but actually are not. For example, in analysis of reinforced concrete beams and columns, since we are only interested in one directional loading, we generally assumed them isotropic. The loading in other directions is either not possible or not of significant interest. In this book, we will be mostly

Table 2.1 Directional Behavior-Based Classification of Materials

Isotropic	• Same behavior in all directions • All directions uncoupled	• Steel • Material used in line type of members (beams, columns, etc.)
Orthotropic	• Different behavior in orthogonal directions • Behavior is uncoupled	Metal deck, reinforced concrete slabs
Anisotropic	• Different behavior in three directions • Behavior is coupled	Soil, some types of rubber

dealing with isotropic materials used in cross-sections of line-type members.

On the other hand, orthotropic materials have different behaviors in orthogonal directions. The behavior is still uncoupled, i.e., properties in one direction are independent of the properties or stress state in other directions. An example of such behavior is concrete slab having different reinforcement in both directions but the behavior is independent. Other examples include steel deck, precast slabs, and wood. In structural timber, the tensile strength along the wooden fibers is very high while it exhibits a very low tensile strength in the direction perpendicular to fibers.

The third category is anisotropic materials which not only have different behavior in all directions, but also the behavior is coupled. Most of the real-life materials are ideally anisotropic, but for the sake of simplification are assumed to be either orthotropic or isotropic (depending upon their intended use and expected actions). A common example of a highly anisotropic material is soil. In fact, concrete also is anisotropic but is generally assumed otherwise for simplicity sake. Unreinforced masonry is also anisotropic and is considered among the difficult materials to deal with.

For most of the practical purposes, for a beam-type member (where only three stresses and strains are of importance), the full relationship between stress and strain components (represented in the form of a 6×6 stiffness matrix, as shown in Fig. 2.12) can be simplified by assuming the material to be isotropic, thereby removing the directional coupling of properties. By introducing this simplification (neglecting the stresses in directions not under consideration, as well as neglecting the higher powers of Poisson's ratio), we can always get the simple relationships between stress and strain involving only E and G. It means no matter how complex and detailed mathematical formulation we have, for the simplest or uniaxial case, we can always simplify back to original form of Hook's law.

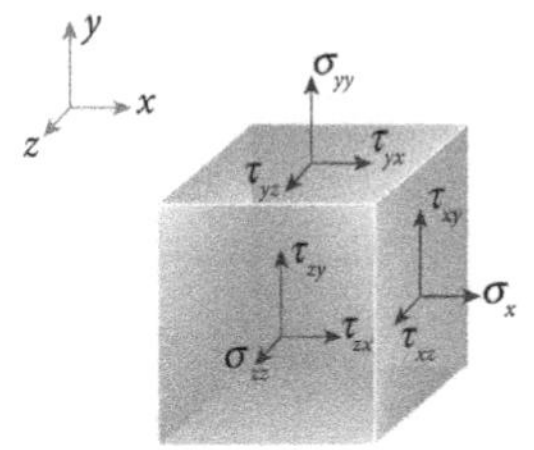

$$\begin{bmatrix} \sigma_x \\ \sigma_y \\ \sigma_z \\ \tau_{xy} \\ \tau_{yz} \\ \tau_{zx} \end{bmatrix} = \frac{E}{(1+v)(1-2v)} \begin{bmatrix} 1-v & v & v & 0 & 0 & 0 \\ v & 1-v & v & 0 & 0 & 0 \\ v & v & 1-v & 0 & 0 & 0 \\ 0 & 0 & 0 & \frac{1-2v}{2} & 0 & 0 \\ 0 & 0 & 0 & 0 & \frac{1-2v}{2} & 0 \\ 0 & 0 & 0 & 0 & 0 & \frac{1-2v}{2} \end{bmatrix} \begin{bmatrix} \varepsilon_x \\ \varepsilon_y \\ \varepsilon_z \\ \gamma_{xy} \\ \gamma_{yz} \\ \gamma_{zx} \end{bmatrix}$$

Figure 2.12 The relationship between stress and strain in a three-dimensional isotropic material (generalized form of Hook's law with a constant E and v in all directions).

Is it Easy to Determine and Define Material Properties?

At a uniaxial material level, Hook's law states that the elastic modulus can simply be determined by dividing the stress with strain value at a particular loading level resulting in a single constant numerical value. However, for an actual three-dimensional block of any anisotropic material, elastic modulus is not just a single constant number (unlike Fig. 2.12). For such a general case, there would be many stress and strain components (corresponding to each direction or face of the block) and a single value of E is by no means expected to accurately relate all those stress and strain components.

In fact, to accurately describe the complete stress state of a three-dimensional material block, we need at least three axial stress and three shear stress components along with their corresponding strain components. Therefore, we need to have six elastic moduli values (corresponding to three normal planes and three shear planes) to relate these six stress components with their corresponding strains. The generalized stress–strain matrix equation with a constant E (Fig. 2.12) will no longer be applicable in such cases. It is also important to note that these stress components are also not independent of each other, owing to Poisson's effect. Stress in one direction also produces strain in perpendicular direction making the behavior even more complicated. The ratio of strain in transverse direction to the longitudinal strain (or in the direction under consideration) is termed as "Poisson's ratio." The modulus of elasticity for a shearing or torsional force is referred to as shear modulus or modulus of rigidity (often denoted by G). On similar lines to axial stress/strain, it is defined as the ratio of shear stress to the shear strain. In short, for real-life analysis cases, we may have to deal with a number of material properties (not just a constant E) and stress/strain components.

Even for simple material behavior predominantly in one direction, the second part of difficulty lies in the definition of the material properties due to the fact that (in a particular direction) actual relation between stress and strain is not linear elastic throughout the loading history. With increasing strain, the stress starts deviating from original straight line and upon unloading it follows a different trend as compared to loading. Linear relationship lasts only up to a certain level of strain and the complete relationship between stress and strain depends on basic material composition, initial conditions, direction and history of strain, time since initial strain, temperature, rate of strain change, and velocity and acceleration of structural material. The value of Young's modulus changes with variation in any of these factors. In fact, if the structure is assumed linear (constant E), analysis can never predict its collapse or failure and will provide the deformations corresponding to all input levels of applied loading (without entertaining the fact that the real structure might not be able to sustain these loads).

We can test the real members or material samples in laboratory to determine actual relation between stress and strain. However, it is not practically feasible and realistic to perform actual testing every time we have to analyze a structure. We need to be able to standardize and predict the material stress–strain behavior which serves as an input for stiffness calculation at higher analysis levels.

In short, the primary material inputs for structural analysis include the modulus of elasticity at different levels of strain and for each stress component, Poison's ratio, shear modulus, and coefficient of thermal expansion. We may also require some specific properties like relaxation, creep, and fatigue. Relaxation is a gradual change of stress without change in strain. For example, stresses in prestressing wires decrease gradually due to relaxation phenomenon. In concrete, a rather opposite phenomenon (in compression) is very common, referred to as creep. Similarly, concrete shrinkage (both early and long term) may also govern the important response parameters of structure and may need to be known prior to analysis. Fatigue resistance is also another important property which may govern the response and capacity of structures subjected to repeated cyclic loading. After each cycle, the stiffness and/or strength of material tends to degrade significantly. Consideration of all these properties makes the material modeling one of the most cumbersome processes in structural idealization. For design, we may also need the failure criteria, yield and failure stresses, as well as the yield and failure strains in each direction.

The Concept of Specific Length

Can the size of a structure be increased indefinitely for it to be able to carry its own weight? How long can a bar of uniform cross-section be, before it breaks due to its own weight? The first person who attempted to answer this question was Galileo Galilei (AD 1654—42) who introduced the concept of "specific strength" of a material. It simply states that there exists an absolute limit to the length that a bar can attain without breaking under its own weight. Generally, the larger the sizes of a members in a structure are, the larger is the proportion of its own weight to the total load which it can carry. For structures subjected to any loading (other than self-weight), as the size of cross-section increases, its strength increases with the square of the ruling dimensions, while the weight increases with its cube. So depending upon the type of structure, there is a maximum possible size beyond which it cannot carry even its own weight. The specific length of a bar can easily be determined by equating the weight of the bar to its tensile strength (see Fig. 2.13). This concept makes it impossible to construct structures of enormous size. Natural structures (trees, mounds, animals, etc.) also have a size limit. An interesting observation is that proportions of aquatic animals are almost unaffected by their size because their weight is almost entirely supported by buoyancy. An important application of specific strength concept is in the aircraft manufacturing industry. Aircrafts, in addition to external loads, must be capable of being raised into the air under their own weight, requiring the usage of materials with high strength-to-weight ratios. Wood has a relatively higher strength-to-weight ratio especially in tension and was extensively used in early planes. Various aero-modeling clubs use Balsa wood (because of its high strength-to-weight ratio) to make model planes. Similarly, in wind tunnel testing, building models are often made using lightweight Balsa wood to minimize the effects of mass in the model.

The design of long span bridges (with span length > 50 m) are mostly proportioned to carry their own weight. The live load (vehicle loadings) is only a small proportion compared to the self-weight of bridge. As the structural engineering profession is advancing toward more efficient structural systems, engineers are looking for materials that are very light yet can withstand large amount of applied loading. Example of such ultra-strong materials include carbon fibers and carbon nanotubes.

In calculation of specific length, it can be observed that cross-sectional area and weight are related to each other. It means increasing the cross-sectional area does not always help in making the structure strong. An important analogy of this case can be given to punching failure of concrete slabs. For example, in flat slab design, engineers usually have problem with punching shear and the general solution to this issue is often recommended in terms of increasing the slab thickness. The direct relation between the size and weight suggests that increasing the slab thickness may not always help in avoiding punching shear failure (as the shear force is also increased). The overall effect may not be as useful and effective as it seems.

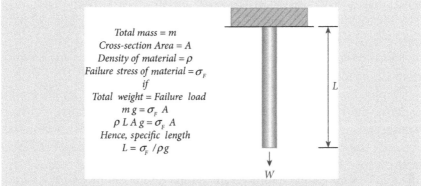

$$Total\ mass = m$$
$$Cross\text{-}section\ Area = A$$
$$Density\ of\ material = \rho$$
$$Failure\ stress\ of\ material = \sigma_F$$
$$if$$
$$Total\ weight = Failure\ load$$
$$m\,g = \sigma_F\,A$$
$$\rho\,L\,A\,g = \sigma_F\,A$$
$$Hence,\ specific\ length$$
$$L = \sigma_F\,/\rho g$$

Figure 2.13 Some concept of specific strength and its implications.

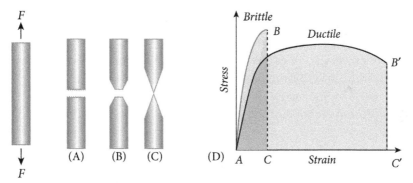

Figure 2.14 Some brittle and ductile material behavior. (A) brittle fracture; (B) ductile fracture; (C) completely ductile failure; (D) typical stress—strain curves of brittle (ABC) and ductile (AB′C′) materials.

Brittle and Ductile Materials

Another classification of materials is possible based on their ability to deform before complete failure. The materials which can undergo strain levels significantly higher than yield strains without complete failure are referred to as ductile. This is an important property in terms of safety and warning before failure. Mathematically, ductility is defined as the ratio of ultimate to yield strain. Steel and some other metals are common examples of ductile materials. On the other hand, those materials which fail suddenly (without any warning) right after their yield limit are termed as brittle materials. Fig. 2.14 shows the typical fracture and corresponding stress—strain relationship of brittle and ductile materials. The design rules as well as behavioral expectations for both type of materials are different.

Unreinforced concrete is a common example of a brittle material. The ductility value for steel may go up to 25 while for concrete it is usually in the range of 1—2. It is important to note that ductility is also temperature-dependent. At low temperatures, steel becomes harder and relatively brittle compared to higher temperatures.

As discussed earlier, for a three-dimensional block of a real-world material, the behavior for each direction (or each stress component) will be different. It means we will require a different stress—strain curve relating each stress component with corresponding strain component. Some of them may be brittle, while others may be ductile; some may exhibit linear behavior, while some can be nonlinear inelastic depending upon the direction of loading; some may exhibit good cyclic behavior (upon repeated loading and unloading cycles), while some may perform poor. So the same material block will have a variety of behaviors depending upon the direction and type of applied action and the deformation. Moreover, properties in one direction may be affected by the strains/ properties from the other direction. For example, a material which is ductile otherwise may become brittle for a certain direction of loading and vice versa. This interaction of stress and strain components also provide us an opportunity to modify the properties in a certain desired direction by changing some factors on perpendicular directions. For example, concrete which is a brittle material otherwise can be forced to behave as a ductile material by restraining the stresses in perpendicular direction. This idea (referred to as concrete confinement) will be discussed later in detail.

Does being brittle or ductile have anything to do with the strength of material? Does ductile material mean low in strength (because by definition, it can deform more)? Does brittle material mean high strength (as it does not deform before failure)? The answer to all of these is, No, the strength and ductility of a material are two different properties which cannot be related and should not be confused with each other. A ductile material may have a low strength but is expected to sustain large deformations before collapsing, "warning" the people of the possible failure. A brittle material may have much higher strength but cannot provide any failure indication even a short while before actual collapse. However, it is possible to use a brittle material and yet make the cross-section ductile. Similarly, it is possible to use a brittle cross-section and yet make the overall structure ductile. This is possible because ductility can be added at higher levels contributing to global stiffness. The key question is where to add ductility (and in what form)? However, it should be noted that the benefit of having ductile materials at first hand cannot be undermined.

It is always best to start with ductile materials if the ultimate goal is to achieve ductile performance against applied actions.

The biggest advantage of using ductile materials is that they provide the opportunity for redistribution of forces. Most statically indeterminate structures are designed to respond in the elastic range under service loads but, given the uncertainties in real strength of material, structural type, magnitude of loading, and accidental actions, the structure can be subjected to inelastic deformations. In such case, the ductile materials will allow for redistribution of stresses. This provides both increased resistance to failure as well as warning before collapse, which is not possible in the case of brittle materials.

Classification Based on Stress—Strain Behavior

Depending upon the exhibited stress—strain behavior, any material can be classified into one of the four major categories including linear elastic, linear inelastic, nonlinear elastic, and nonlinear inelastic. Remember that the Hooke's law states that within the elastic limits, the stress is proportional to the strain.

Some important behaviors are described below.

1. *Linear elastic material:* A linear elastic material is one in which the strain is proportional to stress forever as shown in Fig. 2.15A. Both "loading" and "unloading" curves are same (straight lines).
2. *Linear inelastic material:* A linear inelastic material is one in which the strain is proportional to stress as shown in Fig. 2.15B. However, "loading" and "unloading" curves are not same (although straight lines). The unloading line does not come back to initial loading point and hence exhibit some residual deformation.
3. *Nonlinear elastic material:* For a nonlinear elastic material, strain is not proportional to stress as shown in Fig. 2.15C. Both "loading" and "unloading" curves are same but are not straight lines. Such materials return back to the same initial dimensions (following nonlinear behavior) without any residual strains. A good example of such material is rubber which cannot be easily forced to fail just by repeated twisting. Another important example of such materials are shape-memory alloys.
4. *Nonlinear inelastic material:* For a nonlinear inelastic material, strain is not proportional to stress as shown in Fig. 2.15D. At the same time, "loading" and "unloading" curves are not same in this case (exhibiting some permanent deformation). This category of materials is the most difficult to deal with. A common example of such material is reinforced concrete.

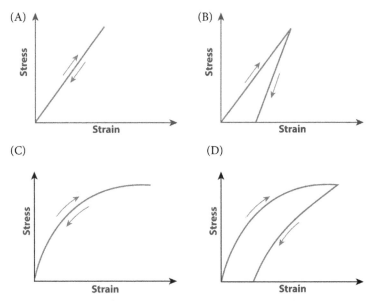

Figure 2.15 Material classification based on stress—strain relations. (A) linear elastic material, (B) linear inelastic material, (C) nonlinear elastic material, (D) nonlinear inelastic material.

Generalized Stress—Strain Curves

Since for most materials, it is difficult to describe the entire stress—strain curve with simple mathematical expressions, the behavior of the materials is generally represented by an idealized stress—strain curve. The stress—strain curve is generally used as the basis for the generation of capacity interaction curves, action deformation curves, and actual stress profiles of cross-sections for various loading or deformation conditions. A general stress—strain curve encompasses several material properties and provides a detailed account of material response under the applied loading in considered direction. The following properties can be inferred from any general stress—strain curve:

- The modulus of elasticity at various levels of strain or stress;
- The characteristic strength;
- The limiting strain values;
- The ductility ratio.

It is quite possible that the basic material of which structural elements are made is linear elastic but the overall structure behaves nonlinearly because the nonlinearity from other components of global stiffness may come into play. In such case, the linear elastic material behavior will contribute from the starting level, but at cross-section and member levels, behavior may change depending upon other factors. For example, the cross-section may crack making the behavior nonlinear and the member distribution or overall structural configuration may impart nonlinearity to global behavior. The only way full linear elastic structural behavior can be ensured is to make sure that all of these levels will exhibit linear elastic behavior. In Chapter 1, Structures and Structural Design, the overall hierarchy of structural stiffness, its components, and levels were discussed. Here, a brief introduction of some of stress—strain relationships for two most important engineering materials will be presented, with specific application to beam- and column-type members.

Idealized Stress—Strain Curves for Steel

The stress—strain curve for steel is fairly complex when considered in its entirety. It may have at least three regions: the elastic region, the plastic region, and the strain hardening region. However, in most cases, a commonly used idealization is elastic—perfectly plastic (nonstrain hardening). In fact, a lot of materials can be represented using this simplified behavior which consists of two straight lines, the first showing the elastic behavior and the second (horizontal) line representing perfectly plastic post-yield behavior for infinite strain (and constant stress). The stress—strain curves for mild and prestressing steels and typically used elasto-plastic behavior are shown in Figs. 2.16 and 2.17.

Idealized Stress—Strain Curves for Unconfined Concrete

The determination of concrete properties has been a focus of intensive research for a considerable time. The stress—strain curves for concrete has been developed and proposed by numerous researchers, design codes, and technical committees. Almost every concrete design code proposes the

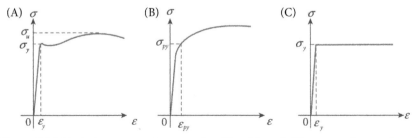

Figure 2.16 Typical stress—strain curves for (A) mild steel, (B) prestressing steel, and (C) simplified curve.

use of a certain stress—strain model to be used in the axial—flexural analysis and design. It has been known for a while that the stress—strain curve for concrete follows a combination of straight lines and parabolas of various order. The stress—strain curve directly translates to the stress diagram when linear strain distribution is assumed. Figs. 2.18 and 2.19 show the typical stress—strain curve for unconfined concrete and stress diagrams across the depth of the cross-section, respectively.

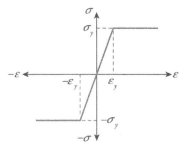

Figure 2.17 An idealized complete elasto-plastic stress—strain curve for steel.

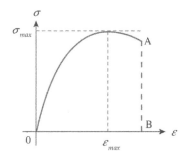

Figure 2.18 Typical stress—strain curve for concrete.

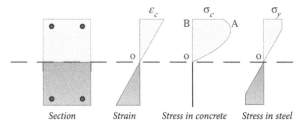

Figure 2.19 The stress—strain curve of material translates to stress diagram on the cross-section.

A large number of stress—strain curves can be found in literature for concrete. In fact, almost every major design code has adopted its own specific stress—strain curve. There are also several specialized stress—strain curves for unconfined concrete, semiconfined concrete, or fully confined. A symbolic comparison is shown in Fig. 2.20. The tension capacity of concrete is generally ignored and not represented in stress—strain curves.

Due to difficulties in the manual computation of area and centroid of the stress block on the compressed portion of cross-section, several simplifications are made to the basic stress—strain curves. The most popular is the conversion of parabolic stress—strain curve to a rectangular block, used in the American codes. This translates into a rectangular stress block on the compressed portion of the section. This conversion is based on the assumption that if the area of the rectangular stress block and its centroid is the same as the original parabolic stress block, then the value of the stress resultants, and hence, the interaction curves would not be affected.

It is obvious from Fig. 2.21 that for the actual compressed area which is not rectangular, the rectangular stress block approximation may not provide the same location of resultant force. These days, most of the computation for cross-section capacity is done by computer programs; it therefore makes sense to use the more refined stress—strain curves, derived from the test directly rather than the simplified versions. The detailed discussion on the development of general stress—strain curve for concrete, together with the short- and long-term effects can be found in several available textbooks on design of reinforced concrete structures (MacGregor, Wight, Teng, & Irawan, 1997; Nawy, 2005; Nilson, Darwin, & Dolan, 2006).

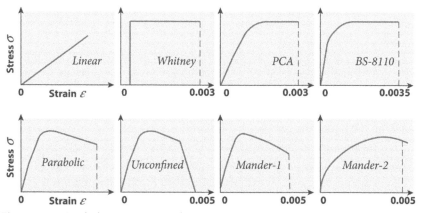

Figure 2.20 Symbolic comparison of various stress—strain curves for concrete.

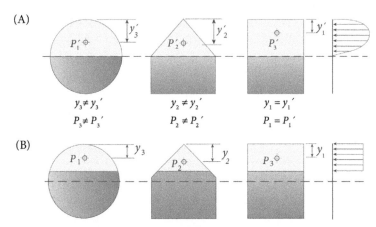

Figure 2.21 The simplified stress may not give the same results as original stress block: (A) original stress block; (B) simplified stress block.

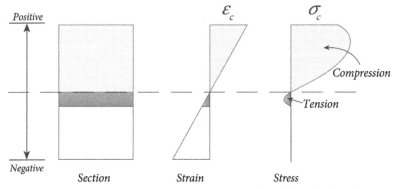

Figure 2.22 Tension capacity of concrete can be considered by including the tension part of the stress−strain curve.

Generally, the tension capacity of concrete is ignored in computing the axial−flexural capacity of reinforced concrete or composite cross-sections. However, in the general formulation for computing the axial−flexural capacity, it is easy to include the effect of tension capacity of concrete. The tension part of the concrete stress−strain curve is included in the computations, and the usual calculations are performed. Fig. 2.22 shows the stress diagram for concrete including the tension part.

The discussion about the materials and their properties presented in this chapter is neither comprehensive nor exhaustive, and there are several well-written texts (Ashby & Jones, 2012; Mehta, 1986; Neville, 1996, 2006) on this subject. The information presented here is just enough to develop an understanding of cross-section response discussed in subsequent parts of the book.

CROSS-SECTION PROPERTIES—AN OVERVIEW

The member cross-section properties are involved in almost all relationships involving actions, deformations, stresses, and strains. They are used for determining the stiffness properties when relating actions with deformations. These relationships are the basis of the stiffness matrix method and the finite element method (FEM) for computer-aided structural analysis and in determining deformations directly from actions. These relationships require basic or intrinsic cross-section properties. The properties are used in two contexts, first to determine the stresses from the actions and second to determine the stresses resultants themselves. The simplest case is determining axial stresses from axial load by dividing it with the cross-sectional area. More complex applications include stress interaction, and shear stress due to shear and torsion.

Difference Between Geometric and Cross-Sectional Properties

Apparently, the geometric and cross-section properties may seem to refer to the same thing. However, there is a basic difference. The geometric properties refer to the properties computed only on the basis of the shape geometry. The shape does not have to be a cross-section of a structural member. For example, a rectangular or Tee shape has an area property; this is its *geometric property*. However, when the shape or combination of shapes is used to define a cross-section of a structural member, then the properties for those shapes become cross-section properties. Cross-section properties must include the physical meaning of the shape(s) and must take into account the material behavior and material properties. For example, if an I-shape is placed inside a rectangular shape, when the rectangular shape is made from concrete and I shape is made from steel, the geometric properties and the cross-section properties are no longer the same. The cross-section properties must now be computed in equivalent terms with reference to a base material, which could either be steel or concrete. In each case, the numerical values as well as the meaning of the values will be different from the simple geometric properties.

Similarly, if a few small circles are placed inside a rectangular shape, unique geometric properties can be computed based on the dimensions of the shape, diameter of circles, and their location. However, if the same shape refers to a reinforced concrete section, with circles representing rebars, the cross-section properties cannot be computed without specifying the level of cracking and the material stiffness. The cross-section properties may either be computed as gross, equivalent, cracked transformed, or effective properties. Each of these will have a different numerical value, as well as meaning.

Role of Cross-Section Properties in Section Stiffness

As briefly introduced in Chapter 1, Structures and Structural Design, the action along each DOF is related to the corresponding deformation by the member stiffness, which in turn, depends on the cross-section stiffness.

Therefore, there is a particular cross-section property corresponding to member stiffness for each DOF. Hence, for the seven DOFs shown in Fig. 2.23, the related cross-section properties are:

- u_z ⇨ cross-sectional area, A_z
- u_x ⇨ shear area along x, SA_x
- u_y ⇨ shear area along y, SA_y
- r_z ⇨ torsional constant, J

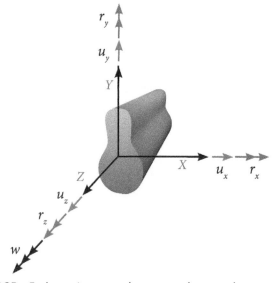

Figure 2.23 DOFs. Each section on a beam member can have seven DOFs with respect to its local axis.

- $r_x \Rightarrow$ moment of inertia, I_x
- $r_y \Rightarrow$ moment of inertia, I_y
- $w_z \Rightarrow$ warping constant, W_z or C_w

Some relationships between actions and deformations involving the material and section properties are shown in Tables 1.3 and 1.4.

Classification of Cross-Sectional Properties

Cross-sectional properties can be categorized in many ways. From the computational point of view, we can look at the properties in terms of
- basic or intrinsic properties
- derived properties
- specific properties for reinforced concrete sections
- specific properties for steel sections

Basic properties are those which are computed directly from the cross-section geometry and material properties. These include:
- the area of the cross-section, A_x
- the first moment of area about a given axis ($A.y$ or $A.x$, etc.)
- the second moment of area about a given axis ($A.y^2$ or $A.x^2$, etc.)
- the moment of inertia about a given axis, I_x
- the polar moment of inertia, I_{xy}
- the shear area along a given axis, SA
- the torsional constant about an axis, J
- the warping constant about an axis, Wz or Cw
- the plastic section modulus about a given axis, Z_P
- the shear center, SC

On the other hand, derived quantities are those which are computed from the basic properties. They include but are not limited to
- the geometric center with reference to the given axis, x_0, y_0
- the plastic center with reference to the given axis, x_p, y_p
- the elastic section modulus with reference to the given axis, s_x, s_y
- the radius of gyration with reference to the given axis, r_x, r_y
- moment of inertia about the principal axis of bending, I_{11}, I_{22}
- the orientation of the principal axis of bending, θ

For reinforced concrete sections, due to special behavior of combined materials for various actions, additional properties may be needed for design. These include
- the cracked section neutral axis depth, c
- the cracked moment of inertia, I_{cr}

- the cracked transformed moment of inertia, I'_{cr}
- the effective moment of inertia, I_{eff}
- the torsional stiffness parameters, C_t
- torsional stress flow parameters, A_{cp}, P_{cp}

The prestressed concrete sections behave differently than the reinforced sections, especially when subjected to bending moments. The prestressing modifies the cracked section properties and also the effective/transformed properties during various stages of casting, stressing, etc. The additional considerations include

- stage-wise and incremental section modulus
- location of force resultant

The steel cross-sections, due to their slender nature, are often designed for local buckling consideration. This requires the computation of additional properties such as

- the lateral torsional buckling constant, r_T
- the b/t_f (width to thickness) ratio for flanges
- the h/t_w (height to thickness) ratio for webs
- the warping constant, C_w
- effective compressed zones

Reference Axis

Before we begin a detailed discussion of the cross-section properties, we need to define a frame of reference. The cross-section dimensions as well as properties are generally defined with respect to a local axis system.

The entire structural definition, modeling, analysis, and result interpretation in computer-added systems are carried out with reference to some coordinate system. In fact, several reference coordinate systems are used at various levels and for various purposes of analysis and design. Before proceeding with the calculation procedures for computing cross-sectional properties, it is important to review coordinate systems relevant to cross-sections and members.

The Global Axes

This is the main coordinate system used to define the overall structure. It is generally a right-hand orthogonal system consisting of $X, Y,$ and Z axes. Sometimes the Z-axis is used as global up and sometimes Y is used as the up axis. In this book, the Z-axis will be used as up, with the XY-plane defining the "plan" of the structure. The structural members are located and referenced with respect to the global axis and global origin.

The Local Member Axes

Each member in the structural model is defined with respect to its own local coordinate axis system. For one-dimensional, or beam-type members, this is again a right-hand, orthogonal system. For the purpose of defining the DOFs, stress, strain, and actions, the x, y, z-coordinate notation are used.

These x, y, z-axes are directional axes, with proper sense of positive and negative direction and are used for the reference and definition of vector quantities. However, for the purpose of computing and defining the cross-section properties, which are basically scalar in nature, a secondary, nondirectional axis system can be defined. This system is referred to as a 1, 2, and 3 system located at the member cross-section centroid, where 1 is along the member length and 2 and 3 are on the plane defining the cross-section. All cross-section properties are referenced with respect to this axis system, using 1 in direction out of the section.

The Cross-Section Coordinate Axes

This is a "temporary" two-dimensional coordinate system used to define the coordinates of the cross-section for the purpose of computing cross-section properties. In this book, for computational convenience, this system will be referred to as the $x-y$ system and located at the extreme left-bottom position so that all coordinates in the cross-section are positive and are in the first quadrant.

The intersection of the $x-y$ axes serves as the coordinate origin, which will be used to refer to the geometric, plastic, and shear centers (Figs. 2.24–2.26).

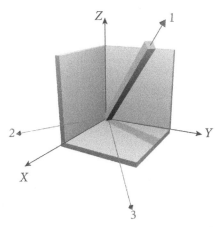

Figure 2.24 Global structure coordinate axis and member local axis.

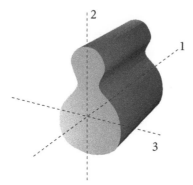

Figure 2.25 For the purpose of defining materials and computing cross-sectional properties, a nondirectional axis system is defined on the local member axis.

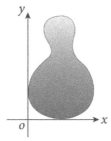

Figure 2.26 The temporary reference cross-section coordinate axes for computation purposes.

BASIC CROSS-SECTION PROPERTIES
Computation of Section Properties

Cross-sectional properties of beam-type members or of one-dimensional elements are the basic information required for structural analysis and member design. Most software for structural analysis and design have the capability to compute these properties automatically for standard shapes. Some programs also have the additional functionality to read-in the properties from standard section databases, especially hot-rolled steel shapes. However, several structures demand special sections not defined in the programs and/or not available in standard databases. This is especially true for odd concrete shapes used as columns in buildings, in bridge piers, for composite sections in bridge decks, and built-up sections in steel structures. Special techniques and tools are needed to compute the properties of these cross-sections.

Several special-purpose "section builder" or "section designer" software (e.g., CSI Section Designer, CSiCOL, ACECOMS Gear, RISASection, Response 2000, ENGISSOL, etc.) are available to compute the properties of general cross-sections. Chapter 7, Retrofitting of Cross-Sections, presents an overview and comparison of some of these software. In this section, a consistent approach will be presented to compute all the cross-sectional properties needed for analysis as well as for design and the necessary mathematical equations and computing techniques.

The properties of "standard" or "regular" shapes can be computed based on their dimensions. Important properties with formulas derived from general equations for some of the most commonly used shapes are listed in Annexure A of this book. For more complex shapes and sections, for which direct formulae cannot be derived conveniently, two numerical methods will be presented later.

A summary of equations and mathematical representation of the basic properties is given here so that the discussion about properties can be self-contained. More detailed information and derivations of the equations can be found in many references (Beer, Johnston, & Dewolf, 2004; Gere & Goodno, 2012). These equations can be used to compute properties directly or implemented into computer programs. It is worth mentioning here that most of the integration shown here can be easily converted into summations and can be implemented in computer applications for convenient computations.

The Modular Ratio

Definition

Modular ratio (n) is the ratio of the elastic modulus of a particular material in a cross-section to the elastic modulus of the "base" or the reference material.

Significance

The concept of modular ratio is very important in the computation of properties of reinforced, prestressed, jacketed, encased, and composite cross-sections. The properties of each component of the cross-section are scaled by the modular ratio of the corresponding material. This is necessary so that the final properties can be multiplied by the modulus of elasticity (E or G) in determining the total cross-section stiffness.

Mathematical Computation

For homogeneous section, the modular ratio is 1 and therefore not used in computations. In a composite section, the modular ratio of a material A with respect to a base material B can be simply determined as

$$n_A = \frac{E_A}{E_B} \tag{2.1}$$

Cross-Section Area

Definition

This is most basic and commonly used property of a shape or a cross-section. It represents the area within the bounds of the geometry of the shape, excluding any holes or openings.

Significance

The cross-sectional area is the simplest of all properties, which simply measures the region enclosed by a boundary. For structural members, this property is a measure of axial stiffness and is involved in all expressions related to mass or weight of the members, axial and flexural actions, deformations, strains, and stresses.

Mathematical Computation

The cross-sectional area of a homogenous section is simply an integration of all the small areas within the shape boundaries (Fig. 2.27):

$$A = \int_A dA \tag{2.2}$$

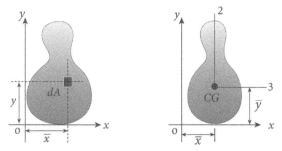

Figure 2.27 Computation of basic properties (left) about the $x-y$ coordinate system and transformation to local 2-3 system (right).

For a composite section, each area components should be multiplied with the modular ratio of the material at that area for determination of other properties.

Moments of Area and Moments of Inertia

Definition
The first moment of area is determined by multiplying the portion of cross-sectional area by the distance of its centroid from any specified axes. It can be taken about the x-axis or about the y-axis or about any arbitrary axis. The second moment of area is simply the product of area and square of distance of centroid of area to an axis. The moment of inertia is the second moment of mass about an axis.

Significance
The moment of area is a very useful property, which is also used to determine other properties. The second moment of area is used to determine the moment of inertia. Generally, the moment of inertia is computed with respect to the local axis of the cross-section, whereas the second moment of area can be computed with reference to any axis. When the area is multiplied by distance of its centroid from two orthogonal axes, this is often called product moment of inertia. The product moment of inertia is zero for shapes that are symmetrical about at least one reference axis whose inertia is being determined.

For member design, the area of cross-section and the moment of areas are involved in the following computations:
- Determining reinforcement ratio;
- Determining basic axial stress using combined stress equation;
- Determining elastic stiffness properties (EI) used in computing elastic buckling load, elastic deformations, effective length factors, frame element stiffness matrix, cracked section properties, etc.

Mathematical Computation
First moment of area about x and y axes:

$$A \cdot x = x \int_A dA \tag{2.3}$$

$$A \cdot y = y \int_A dA \tag{2.4}$$

Second moment of area about x and y axes:

$$A \cdot x^2 = x^2 \int_A dA \tag{2.5}$$

$$A \cdot y^2 = y^2 \int_A dA \tag{2.6}$$

Moment of inertia about x and y axes:

$$I_x = \int_A x^2 \, dA \tag{2.7}$$

$$I_y = \int_A y^2 \, dA \tag{2.8}$$

Product of inertia about xy axes:

$$I_{xy} = \int_A xy \, dA \tag{2.9}$$

Moment of inertia about the local 3−3 axis:

$$I_{33} = I_{xx} - A\bar{y}^2 \tag{2.10}$$

Moment of inertia about the local 2-2 axis:

$$I_{22} = I_{yy} - A\bar{x}^2 \tag{2.11}$$

Product of inertia:

$$I_{32} = I_{xy} - A\bar{x}\bar{y} \tag{2.12}$$

Shear Areas

Definition

Shear area is the area of cross-section parallel to the applied shear force vector (i.e., the area in which primary shear stress is developed).

Significance

The shear area is needed mostly for analysis programs that include the shear deformation in the calculation of member stiffness and for computing average shearing stress. Shear area is often less than the cross-sectional area, and the ratio between the shear area and the cross-sectional area is sometimes referred to as the "shape factor." The shear area takes into account the distribution of shear stresses across the cross-section and may not be the same along different axes if the shape is not symmetrical. The determination of shear area is not a trivial task. The expressions used to compute the shear area for a section are discussed next.

Mathematical Computation

The shear area is computed using double integral as shown in equations below. The manual solution of integration of shear stress distribution described by the shear area equation is difficult for general shapes and cross-sections. Therefore, a numerical integration is often used over the depth of the section along the axis being considered. One way to compute the shear area by numerical integration is to the divide the cross-section into narrow strips and computes the width, area, the first moment of area, and the second moment of area of each strip about any arbitrary axis parallel to the strips. The summation of these properties is then carried out about the shear plane (called Q), which in turn is moved from the top of the section to the bottom. This summation at each level is a measure of the shear stress intensity. The moment of inertia is then divided by the total summation of Q to obtain the shear area. Shear stress distribution for some common cross-section shapes is shown in Fig. 2.28.

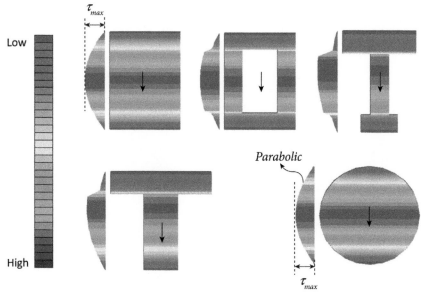

Figure 2.28 Shear stress distribution due to shear along the y-axis in some common cross-section shapes.

Shear area about x and y axes can be computed as

$$SA_3 = \frac{I_{33}^2}{\displaystyle\int_{y_b}^{y_t} Q^2(y)dy/b_y} \tag{2.13}$$

$$SA_2 = \frac{I_{22}^2}{\displaystyle\int_{x_b}^{x_t} Q^2(x)dx/b_x} \tag{2.14}$$

where

$$Q(y) = \int_{y}^{y_t} n_y b_y \, dn_y \tag{2.15}$$

$$Q(x) = \int_{x}^{x_t} n_x b_x \, dn_x \tag{2.16}$$

The notation for these equations are shown in Fig. 2.29.

Torsional and Warping Properties

Definition

The torsional constant is generally needed to determine the torsional stiffness of a member, and in some cases shear stresses due to torsion. Strictly speaking, the torsional constant computed by the elastic theory is valid only for steel sections or other homogenous sections. In reinforced

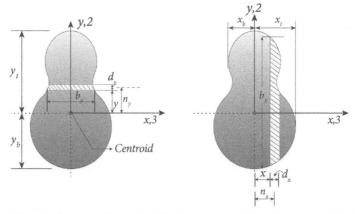

Figure 2.29 The shear area is obtained by integration of the shear stress distribution. The shear stress distribution is obtained by integration of static moment of area above the location on which shear stress is being determined.

concrete, the validity of the elastic torsional constant is limited only to the case when the section is uncracked and is uniformly reinforced around the perimeter. Once the section cracks, either due to axial tension stresses or due to tension produced by shear stresses, the elastic torsional constant become invalid. For example, for an unreinforced concrete section subjected to torsion, as soon as the tensile capacity of concrete is reached it will fail in a brittle manner and the shear plane will be broken. The torsional constant in this case is only valid just before cracking. However, if the beam section is adequately and properly reinforced against shear and axial tension, then after cracking the shear flow will be transferred to the reinforcement, and the section will again possess a torsional stiffness. The actual transition from elastic behavior to post-elastic behavior in this case will depend on several factors, including the presence of shear force, bending moment, axial force, the cross-section shape, the amount and distribution of reinforcement, etc.

For practical applications, most design codes use a space truss analogy to develop the design procedures for torsion. This basically means that after cracking, all concrete in tension is ignored and tension is entirely taken by steel ties, whereas all compression is provided by concrete struts. The detailed discussion of this tie-strut model is beyond the scope of this book. However, the cross-sectional properties needed to determine the torsional stiffness in this case are based on a thin-walled tubular model of the cross-section. The tube is assumed to be centered on the transverse reinforcement cage. The torsional properties of this equivalent tube can then be used to determine the torsional constant. Therefore, it is important that the reinforcement arrangement is decided before the torsional stiffness is computed.

Torsional Constant for Thin-Walled Open Shapes

For thin-walled open shapes, such as hot-rolled steel sections, J can be computed simply as

$$J = \sum \frac{bt^3}{3} \qquad (2.17)$$

where each flange or part is considered as a thin rectangle (see Fig. 2.30).

Significance
The Torsional Constant

Torsional constant is actually a kind of moment of inertia and is needed to compute the torsional stiffness of the cross-sections. The torsional constant is

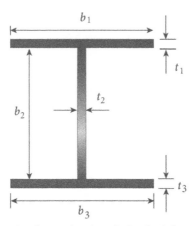

Figure 2.30 Various notation for torsional analysis of reinforced concrete beams.

a very elusive property to determine and to explain. There are several procedures available for computing torsional constant of different shapes and cross-sections. Similar to shear area, the torsional constant also depends upon the distribution of shear stresses. However, unlike the shear area, where the shear stress distribution is defined for different shapes by a single equation, the shear stress distribution due to torsion is highly dependent upon the cross-section shape. In general, if the shear stress distribution function is available, it can be integrated over the cross-section to obtain the torsional constant for a unit rotation. However, the shear stress distribution due to torsion is completely different in the solid sections, in thin-walled open sections and in thin-walled closed sections. No single equation or formula is available to cover all these cases, although there are several numerical methods available to determine the torsional constant. The stress distribution (which is the basis for determining the torsional constant) due to torsion for several types of different shapes and cross-sections are shown in Fig. 2.31.

The Warping Constant
Like the torsional constant, the warping constant is also related to the torsional stiffness of the beam-type members. It is mostly relevant for the members that have small resistance to torsion and whose cross-sections are prevented at the ends from warping. For example, if an I-section is fixed at the ends in such a way that no part of the cross-section can move in any direction, then the stresses due to the prevention of warping will be significant and the warping stiffness needs to be considered. On the other hand, if the beam is supported in such a way that it is prevented

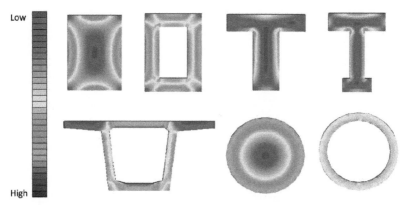

Figure 2.31 Shear stress distribution due to torsion in some common cross-section shapes.

from rotation, but not prevented from out-of-plane cross-section deformation, then warping stiffness is negligible and can be ignored. On the other hand, for the cross-sections that have high warping resistance (large value of warping constant), such as solid circular section or tubular sections, the warping stresses are low in comparison with other stresses and are often ignored. So warping constant is important for members that are restrained against warping and have low warping resistance. Generally, all open and thin-wall walled sections have low warping (and torsional) resistance and if they are restrained against warping, then the stresses and deformation due to warping should be considered.

Mathematical Computation
The General Torsional Equations
Torsional stiffness = Saint Venant's torsional stiffness + warping stiffness

$$S_t = GJ + EC_w \tag{2.18}$$

$$T = G\phi J \tag{2.19}$$

$$\phi = \frac{T}{GJ} \tag{2.20}$$

$$q = \frac{\rho T}{J} \tag{2.21}$$

Solid shapes are the most difficult category of shapes for computing of torsional constant, as their shape can be a circle, ellipse, rectangle, or any other irregular polygon. Approximate formulas are available for rectangular sections and exact formulas for circular sections.

For circular sections:

$$J = \frac{\pi r^3}{2} \qquad (2.22)$$

For rectangular sections, J depends on the ratio of the two dimensions. In general,

$$J = kxy^3 \qquad (2.23)$$

where x is the longer dimension. The values of k for different aspect ratios are listed in Table 2.2.

Sometimes the complex shapes can be divided into a combination of simple shapes for convenient computations. For example, a T-shaped section may be divided into several rectangles. For each rectangle, J is calculated individually using the above equation. The total J of the shape is calculated by summing up J for individual shapes. Since the flanged shapes can be divided in several ways, the lowest value of J should be used.

In the example above in Fig. 2.32, the T-shaped cross-section can be divided into sum of rectangles in two possible ways shown.

Table 2.2 The Values of k for Various Dimensional Ratios of Rectangular Shape

x/y	k
1.0	0.141
1.5	0.196
2.0	0.229
2.5	0.249
3.0	0.263
4.0	0.281
5.0	0.291
6.0	0.299
10.0	0.312
∞	0.333

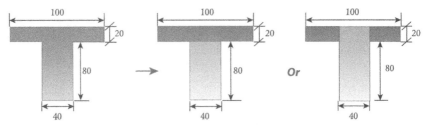

Figure 2.32 The torsional constant of flanged shapes can be approximated by dividing the shape into two or three rectangles, computing the torsional constant for each rectangle, and summing it up. If several options are possible, then the lowest J is taken.

For the first case,

$$J \quad = \quad [0.291(100 \times 20^3)] + [0.229(80 \times 40^3)]$$
$$= \quad 232,800 + 1,172,480$$
$$= \quad 1.41 \times 10^6 \text{ mm}^4$$

For the second case,

$$J \quad = \quad 2[0.196(30 \times 20^3)] + [0.229(80 \times 40^3)]$$
$$= \quad 94,080 + 1,172,480$$
$$= \quad 1.27 \times 10^6 \text{ mm}^4$$

The lowest value of $1.27 \times 10^6 \text{ mm}^4$ should be used.

Finite Element Solution for Torsional Constant

The FEM can also be used for the analysis of cross-sections and determining torsional constant and other section properties. The solution using this approach requires the section to be modeled using finite elements in a computer analysis package. There are a number of specialized software tools available to determine the torsion constant using FEM. These include "Shape Designer" by Mechatools Technologies, "Shape Builder" by IES, Inc., "STAAD Section Wizard" by Bentley, "Section Analyzer" by Fornamagic Ltd., "Strand7 BXS Generator" by Strand7 Ltd., etc.

Plastic Section Moduli

Definition

If we assume that due to bending, half of the section has completely yielded in compression and the other half is completely yielded in tension. The distance between the centroid of the upper half and the centroid of the lower half area (which are also the centroid of the internal stress resultants in this case) is multiplied by the area itself and the product is denoted as the plastic section modulus about the x-axis.

Significance

The plastic section modulus is mostly used in determining the plastic moment capacity of a very compact or thick steel section.

Mathematical computation

The plastic section modulus of a rectangular section with width b and height h about the x- and y-axes can be determined using the following expressions:

$$Z_{Px} = \frac{bh^2}{4} \qquad (2.24)$$

$$Z_{Py} = \frac{hb^2}{4} \qquad (2.25)$$

DERIVED CROSS-SECTION PROPERTIES

Overview

Several cross-section properties can be derived and computed based on the basic or the intrinsic properties, such as the area, moments of area, etc. The derived properties are often needed for specific analysis or design applications.

The Geometric, Elastic, Plastic, and Shear Centers

Definitions

The geometric center (GC) is the center of cross-section without regard to the material properties of its components. This can be viewed as the center of gravity of a cardboard model of cross-section with constant thickness. We can "balance" the cross-section geometry around this point.

The elastic center (EC) or sometimes called transformed center is the center of cross-section with due regard to the material elastic modulus properties of the cross-section components. In other words, this is the geometric center of cross-section, when it is transformed into a single base or reference material using the relevant modular ratio (n) of each component of the cross-section. This can be viewed as the center of gravity of a cardboard model of the cross-section where the thickness of each component corresponds to the n value of the component. For cross-sections made up of a single material, the elastic center is the same as the geometric center.

The plastic center (PC) is the center of cross-section where the area on all sides is equal (irrespective of the distribution or location of the area), with due regard to the material yield properties of the cross-section components. This basically refers to location at which the plastic or yield stress of the cross-section components will change sign to maintain net zero axial stress state.

The shear center (SC) is a point through which, if a force is applied in the plane of the cross-section, does not produce any rotation of the cross-section. Shear center, as the name implies, is related to the center of shear stress resultant on the cross-section. This point may or may not lie within the bounds of the cross-section. For example, for C shape sections, the shear center is outside the shape bounds. The shear center is in line with the plastic center for the axis about which the cross-section is either symmetrical or is asymmetrical. For example, for a uniform rectangular section, the shear center is the same as the plastic center. The shear center for T sections is in line with the plastic center along the y-axis, but away from the plastic center along the x-axis. For the Z section, the shear center is in line with the plastic center along both axes, even though the shape is not symmetrical about either axis (see Fig. 2.33).

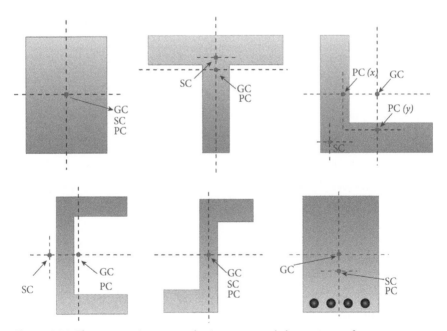

Figure 2.33 The geometric center, plastic center, and shear center of some common shapes.

Significance

The choice of centroid is critical in the design of members, especially when computing the internal stress resultants, and comparing them to the applied actions. The internal stress resultants or capacity and the external actions must be considered with respect to the same centroid; otherwise, the misalignment of the external and internal axial load will produce additional moments, and the direct comparison of the capacity and applied loads would not be consistent. For columns, this behavior is shown in Fig. 2.34. It is obvious that an additional moment M_u must be considered in case (B).

Mathematical Computation

Geometric and elastic centroid:

$$\bar{x} = \frac{\sum A_i x_i}{\sum A_i} \qquad (2.28)$$

$$\bar{y} = \frac{\sum A_i y_i}{\sum A_i} \qquad (2.29)$$

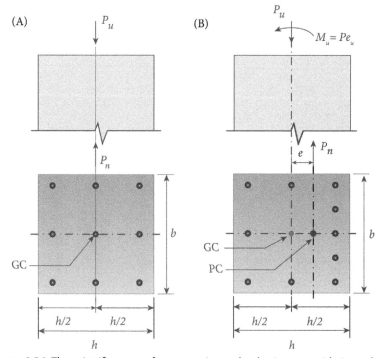

Figure 2.34 The significance of geometric and plastic centroid in columns. (A) Symmetric rebar arrangement; (B) unsymmetric rebar arrangement.

Plastic center:

The plastic center can be calculated by applying force equilibrium conditions to the stresses in various cross-sectional components. Usually it is described in terms of the distance from top (or bottom) of the section to the point at which stresses change their sign to maintain net zero axial stress state.

Shear center:

The shear center is determined by equating the internal moment about centroid caused by shear flow (developed due to applied loading) with the moment due to applied shear force. The moment arm of applied shear force is the location of shear center with respect to centroid.

Elastic Section Moduli

Definition

The elastic section modulus is defined as the ratio of the second moment of area (or moment of inertia) and the distance from the neutral axis to any given (or extreme) fiber. It can also be defined in terms of the first moment of area.

Significance

It is also often used to determine the extreme fiber elastic stresses as well as the yield moment (M_y) of elastic sections such that $M_y = S \times \sigma_y$, where σ_y is the yield strength of the material. It is a commonly used property in the design of flexural members of steel, post-tensioned concrete, and timber sections.

Mathematical Computation

Section moduli of a section about 2 and 3 axes (Fig. 2.27) can be determined using the following expressions:

$$S_{3\,(top)} = \frac{I_{33}}{\overline{y}_{top}} \tag{2.30}$$

$$S_{3\,(bot)} = \frac{I_{33}}{\overline{y}_{bot}} \tag{2.31}$$

$$S_{2\,(left)} = \frac{I_{22}}{\overline{x}_{left}} \tag{2.32}$$

$$S_{2\,(right)} = \frac{I_{22}}{\overline{x}_{right}} \tag{2.33}$$

Radii of Gyration
Definition
The radius of gyration is a measure of the elastic stability of a cross-section against buckling. In reality, there are several values of the radius of gyration for a cross-section, depending on which axis is being considered.

Significance
While considering the elastic buckling of columns, the radius of gyration is used in computing the kl/r (the so-called slenderness ratio). The direction about which the r needs to be considered is important, and often the least value will govern the determination of the buckling capacity. Sometimes the least r may be about the principal axes as will be discussed later.

Mathematical Computation
In a circular section, this is a single value. The radius of gyration about any axis can be computed as

$$r = \sqrt{\frac{I}{A}} \tag{2.34}$$

For other sections, the radius of gyration about 3 and 2 axes can be computed as

$$r_3 = \sqrt{\frac{I_{33}}{A}} \tag{2.35}$$

$$r_2 = \sqrt{\frac{I_{22}}{A}} \tag{2.36}$$

Principal Properties
Definition
Each section has a major and minor axis of bending, passing through the plastic center. It means that if a force is applied along any one of these axes in the plane of the cross-section, it will only cause bending along that axis. For example, if the load is applied to a rectangular cross-section along the y-axis through the plastic center, then the cross-section will only move along the y-axis, causing no bending component along the x-axis. In this case, the x–y axes are the principal axes. Alternatively, we can say that the principal axes are aligned along the x–y axes.

On the other hand, for an angle section, if we apply a force along the y-axis through the plastic center, the section will move, both along

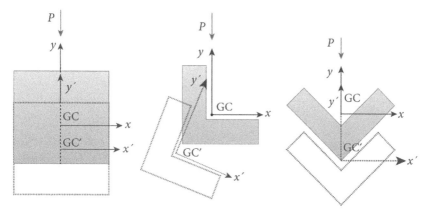

Figure 2.35 Principal axes and section deformation.

the y as well as along the x-axis, causing bending about both axes. In this case, the $x-y$ axes are not the principal axes of the cross-section. This behavior is illustrated in Fig. 2.35. One way to determine whether the principal axes are aligned with the $x-y$ axes is to check the value of the product of moment of inertia I_{xy} about these axes. If I_{xy} is 0, then the principal axes are aligned with the $x-y$ axes, otherwise not. If the principal axes do not correspond to the local cross-section axis, the transformed properties with respect to the principal axes should be calculated and used in analysis and design instead of the local axis values. The properties that need to be computed along the principal axes are:

- moment of inertia,
- the elastic section modulus,
- the radius of gyration, and
- plastic section modulus.

Significance

The issue relating to the principal bending direction is important in the design of columns, especially when slenderness effects are being considered. The column will buckle, or will undergo moment magnification in the direction of least stiffness, or along the weakest axis. In the case of unsymmetrical sections, this axis may not be aligned with the global $X-Y$ axes of the structure or may not match with the framing or loading direction. This aspect will be further discussed in Chapter 4, Response and Design for Shear and Torsion; and Chapter 5, Response and Design of Column Cross-Sections.

Mathematical Computation

The principal properties are computed from a transformation equation and the direction is computed as an angle from local 3—3 or x-axis. The notation for the principal axes are shown in Fig. 2.36.

Moment of inertia about the principal axis uu:

$$I_{uu} = \frac{I_{33} + I_{22}}{2} + \frac{I_{33} - I_{22}}{2} \cos 2\theta_p - I_{32} \sin 2\theta_p \qquad (2.37)$$

Moment of inertia about the principal axis vv:

$$I_{vv} = \frac{I_{33} + I_{22}}{2} - \frac{I_{33} - I_{22}}{2} \cos 2\theta_p + I_{32} \sin 2\theta_p \qquad (2.38)$$

Principal axis angle:

$$\tan 2\theta_p = \frac{2I_{32}}{I_{33} - I_{22}} \qquad (2.39)$$

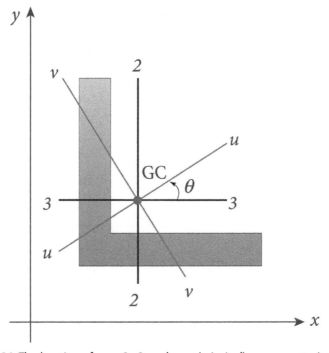

Figure 2.36 The location of x—y, 2—3, and u—v (principal) axes on a typical unsymmetrical section.

Radius of gyration about the principal axis:

$$r_u = \sqrt{\frac{I_{uu}}{A}} \text{ and } r_v = \sqrt{\frac{I_{vv}}{A}} \tag{2.40}$$

SPECIFIC PROPERTIES OF RC SECTIONS
Overview

Being composed of two materials with widely different properties, reinforced concrete sections require some specific properties for their design. These properties may change during the deformation history and the level of applied loading and are discussed below.

Equivalent Transformed Properties
Definition

One important assumption in elastic flexure stress theory is that the material is homogeneous and has a constant elastic modulus value. An RC section does not satisfy this requirement and thus the elementary bending stress theory is not directly applicable. However, we can still apply it within a certain range of loads and conditions by converting RC section into an equivalent concrete section with additional concrete area (replacing steel). This section is called transformed section.

Significance

The concept of transformed section is not only specific to RC sections. For all composite sections, we can invoke this concept for convenient determination of properties by converting composite section in an equivalent section made of only one base material.

Mathematical Computation

Since within the elastic range, the transformed RC section is required to carry the same load as the original section, we have to replace steel bars with equivalent amount of concrete such that the load carried by the section is unaltered. Thus, if the area of steel is A_s, the equivalent area of concrete will be nA_s, where n is the modular ratio. The transformation of a singly reinforced concrete beam section is shown in Fig. 2.37. The transformed section can be based on nA_s, which will create a small

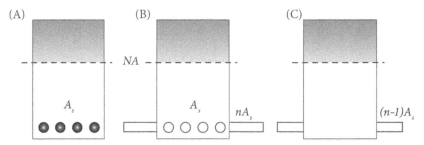

Figure 2.37 Transformation of a singly reinforced beam section. (A) Original section, (B) transformed section, with overlap $A_t = A_c + nA_s$, and (C) transformed section, with no overlap $A_t = A_g + (n-1)A_s$.

concrete and steel overlap where rebars exist. A more accurate transformed section can be defined using $(n-1)A_s$ so that there is no duplication or overlay between any areas.

Cracked Properties
Definition
Cross-sections made of certain materials may experience cracking when subjected to loadings higher than their capacity. The most common example of this is reinforced concrete sections. The effect of cracking on cross-sectional properties and stiffness is a major consideration in designing of these sections.

Significance
The properties of reinforced concrete sections change during the deformation history. For example, let us consider a beam under the influence of bending moment due to some loads. When the moment, the curvature, and the axial strains are small, the corresponding stress in concrete can then be computed on the basis of "equivalent transformed" moment of inertia and also computed on the basis of concrete dimensions and reinforcing bars. When the moment is such that the stress in concrete on the extreme tension fiber exceeds the tensile capacity, the concrete fails in tension. The cross-section size reduces up to the level where the section is uncracked. This reduction in section depth reduces the moment of inertia, which in turn increases the extreme fiber stress in the cracked concrete (which is already beyond its capacity). Further cracking again reduces the moment of inertia and this continues until an equilibrium is reached, aided by increased stresses in the reinforcement placed in the

tension zone. The properties of this cracked transformed section are different from the original section and need to be determined.

Mathematical Computation

The actual level of cracking and the real moment of inertia is very difficult to assess because of several other elements such as aggregate interlock, tensile capacity of concrete, distribution of cracks, curvature distribution, etc. We assume that all concrete in the tension zone can be ignored and that the moment is balanced by the internal stress resultant formed by the compressive force in concrete and the tensile force in the reinforcement. If we further assume linear distribution of strains and stresses, then we can compute the properties for this special condition and the corresponding internal moment resultant. This behavior is very important for the large deformations and for computing ductility of columns, especially near joint regions in frames subjected to seismic load. A detailed discussion of this behavior, together with the significance of moment curvature diagram will be presented in Chapter 3, Axial-Flexural Response of Cross-Sections; and Chapter 5, Response and Design of Column Cross-Sections.

To compute the cracked section properties of reinforced concrete, prestressed concrete, or for a composite section, the following general procedure can be used:

- Assume a certain neutral axis state after cracking. Transform all parts of the section, including rebars and steel sections in terms of concrete properties.
- Assume that the parts of the concrete on the tension side can be ignored and the tension is only provided by the reinforcement or by steel shapes.
- Tension and compression are in equilibrium with the external axial action. However, if the external action (axial load) is zero, then tension and compression must be equal.
- Based on these assumptions, determine the location of the neutral axis. This is often done in an iterative manner.
- Using this neutral axis, compute the moment of inertia of the cracked transformed section.

The most important and difficult step is the determination of the neutral axis location. For a rectangular section, with reinforcement at the top and bottom, a closed-form solution can be derived for the neutral axis depth, as well as for cracked properties. For the general section, however, this can be a fairly difficult problem if solved manually. A simple computer-based solution is of course possible using an iterative process, in which a neutral axis location is assumed in each cycle and the above steps are repeated until the equilibrium is satisfied.

For a rectangular, singly reinforced section (Fig. 2.38), the location of the neutral axis of the cracked section, and the corresponding cracked moment of inertia can be determined as follows:

$$I_{cr} = \frac{b}{3}(kd)^3 + nA_s(d-kd)^2 \tag{2.41}$$

$$y_t = d - kd \tag{2.42}$$

By equating the moments of compression and tension areas about the neutral axis, we can get the value of k as follows:

$$k = \sqrt{(n\rho)^2 + 2n\rho} + n\rho \tag{2.43}$$

where $\rho = \dfrac{A_s}{bd}$.

The expression for effective moment of inertia proposed by ACI 318–14 provides a transition between the upper and lower bounds of gross

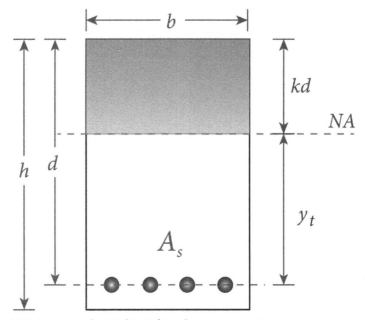

Figure 2.38 A rectangular singly reinforced concrete section.

inertia I_g and cracked inertia I_{cr} as a function of the ratio M_{cr}/M_a. For most cases, I_e is less than I_g.

$$I_e = \left(\frac{M_{cr}}{M_a}\right)^3 I_g + \left[1 - \frac{M_{cr}}{M_a}\right]^3 I_{cr} \qquad (2.44)$$

where, the cracking moment is

$$M_{cr} = \frac{f_r I_g}{y_t} \qquad (2.45)$$

and

$$f_r = 0.62\lambda\sqrt{f'_c} \qquad (2.46)$$

I_{cr} = moment of inertia of cracked section transformed to concrete, mm^4
I_e = effective moment of inertia for computation of deflection, mm^4
I_g = moment of inertia of gross concrete section about centroidal axis, neglecting reinforcement, mm^4
M_{cr} = cracking moment, N mm
M_a = applied moment, N mm
f'_c = compressive strength of concrete, MPa
f_r = modulus of rupture of concrete (flexural strength), MPa
λ = factor for lightweight aggregate concrete
y_t = distance from centroidal axis of gross section, neglecting reinforcement, to tension face, mm

SPECIFIC PROPERTIES OF STEEL SECTIONS

Overview

The inherent high strength-to-weight ratio of steel can be exploited to efficiently resist loads/actions in various structures. Small section sizes and long-span steel solutions not only create open and column-free space, but can also allow additional floors to be constructed in multistory buildings. As discussed earlier, a steel structural member can be a rolled shape or can be built up from two or more rolled shapes or plates, connected by welds or bolts. The most commonly used steel sections in building construction include wide flange (W), angle (L), channel (C), and tee (WT). The basic dimensions of these sections are directly used as their designation as shown in Table 2.3.

Table 2.3 Designations of Some Common Steel Section Shapes as per AISC LRFD Specification

Type of Shape	Cross-Section	Example of Designation	What Designation Means
W (wide flange)		W14 × 90	Nominal depth = 14 in. Weight = 90 lb/ft
C (channel)		C12 × 30	Depth = 12 in. Weight = 30 lb/ft
L (angle)		L4 × 3 × 1/4	Long leg = 4 in. Short leg = 3 in. Thickness = ¼ in.
WT (structural tee cut from W shape)		WT7 × 45	Nominal depth = 7 in. Weight = 45 lb/ft

Due to the small thickness of these sections, the local stability of parts of the section becomes a major consideration in design. Additional properties are, therefore, needed that are specific to steel structures.

Net Area and Effective Net Area

Most steel members are connected to each other through bolted or riveted connections. The gross area of a member at any point is the total area of the cross-section, with no deductions for holes made for bolts and rivets. The net area is the gross area minus the area of the holes. In computing the net area for tension, the width of a hole is often taken as 1/16 in. (3/2 mm) greater than its specified dimension. Since tolerances require that a bolt hole be greater than the diameter of the bolt, for design purposes the width of any hole is assumed to be twice of this (i.e., 1/8 in. or 3 mm) greater than the diameter of the bolt. The net area of an element is its net width multiplied by its thickness. For holes running perpendicular to the axis of the member, the net width is the gross width minus the sum of the widths of the holes. However, if a chain of holes extends across a part in a diagonal fashion, the net width is the gross width minus the sum of the hole dimensions plus the quantity $s^2/4g$ for each gage space in the chain, where

s = pitch (the longitudinal center-to-center spacing of any two consecutive holes)

g = gage (the transverse center-to-center spacing between fastener gage lines)

It may be necessary to examine several chains to determine which chain has the least net width (see Fig. 2.39).

The concept of effective net area accounts for shear lag in the vicinity of connections. When the member end connection transmits tension directly to all cross-sectional elements of the member, A_e and A_n are same. However, if the end connection transmits tension through some, but not all, of the cross-sectional elements, a reduced effective net area is used instead.

For bolted and riveted members,

$$A_e = UA_n \tag{2.47}$$

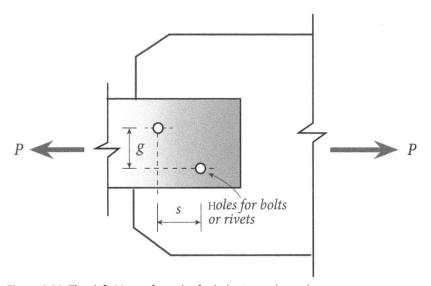

Figure 2.39 The definitions of s and g for holes in steel members.

For welded members,

$$A_e = UA_g \qquad (2.48)$$

Design values for this reduction factor U and A_e are given in AISC LRFD or similar specifications.

Width-to-Thickness Ratio

Definition

The cross-sections of structural steel members are classified as either compact, noncompact, or slender-element sections, depending on the width-to-thickness ratios ($\lambda = b/t$) of their elements. In compact sections, the section is expected to be safe from local buckling of flanges or web, which can otherwise prevent the achievement of full yield strength of section. It means that the possibility of a local failure (before global failure) is not considered in compact sections. Various codes provide the limiting width-to-thickness ratios as a function of tensile yield strength to define if a section is compact, noncompact, or slender. These limiting ratios are generally denoted as λ_p and λ_r in literature. As an example, AISE 360 specifications prescribe that a section is compact if the flanges are continuously connected to the web and the width-to-thickness ratios of all its compression elements are less than λ_p. A section is noncompact if the width-to-thickness ratio of at least one element is greater than λ_p, provided the width-to-thickness ratios of all compression elements are less than λ_r. If the width-to-thickness ratio of a compression element is greater than λ_r, that element is a slender compression element and the cross-section is called a slender-element section. The width-to-thickness ratios for columns and beams, and the corresponding limiting values of λ_r and λ_p can be seen various AISC LRFD specifications. As an example, Table 2.4 provides the limiting width-to-thickness ratios taken from AISC 360.

Significance

Steel members with compact sections can develop their full compressive strength without local instability. Noncompact shapes can be stressed to initial yielding before local buckling occurs. In members with slender elements, elastic local buckling is the limitation on their strength. Hence, the width-to-thickness ratio is very vital factor controlling the design method for steel sections.

Table 2.4 Limiting Width-to-Thickness Ratios to Check if a Section is Compact (Table B4.1, AISC 360)

Description	Parameter (λ)	λ_p	λ_r
Flexure in flanges of rolled I-shaped sections and channels	Width-to-thickness ratios (b/t)	$0.38\sqrt{\dfrac{E}{f_y}}$	$1.0\sqrt{\dfrac{E}{f_y}}$
Flexure in webs of rolled I-shaped sections and channels	Height-to-thickness ratios (h/t_w)	$3.76\sqrt{\dfrac{E}{f_y}}$	$5.70\sqrt{\dfrac{E}{f_y}}$

where
b = width of the I-shape or channel's flange
t = thickness of the flange
h = height of the web (does not include the flange thickness)
t_w = thickness of the web
E = modulus of elasticity of steel
f_y = yield strength of the steel

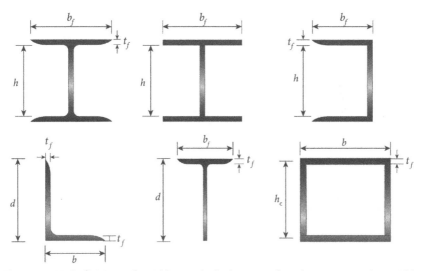

Figure 2.40 Definition of widths and thicknesses for determining the width-to-thickness ratio of various steel sections.

Mathematical Computation
The definition of different widths for various sections is shown in Fig. 2.40.

Height-to-Web Thickness Ratio
Definition
Another important parameter required to determine the design shear strength of compact beam sections is the height-to–web thickness ratio. Although flexural strength usually controls the selection of rolled beams,

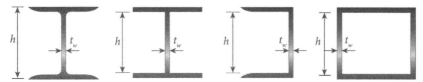

Figure 2.41 Definition of *h* for determining height-to-thickness for various steel sections.

shear strength may occasionally govern, particularly for short-span members or those supporting concentrated loads. Therefore, the shear strength of beams should always be checked.

Significance
In built-up members, the thickness of the web plate is often determined by shear. For rolled shapes and built-up members without web stiffeners, AISC LRFD specifications provide the equations to determine design shear strength based on ranges of the height-to-web thickness ratio (Table 2.4).

Mathematical Computation
The definitions of height for determination of height-to-web thickness ratio of various steel sections is shown in Fig. 2.41.

NUMERICAL COMPUTATIONS OF SECTION PROPERTIES
Overview—The Point, Polyline, and Polygon Method

The basic geometric properties for general solid shapes can be computed by using the polygon method. In this method, all shapes are represented by closed polygons. Curvilinear and circular shapes are represented by several straight-line segments. The properties of the overall shape are computed by geometric summation of the properties of a trapezoid defined by a projection of two consecutive points of the polygon on to the *x*- and *y*-axes. All concrete and hot-rolled steel shapes can be handled in this manner.

The geometric properties of thin-walled open shapes can also be computed using the polyline method. Thin-walled open shapes are represented by polylines with specified thickness. Curvilinear and circular

(A)

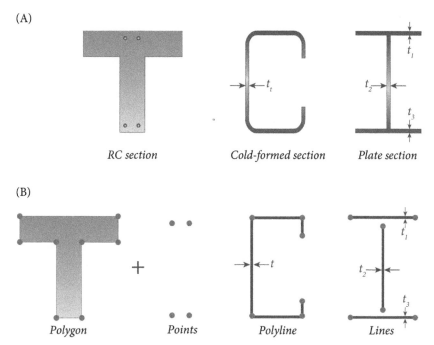

(B)

Figure 2.42 Section representation by geometric entities. Polygons can be used to represent almost any shape, even curved. (A) Original section, (B) representation by polygons, polylines, lines, and points.

shapes or edges are represented by several straight-line segments and the properties of overall shape are computed by geometric summation of the properties of a line segment defined by the projection of two consecutive points of the polyline onto the x- and y-axes.

The properties of small portions of section, such as rebars, in a reinforced concrete section can be computed by representing them by points. Fig. 2.42 shows this concept symbolically.

The basic properties computed in this manner are:
- Total area of cross-section
- First moment of area
- Second moment of area about the x- and y-axes
- Shear areas about 3−2 coordinate system

The derived properties can then be computed using the standard mathematical equations presented earlier in the chapter and may include
- geometric centroid of shapes,
- polar moment of inertia about the origin,
- modulus of section at top, bottom, left, and right extreme of the shape,

- radius of gyration about x- and y-axes,
- plastic modulus about 3−2 coordinate system,
- orientation of the principal axis, and
- principal moment of inertia.

The total properties of the cross-section can be obtained by adding the properties of each polygon/polyline/point multiplied by the corresponding modular ratio. For each polygon, the properties are computed using the area under the trapezoid concept, as demonstrated in Figs. 2.43 and 2.44.

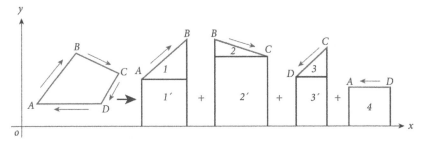

Figure 2.43 The properties of polygon about a given axis can be computed by projecting each line in the polygon on to the axis and by summation of all properties.

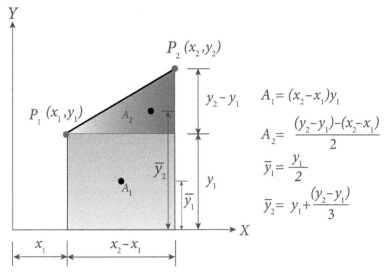

$$A_1 = (x_2 - x_1)y_1$$

$$A_2 = \frac{(y_2 - y_1)-(x_2 - x_1)}{2}$$

$$\bar{y}_1 = \frac{y_1}{2}$$

$$\bar{y}_2 = y_1 + \frac{(y_2 - y_1)}{3}$$

Figure 2.44 Proceeding clockwise, each line in the polygon is projected on to the x-axis (or y-axis) and the resulting trapezoid is divided into a rectangle and a triangle. The properties of these basic shapes are computed for the x- and y-axes. The total properties of the polygon are the sum of properties for each line. Other properties are derived from the basic properties.

For the thin-walled sections, the properties may be reasonably estimated by representing each part of the section by a line segment, with a
thickness assigned to it. In this case, the local properties of the segment
can be computed using a rectangular segment, about its middepth,
and then global properties by using the moments of area as shown in
Figs. 2.45 and 2.46.

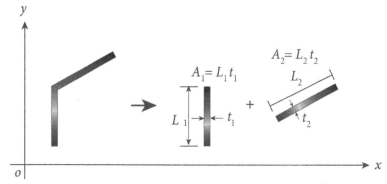

Figure 2.45 The properties of plates and thin wall sections can be computed using
summation of all segments.

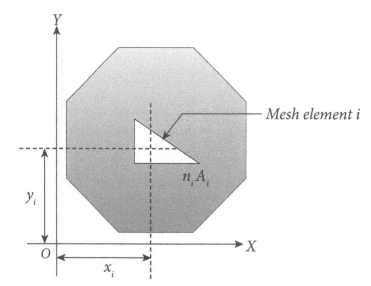

Figure 2.46 Using meshing to compute properties.

Accuracy of the Polygon and Polyline Methods

The polygon and polyline methods for computation of geometric properties are generally adequate for shapes of usual dimensions and proportions. The methods may lose some of the accuracy for very oblong shapes, highly curvilinear shapes, and for shapes with very small or very large dimensions. Accuracy may also be lost if the shape is too far from the coordinate origin. These errors are introduced due to numerical round off and may be more pronounced in the results for second moment of area. The errors can be significantly reduced if the section is located close to the origin, and the dimension units are such that the values are not in thousands or not in thousandths. For example, for a typical column in a building, good results will be obtained by using foot, inch, millimeter, or centimeter units. For large sections, however, such as in bridge piers, it may be best to use foot or meter to define the section for calculation of properties using the above methods.

In the polygon method, the properties of each shape are computed independently and then added up to determine the properties of the entire cross-section. However, if the cross-section is made up of several shapes of different material properties, which are either partially overlapping, completely encased, or enclosed inside other shapes, it is difficult to determine the properties of the cross-section using this method. Also, the polygon method cannot be used directly to compute properties like shear area, the plastic section modulus, and the torsional constant.

Using Meshing to Compute Properties

As mentioned above, the polygon and polyline method is difficult to use when dealing with complex sections with holes and composites made from several materials. Polygon method also cannot be used to compute some of the section properties.

The main purpose of meshing is to divide complex shape polygons into simpler quadrilaterals, trapezoids, and ultimately triangles. This subdivision allows for several operations to be performed on these shapes that would not be otherwise possible. These include:

- The determination of the overlapping area.
- Combining, adding, and merging shapes. Determining cross-section properties using a consistent and accurate procedure using exact solution of triangles.

- Computing section stress resultants using an exact solution of stress integration for triangles.
- Drawing and rendering of concave and complex polygons, which would otherwise be difficult.
- Finite element discretization of general polygons and cross-sections for the purpose of computing the torsional constant.
- Calculating and plotting of stress and deformation contours.
- Computing the shear area by using the strip or quad mesh.

Once all the polygons in the cross-section have been meshed, the overlapping meshes are removed, retaining the mesh with the higher modular ratio. Finally, only a single mesh element occupies any particular region in the cross-section. The mesh is then divided into triangles. The properties of each triangle are computed and then summed up using the transformation rules. For example, area and moment of inertia are computed as

$$A = \sum_{i=1}^{m} A_i \, n_i \tag{2.49}$$

$$I_{xx} = \sum_{i=1}^{m} \left(I_{x_i} \, n_i + A_i \, \bar{y}^2 \, n_i \right) \tag{2.50}$$

$$I_{yy} = \sum_{i=1}^{m} \left(I_{y_i} n_i + A_i \bar{x}^2 n_i \right) \tag{2.51}$$

Meshing of Sections Made Up from Polygon Shapes

The automatic meshing of regions or enclosed areas, surfaces, and volumes is a topic of significant research and development. The greatest application of meshing is in finite element analysis where complex regions need to be subdivided into simpler but connected elements for the purpose of stiffness matrix formulation. Several methods and techniques have been developed to handle automatic meshing problems. One method is applicable to two-dimensional or planar regions made up from single or multiple polygons. This method is ideally suited to mesh the complex cross-section to simpler quadrilaterals and triangles. This book will cover only the general principles, without going into the implementation details which require a significant discussion of analytical geometry, vector geometry, and computer graphics, all of which are outside the scope of this book.

Meshing of a Single Polygon

Let us start with a simpler case of a single polygon that needs to be subdivided or meshed into simpler shapes, as shown in Fig. 2.47.

Step-1: Pass horizontal lines at all vertices and find the intersection of these lines with edges of the polygon. The resulting subregions will be either triangular or trapezoids/quads. Additional horizontal grid lines may also be added to refine the mesh.

Step-2: Pass vertical lines either through all vertices and at all newly formed intersections of polygon edges and horizontal lines, or at any arbitrary location. In the first case, all resulting mesh elements will be simple quad or triangles. In the second case, some mesh elements will be triangle, some will be quads and some will have five or six sides. The five or more sided polygons can be subdivided into triangles at a later stage.

The method described for a single polygon can be extended to multiple polygons with overlapping or nonoverlapping regions. If the polygons do not overlap, then each polygon can be meshed independently. If the polygons overlap or intersect or touch each other, additional lines and intersection points must be determined before proceeding with meshing of each polygon. The portion of each polygon within the other polygon is determined and then meshing of each polygon, with the region of the other polygon inside is carried out using the previous procedure.

The above description simply outlines the overall methodology. There are several implementation issues and tolerance problems that need to be tackled before such a method can be used reliably for actual applications. For example, even the apparently simple problem of finding which part of a polygon lies within the other polygon becomes a significant programming challenge. In fact, it is fairly difficult to check whether a point lies inside or outside a general complex polygon. Even the best of algorithms fail in certain special situations.

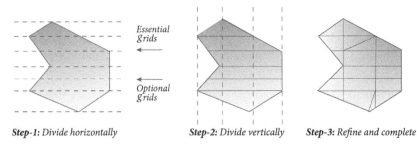

Step-1: Divide horizontally   *Step-2: Divide vertically*   *Step-3: Refine and complete*

Figure 2.47 Meshing of a polygon can be carried out in three steps. The final result is a set of interconnected quads or triangles.

Once the polygons have been meshed into smaller elements, the elements are converted into small polygons themselves or into triangles if needed. The order of vertices in the parent polygons and the meshed polygons is crucial for the meshing and subsequent algorithms to work. In general, all parent polygons must be defined in clockwise direction. The resulting polygon mesh can either be clockwise or counterclockwise depending on the intended application. The property calculation procedures generally use clockwise direction, whereas the finite element algorithms generally use counterclockwise direction (see Fig. 2.48).

Merging of Shapes

The merging of shapes can be used to create more complex shapes and cross-sections. Once two shapes are merged, a new shape is created, which in turn can be merged with other shapes. Merging can be performed for adjoining or for overlapping shapes. Two algorithms are mentioned here (Fig. 2.49).

- Merging using meshing: This method works by removing the duplicate mesh elements and finding the boundary of the final mesh. This algorithm is most suitable for rectangular and adjoining shapes. It may not produce good results for circular or complex overlapping shapes.

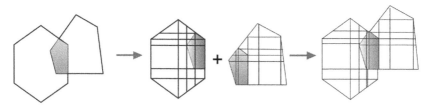

Figure 2.48 Overlapping polygons are first broken into individual polygons considering the overlapping part, meshed independently and then connected together.

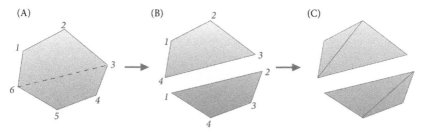

Figure 2.49 Meshing of polygons. (A) Polygon, (B) two quads, and (C) four triangles.

• Merging using boundary intersection: This method works by finding the intersection of shapes' boundaries and removing line segments lying inside the polygon region. This method is more suitable for overlapping and nonrectangular shapes. It may not produce desired results for adjoining shapes with no overlaps or partial overlaps. Generally, this method is faster than the method based on meshing. After merging of the simple shapes, the complex shapes need to be meshed again for computing the properties.

SOLVED EXAMPLES
Cross-Sectional Area and Bearing Area

Question 1: Determine the maximum safe load P that can be applied to the shown assembly (see Fig. Ex. 2.1) of two plates joined together using four rivets, if the shear strength (σ_s) of rivet material is 96.5 N/mm^2 and bearing stress (σ_b) is limited to 124 N/mm^2. Assume a uniform load distribution.

Given
Shear strength (σ_s) of rivet material $= 96.5$ N/mm^2
 Limiting bearing stress (σ_b) $= 124$ N/mm^2

Solution
The safe load can be calculated based on two criteria, i.e., shear failure and bearing stress failure of the rivet material.

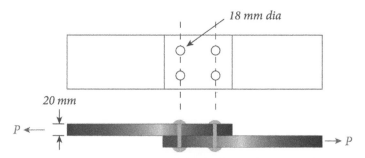

Figure Ex. 2.1 Computing maximum safe load of two plates joined together using rivets.

1. **Based on shear failure:**
    Area under shear stress (parallel to the applied load) $= 4\pi\frac{d^2}{4} = 4\pi\frac{18^2}{4} = 1017.876$ mm^2
    Load $(P) =$ Shear strength $(\sigma_s) \times$ area under shear $(A_s) = 96.5 \times 1017.876 = 98,225.03$ N $= 98.225$ KN
2. **Based on bearing failure:**
    Area of bearing $A_b = 4dt = 4 \times 18 \times 20 = 1440$ mm^2
    Load $(P) =$ Bearing strength $(\sigma_b) \times$ bearing area $(A_b) = 124 \times 1440 = 178,560$ N $= 178.56$ KN

Result
So the safe load that can be applied is minimum of the above two, i.e., $P = 98.225$ KN

Specific Length

Question 2: Find the specific length of a steel rod shown in Fig. Ex. 2.2, if its tensile failure stress is 250 N/mm^2.

Given
Tensile failure stress of steel rod $(\sigma_f) = 250$ N/mm^2

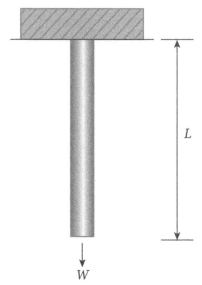

L

W

Figure Ex. 2.2 A hanging rod of steel.

Solution

Let us denote specific length $= L_s$

Equating, total weight $=$ failure load

$$\text{Mass} \times g = failure \text{ stress} \times A$$

$$mg = \sigma_f A$$

The mass of the rod can be written as the product of mass density (ρ) and volume (cross-sectional area, A multiplied with length, L_s). The mass density of steel is assumed to be 7800 kg/m^3.

$$\rho A L_s = \sigma_f A$$

$$L_s = \frac{\sigma_f}{\rho g} = \frac{250 \times 10^6}{7800 \times 9.81} = 3267.21 \text{ m}$$

Result

Specific length of steel rod shown above is 3267.21 m, which is the maximum length it can sustain without failing under its own weight.

Properties of a Transformed Section

Question 3: Find the modular ratio, equivalent area of transformed section, equivalent first moment of area, and equivalent centroid location and neutral axis depth for the RC section shown in Fig. Ex. 2.3.

Given

Dimensions $= 250$ mm $\times 400$ mm

Effective cover $= 40$ mm

Yield strength of steel (f_y) $= 415$ N/mm^2

Compressive strength of concrete (f_c') $= 20$ N/mm^2

Modulus of elasticity of steel (E_s) $= 2 \times 10^5$ N/mm^2

Solution

Gross section area of concrete (A_g) $= 250 \times 400 = 1 \times 10^5$ mm^2

Area of steel (A_s) $= 3\pi \frac{d^2}{4} = 3\pi \frac{12^2}{4} = 339.292$ mm^2

Modulus of elasticity of concrete (E_c) $= 5000\sqrt{f_c'} = 5000\sqrt{20} = 22,360.7$ N/mm^2

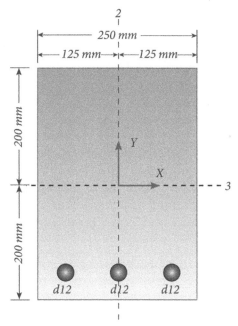

Figure Ex. 2.3 A reinforced concrete section.

Modular ratio (n)

$$n = \frac{E_s}{E_c} = \frac{2 \times 10^5}{22360.7} = 8.944$$

Equivalent area of transformed homogenous material as concrete

$$A_t = A_g + nA_s = 1.03 \times 10^5 \text{ mm}^2$$

Moment of area of concrete about base

$$M_c = \sum A_i y_i = 1 \times 10^5 \times \frac{(400)}{2} = 20 \times 10^6 \text{ mm}^3$$

Moment of area for steel about base

$$M_s = \sum A_i y_i = 339.29 \times 40 = 13,571.6 \text{ mm}^3$$

Equivalent moment of area

$$M_t = M_c + nM_s = 20 \times 106 + 8.944 \times 13,571.6 = 20.12 \times 10^6 \text{ mm}^2$$

Equivalent centroid

$$d = \frac{M_c + nM_s}{A_g + nA_s} = 195.34 \text{ mm}$$

Neutral axis depth

Tension force can be determined as following. A strength reduction factor of 0.87 is applied to tensile strength of steel (f_y).

$$T = 0.87A_{st} = 0.87 \times 415 \times 339.29 = 12.25 \times 10^4 \text{ N}$$

Compressive force can be determined as follows. A strength reduction factor of 0.85 is applied to compressive strength of concrete (f_c').

$$C = 0.85f_c' ba = 0.85 \times 20 \times 250 \times a = 4250a$$

Equating the tensile and compressive forces for equilibrium condition, we get

$$a = \frac{12.25 \times 10^4}{4250} = 28.82 \text{ mm}$$

And neutral axis depth, $d_{NA} = \dfrac{a}{0.85} = 33.91$ mm from top of the section.

Results

Modular ratio (n) = 8.944

Equivalent area of transformed section (A_t) = 1.03×10^5 mm

Equivalent first moment of area (M_t) = 20.12×10^6 mm^3

Equivalent centroid (d) = 195.34 mm

Neutral axis depth (d_{NA}) = 33.91 mm from top

First and Second Moment of Areas

Question 4: Find the moment of area and moment of inertia of the given cross-section about 2 and 3 axes shown in Fig. Ex. 2.4.

Given

Flange width = 50 mm

Web width = 60 mm

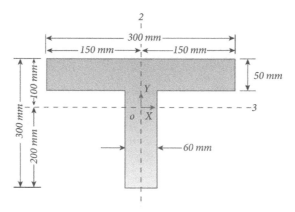

Figure Ex. 2.4 A T-section.

Solution
Geometric center of the section

First we have to find the geometric center of given cross-section. So, using prefix 1 for flange and 2 for web part and taking left-bottom corner as reference axis (origin), we have

$$\bar{x} = \frac{\sum A_i x_i}{\sum A_i} = \frac{300 \times 50 \times \left(\dfrac{300}{2}\right) + 60 \times 250 \times \left(\dfrac{60}{2} + \dfrac{300-60}{2}\right)}{300 \times 50 + 60 \times 250} = 150 \, \text{mm}$$

and

$$\bar{y} = \frac{\sum A_i y_i}{\sum A_i} = \frac{300 \times 50 \times \left(250 + \dfrac{50}{2}\right) + 60 \times 250 \times \left(\dfrac{250}{2}\right)}{300 \times 50 + 60 \times 250} = 200 \, \text{mm}$$

Moment of inertia about 3−3 axis

$$I_{33} = \sum \frac{bd^3}{12} + Ay^2$$

$$I_{33} = \frac{300 \times 50^3}{12} + 300 \times 50 \times (275 - 200)^2 + \frac{60 \times 250^3}{12}$$

$$+ 60 \times 250 \times \left(200 - \frac{250}{2}\right)^2$$

$$= 2.5 \times 10^8 \, \text{mm}^4$$

Moment of inertia about 2-2 axis

$$I_{22} = \sum \frac{bd^3}{12} + Ax^2$$

$$I_{22} = \frac{50 \times 300^3}{12} + \frac{250x60^3}{12} = 1.17 \times 10^8 \text{ mm}^4$$

Moment of area about neutral axis

$$M_{na} = \sum A_i y_i$$

$$M_{na} = 300 \times 50 \times (275 - 200) + 50 \times 60 \times 25 = 1.2 \times 10^6 \text{ mm}^3$$

Moment of area about flange and web interface

$$M_a = \sum A_i y_i$$

$$M_a = 300 \times 50 \times 75 = 1.125 \times 10^6 \text{ mm}^3$$

Results
Moment of inertia about 3-3 axis = 2.5×10^8 mm^4
Moment of inertia about 2-2 axis = 1.17×10^8 mm^4
Moment of area about neutral axis = 1.2×10^6 mm^3
Moment of area about flange and web interface = 1.125×10^6 mm^3

Principal Axes, Moment of Inertia, and Radius of Gyration

Question 5: Find the moment of inertia and radius of gyration about centroidal axis of the channel section shown in Fig. Ex. 2.5.

Given
Flange width = 60 mm
Web Width = 100 mm
Thickness = 2 mm

Solution
Determining the centroid of channel section

$$\bar{x} = \frac{\sum A_i x_i}{\sum A_i} = \frac{\left(60 \times 2 \times \left(\frac{60}{2}\right)\right) + (96 \times 2 \times 1) + \left(60 \times 2 \times \left(\frac{60}{2}\right)\right)}{(60 \times 2) + (96 \times 2) + (60 \times 2)}$$

$$= 17.11 \text{ mm}$$

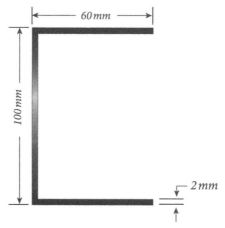

Figure Ex. 2.5 A channel section.

From symmetry,

$$\bar{y} = 50 \text{ mm}$$

Moment of inertia about centroidal axes

$$I_{xx} = \sum \frac{bd^3}{12} + \sum Ah^2$$

$$I_{xx} = 2 \times \left(60 \times \frac{2^3}{12} + 60 \times 2 \times (100-1-50)^2\right) + 2 \times \frac{96^3}{12} = 7.23 \times 10^5 \text{ mm}^4$$

Similarly,

$$I_{yy} = \sum \frac{db^3}{12} + \sum Ah^2$$

$$I_{yy} = 2 \times \left(2 \times \frac{60^3}{12} + 2 \times 60 \times \left(\frac{60}{2} - 17.11\right)^2\right) + 96 \times \frac{2^3}{12}$$

$$+ 96 \times 2 \times (17.11 - 1)^2 = 1.6177 \times 10^5 \text{ mm}^4$$

Radius of gyration

$$R_x = \sqrt{\frac{I_{xx}}{A}} = \sqrt{\frac{7.23 \times 10^5}{432}} = 40.91 \text{ mm}$$

$$R_y = \sqrt{\frac{I_{yy}}{A}} = \sqrt{\frac{1.617 \times 10^5}{432}} = 19.35 \text{ mm}$$

Results

Moment of inertia about centroidal axis $(I_{xx}) = 7.23 \times 10^5$ mm^4

    Moment of inertia about centroidal axis $(I_{yy}) = 1.6177 \times 10^5$ mm^4

    Radius of gyration $(R_x) = 40.91$ mm

    Radius of gyration $(R_y) = 19.35$ mm

Question 6: Find the principal axis plane, moment of inertia about the principal axis, and radius of gyration about the principal axis of the section as shown in Fig. Ex. 2.6. All dimensions are in mm.

Given

Leg 1 $= 10$ mm $\times$ 100 mm

    Leg 2 $= 150$ mm $\times$ 10 mm

Solution

Locating the centroid of given section

$$x = \frac{\sum A_i x_i}{\sum A_i} = \frac{A_1 x_1 + A_2 x_2}{A_1 + A_2} = \frac{10 \times 100 \times 5 + 140 \times 10 \left(10 + \dfrac{140}{2}\right)}{10 \times 100 + 140 \times 10} = 48.75 \, \text{mm}$$

$$\overline{y} = \frac{\sum A_i y_i}{\sum A_i} = \frac{A_1 y_1 + A_2 y_2}{A_1 + A_2} = \frac{10 \times 100 \times \left(\dfrac{100}{2}\right) + 140 \times 10 \times 5}{10 \times 100 + 140 \times 10} = 23.75 \, \text{mm}$$

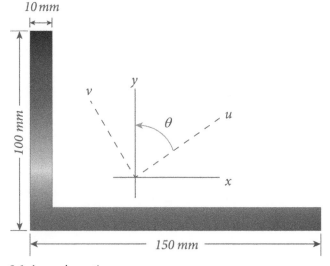

Figure Ex. 2.6 An angle section.

Moment of inertia about 3−3 and 2−2 axes

$$I_{33} = \frac{\sum bd^3}{12} + \sum A\bar{y}^2$$

$$I_{33} = \frac{10 \times 100^3}{12} + 10 \times 100 \times (50-23.75)^2 + 140 \times \frac{10^3}{12}$$

$$+ 140 \times 10 \times \left(23.75 - \frac{10}{2}\right)^2 = 2.02 \times 10^6 \text{ mm}^4$$

$$I_{22} = \frac{\sum bd^3}{12} + \sum A\bar{x}^2$$

$$I_{22} = \frac{100 \times 10^3}{12} + 10 \times 100 \times (48.75-5)^2 + 10 \times \frac{140^3}{12}$$

$$+ 140 \times 10 \times \left(10 + \frac{140}{2} - 48.75\right)^2 = 5.576 \times 10^6 \text{ mm}^4$$

Product moment of inertia

For Leg 1 (10 mm × 100 mm):

$$I_{32} = \iint xy \, dx.dy = \int_0^{10} x \, dx \int_0^{100} y \, dy = \frac{10^2}{2} \times \frac{100^2}{2} = 250 \times 10^3 \text{ mm}^4$$

For Leg 2 (150 mm × 10 mm):

$$I_{32} = \iint xy \, dx.dy = \int_{10}^{150} x \, dx \int_0^{10} y \, dy = \frac{150^2 - 10^2}{2} \times \frac{10^2}{2} = 560 \times 10^3 \text{ mm}^4$$

Now I_{32} for total cross−section is

$$I_{32} = \sum I_{32} - A_{xy}$$

$$I_{32} = 250 \times 10^3 + 560 \times 10^3 - 2400 \times 48.75 \times 23.75 = -1.97 \times 10^6 \text{ mm}^4$$

Principal angle (θ)

$$\tan(2\theta) = \frac{2 \times I_{32}}{I_{33} - I_{22}} = \frac{2 \times (-1.97 \times 10^6)}{(2.02 \times 10^6) - (5.57 \times 10^6)} = 1.11$$

$2\theta = 47.94$ degrees or, $\theta = 23.97$ degrees with respect to x-axis, or $\frac{\pi}{2} - 23.97 = 66.025$ degrees with respect to vertical axis.

Moment of inertia about principal axes

$$I_{uu} = \frac{I_{33} + I_{22}}{2} + \frac{I_{33} - I_{22}}{2} \cos(2\theta) - I_{32} \sin(2\theta) = 6.45 \times 10^6 \text{ mm}^4$$

$$I_{vv} = \frac{I_{33} + I_{22}}{2} - \frac{I_{33} - I_{22}}{2} \cos(2\theta) + I_{32} \sin(2\theta) = 1.15 \times 10^6 \text{ mm}^4$$

Radius of gyration about principal axes

$$R_u = \sqrt{\frac{I_{uu}}{A}} = \sqrt{\frac{6.45 \times 10^6}{2400}} = 51.84 \text{ mm}$$

$$R_v = \sqrt{\frac{I_{vv}}{A}} = \sqrt{\frac{1.15 \times 10^6}{2400}} = 21.84 \text{ mm}$$

Results

Moment of inertia about the principal axis (I_{uu}) = 6.45×10^6 mm^4

Moment of inertia about the principal axis (I_{vv}) = 1.15×10^6 mm^4

Radius of gyration about the principal axis (R_{uu}) = 51.84 mm

Radius of gyration about the principal axis (R_{vv}) = 21.84 mm

Torsional Constant

Question 7: Find the torsional constant (J) of the following sections in Fig. Ex. 2.7.

Given

Radius of solid cylinder = 200 mm

Dimensions of a rectangular bar = 4000 mm $\times$ 250 mm $\times$ 250 mm

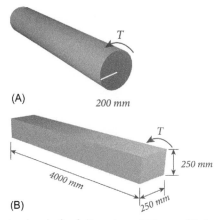

Figure Ex. 2.7 (A) A circular shaft of diameter = 200 mm. (B) A rectangular bar.

Solution
Case (a): A solid cylinder

$$J = \pi \frac{D^4}{32}$$

$$J = \pi \frac{0.2^4}{32} = 1.57 \times 10^{-4} \; \text{m}^4$$

Case (b): A rectangular bar

$$J = 2.25 a^4$$

$$J = 2.25 \times 0.25^4 = 8.789 \times 10^{-3} \; \text{m}^4$$

Angle of Twist and Allowable Torque

Question 8: Determine the ultimate torque that can be applied for the following plain concrete section in Fig. Ex. 2.8.

Given
Diameter of circular section $(D) = 500$ mm
   The compressive strength of plain concrete $(f'_c) = 25 \; \text{N/mm}^2$

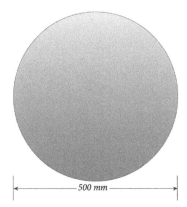

Figure Ex. 2.8 A solid plain concrete section.

Solution
Polar moment of inertia

$$J = \pi \frac{D^4}{32} = \pi \frac{500^4}{32} = 6.13 \text{ mm}^4$$

Shear stress, using 0.8 as strength reduction factor to compressive strength, can be estimated as

$$\tau = \frac{\sqrt{0.8 f_c'}}{6} = 0.745 \text{ N/mm}^2$$

Ultimate torque

$$T = \frac{\tau J}{C} = \frac{0.745}{d/2} J = 0.745 \times 6.13 \times \frac{10^9}{250} = 18.26 \times 10^3 \text{ N m}$$

Result
Ultimate torque for the given unreinforced concrete section $(T) = 18.26 \times 10^3$ N m

Question 9: A solid steel bar of circular cross-section as shown in Fig. Ex. 2.9 has a diameter $D = 50$ mm, length $l = 1.5$ m, and shear modulus elasticity of $G = 80$ GPa. The bar is subjected to torque $T = 350$ N m at the ends. (a) Find the maximum stress in the bar and angle of twist. (b) If maximum angle of twist is $\pi/128$ and maximum stress is 22 MPa, what is the maximum permissible torque?

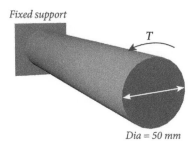

Fixed support

T

Dia = 50 mm

Figure Ex. 2.9 A solid steel cylinder.

Given

Diameter of solid steel bar $(D) = 50$ mm
  Length of solid steel bar $(L) = 1.5$ m
  Shear modulus of elasticity $(G) = 80$ GPa
  Torque $(T) = 350$ N m

Solution (a)
Torsional constant

$$J = \pi \frac{D^4}{32} = 6.136 \times 10^{-7} \text{ m}^4$$

Maximum shear stress

$$\tau_{max} = \frac{T}{J} r = \frac{350 \times 0.025}{6.136 \times 10^{-7}} = 14.26 \text{ MPa}$$

Angle of twist using torsional constant

$$\varphi = \frac{TL}{JG} = \frac{350 \times 1.5}{80 \times 6.136 \times 10^{-7}} = 0.0107 \text{ rad}$$

Solution (b)
Torsional equation

$$\frac{T}{J} = \frac{\tau}{r} = \frac{G\varphi}{L}$$

Maximum possible torque
When the maximum shear stress is 22 MPa,

$$T_{max} = \tau_{max} \frac{J}{r} = 22 \times \pi \frac{D^4}{32} \times \frac{2}{D} = 539.96 \text{ N m}$$

When the maximum angle of twist is $\pi/128$ rad,

$$T_2 = \frac{GJ\varnothing}{L} = \frac{G \times 6.136 \times 10^{-7} \times \pi/128}{1.5} = 803.3 \ N\,m$$

Results

a. The maximum shear stress in the bar $\tau_{max} = 14.26$ MPa with an angle of twist $\varphi = 0.0107$ rad

b. The maximum permissible torque is minimum from two limiting criteria $= 539.96 \ N\,m$

Maximum Bending Stress

Question 10: Calculate the maximum stress in the material of the pipe when full water is flowing through it (see Fig. Ex. 2.10). It is simply supported by 12 m apart.

Given

External diameter $= 250$ mm

    Thickness of pipe $= 15$ mm

Solution

Moment of inertia

$$I = \frac{\pi}{64}\left(D^4 - Di^4\right) = \frac{\pi}{64} \times \left(250^4 - 235^4\right) = 42.04 \times 10^6 \ mm^4$$

Section modulus

$$Z = \frac{I}{y} = \frac{42.04 \times 10^6}{125} = 3.36 \times 10^5 \ mm^3$$

Figure Ex. 2.10 A hollow circular section.

Volume of pipe

$$V = \frac{\pi}{4}\left(D^2 - Di^2\right) = \frac{\pi}{4} \times \left(250^2 - 235^2\right) \times \frac{12}{10^6} = 0.069 \text{ m}^3$$

Wight of pipe

$$w_1 = 0.069 \times 7800 = 534.77 \text{ kg}$$

Volume of water when pipe is full

$$V_w = \frac{\pi}{4} \times 235^2 \times \frac{12}{10^6} = 0.521 \text{ m}^3$$

Weight of water

$$w_2 = 0.52 \times 1000 = 520.48 \text{ kg}$$

Total weight

$$w = w_1 + w_2 = 1055.25 \text{ kg} = 10348.45 \text{ N}$$

Bending Moment

$$M = W.\frac{L}{8} = 10348.45 \times \frac{12}{8} = 15.52 \quad KN \text{ m}$$

Maximum stress when pipe is full

$$\sigma_b = \frac{M}{Z} = 15.52 \times 10^3 \times \frac{1000}{\left(3.36 \times 10^5\right)} = 46.19 \text{ N/mm}^2 = 46.19 \text{ MPa}$$

Results

Maximum stress in the material of pipe when full water is flowing is 49.19 MPa.

Shear Center of a Section

Question 11: Find the shear center of the following section in Fig. Ex. 2.11.

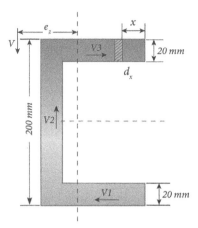

Figure Ex. 2.11 A channel section.

Given
A channel section with thickness = 20 mm

Solution
Shear force loading is parallel to the y-axis. Shear stress oriented along flanges part of channel section (producing couple) can be determined as follows.

$$V_1 = \int_0^{120-20/2} V \frac{Q}{It}\, da$$

where $da = 20\, dx$, $Q = a\bar{y} = xt\left(\dfrac{200}{2} - \dfrac{20}{2}\right)$ and $t = 20$ mm.

Therefore,

$$V_1 = \int_0^{110} \frac{V}{I \times 20} \times 20 \times x \times 20 \times 90\, dx$$

Moment of inertia about horizontal cantorial axis, $I_{xx} = I_{web} + 2I_{flange}$

$$I_{xx} = 20 \times \frac{200^3}{12} + 2\left[100 \times \frac{20^3}{12} + 100 \times 20 \times \left(\frac{200}{2} - \frac{20}{2}\right)^2\right] = 45.73 \times 10^6\, \text{mm}^4$$

$$V_1 = \frac{V}{I_{xx}} \times 1800 \times \left[\frac{110^2}{2}\right] = 0.238\, V$$

Now, V_1 and V_3 are making a couple which will be balanced by applied action, $V \times e_Z$.

$$V \times e_Z = 0.238 V \times (200 - 20)$$

$$e_Z = 42.84 \text{ mm}$$

Results
The position of shear center of the above section $e_Z = 42.84$ mm.

Effect of Cross-Sectional Shape on Moment of Inertia

Question 12: Find the moment of inertia of the iron sheet shown in Fig. Ex. 2.12 and compare when transformed to different shapes (from same steel sheet).

A. A straight steel sheet

Moment of inertia about the x-axis:

$$I_{xx} = \frac{bd^3}{12} = \frac{1037 \times 30^3}{12} = 2.333 \times 10^6 \text{ mm}^4$$

$$I_{yy} = \frac{db^3}{12} = \frac{30 \times 1037^3}{12} = 2.79 \times 10^9 \text{ mm}^4$$

B. A circular shape section

The same steel sheet, when folded into a hollow, circular-shaped cross-section will approximately have an outer dimeter of 345 mm. Therefore, $D_o = 345$ mm, $D_i = 315$ mm.

Moment of inertia of this circular section is

$$I_{xx} = I_{yy} = \frac{\pi}{64} \left(D_o{}^4 - D_i{}^4\right) = \frac{\pi}{64} \times \left(345^4 - 315^4\right) = 2.12 \times 10^8 \text{ mm}^4$$

C. An L-shaped section

The location of centroid of this section is determined as follows:

$$\bar{x} = \sum A_i \frac{x_i}{\sum A_i} = \frac{(600 \times 30 \times 15) + \left(407 \times 30 \times \left(30 + \frac{407}{2}\right)\right)}{(600 \times 30) + (407 \times 30)} = 103.31 \text{ mm}$$

$$\bar{y} = \sum A_i \frac{y_i}{\sum Ai} = \frac{(600 \times 30 \times 300) + \left(407 \times 30 \times \left(\frac{30}{2}\right)\right)}{(600 \times 30) + (407 \times 30)} = 184.81 \text{ mm}$$

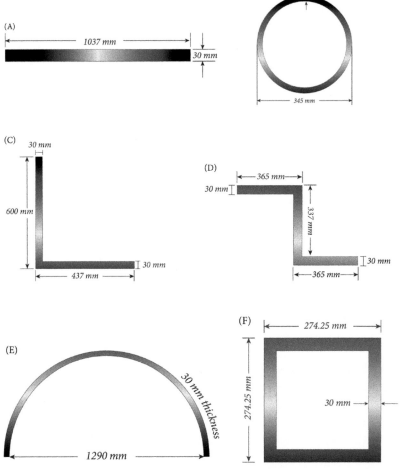

Figure Ex. 2.12 (A) A steel sheet. (B) The same sheet rolled to make a hollow cylinder. (C) The same sheet folded to make an angle section. (D) The same sheet folded to make a Z-section. (E) The same sheet rolled to make a semicylinder. (F) The same sheet folded to make a hollow square section.

Now the moment of inertia about horizontal axis is

$$I_{xx} = \sum \frac{bd^3}{12} + \sum Ay^2$$

$$I_{xx} = 30 \times \frac{600^3}{12} + 30 \times 600 \times \left(\frac{600}{2} - 184.8\right)^2$$

$$+ 407 \times \frac{30^3}{12} + 407 \times 30 \times \left(184.8 - \frac{30}{2}\right)^2 = 1.13 \times 10^9 \text{ mm}^4$$

Similarly,

$$I_{yy} = 5.17 \times 10^8 \text{ mm}^4$$

D. A Z-shaped section

This section is symmetric about both axes, so centroid lies at $\bar{x} = 350$ mm and $\bar{y} = 183.5$ mm.

$$I_{xx} = 2\left[365 \times \frac{30^3}{12} + 365 \times 30 \left(\frac{367}{2} - \frac{30}{2}\right)^2\right] + 30 \times \frac{307^3}{12} = 6.96 \times 10^8 \text{ mm}^4$$

$$I_{yy} = 2 \times \left[30 \times \frac{365^3}{12} + 30 \times 365 \times \left(365 - \frac{30}{2}\right)^2\right] + 307 \times \frac{30^3}{12} = 2.92 \times 10^9 \text{ mm}^4$$

E. An arc-shaped section

The same steel sheet can be rolled in to make an arch section as shown in figure.

$$I_{xx} = 0.3 \ tr_{in}^3 = 0.3 \times 30 \times 645^3 = 2.415 \times 10^9 \text{ mm}^4$$

F. A hollow box-shaped section

For the same steel sheet folded in to a hollow box-shaped section,

$$I_{xx} = I_{yy} = \frac{BD^3}{12} - \frac{bd^3}{12} = \frac{274.25 \times 274.25^3}{12} - \frac{214.25 \times 214.25^3}{12}$$

$$= 2.96 \times 10^8 \text{ mm}^4$$

Results

Comparison of Moments of Inertia of Different Cross-Sectional Shapes

S. No.	Cross-Sectional Shape	I_{xx}	I_{yy}
1	Straight plate	2.33×10^6 mm^4	2.79×10^9 mm^4
2	Hollow circle	2.12×10^8 mm^4	2.12×10^8 mm^4
3	L shape	1.13×10^9 mm^4	5.17×10^8 mm^4
4	Z shape	6.96×10^8 mm^4	2.92×10^9 mm^4
5	Arc shape	2.42×10^9 mm^4	—
6	Hollow box	2.96×10^8 mm^4	2.96×10^8 mm^4

UNSOLVED EXAMPLES

Question 13: Compare the elastic and plastic section moduli of the section shown in Fig. Ex. 2.13.

Question 14: Determine the moments of inertia and the products of inertia about the centroidal axes of the area shown in Fig. Ex. 2.14. Find the centroid polar moment of inertia.

Question 15: Find the elastic and plastic section moduli (in terms of b, d, and b_w) of the section shown in Fig. Ex. 2.15.

Question 16: Determine the torsional constant of the section shown in Fig. Ex. 2.16.

Question 17: Determine the ultimate torque that can be applied to the unreinforced concrete section with compressive strength, $f_c' = 25$ MPa as shown in Fig. Ex. 2.17.

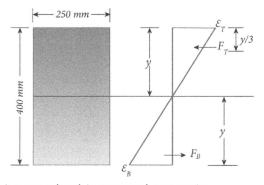

Figure Ex. 2.13 A rectangular plain concrete beam section.

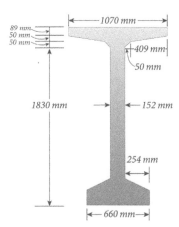

Figure Ex. 2.14 An I-section of plain concrete.

Question 18: Determine the transformed section properties and neutral axis depth of the composite section shown in Fig. Ex. 2.18. Steel section $= 125 \times 75$ mm, area of steel $= 2110$ mm^2, Modular ratio (n) of steel $= 13$.

Question 19: Determine the geometrical properties (first and second moments of areas, elastic section moduli) of the reinforced concrete and steel sections shown in Fig. Ex. 2.19.

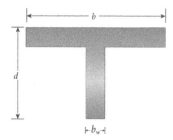

Figure Ex. 2.15 A T-section.

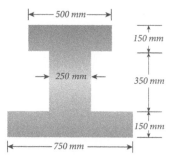

Figure Ex. 2.16 An unsymmetrical I-section.

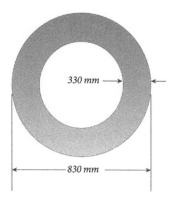

Figure Ex. 2.17 A hollow circular section of unreinforced concrete.

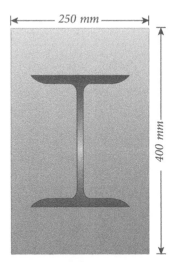

Figure Ex. 2.18 A concrete section with an embedded steel section.

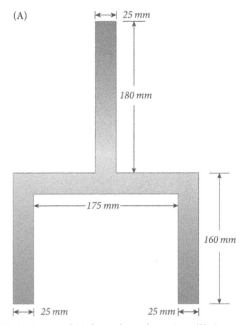

Figure Ex. 2.19 (A) An inverted Y-shaped steel section. (B) A rotated Z-shaped concrete section. (C) A plain concrete girder cross-section. (D) A rectangular concrete section with round corners.

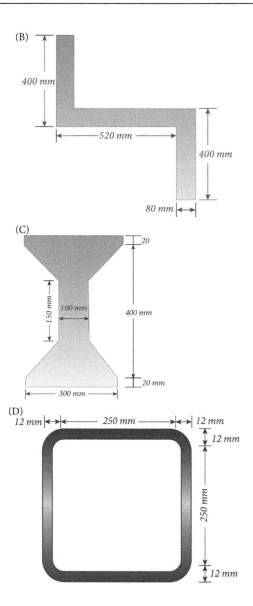

Figure Ex. 2.19 (Continued)

Question 20: Determine the location of neutral axis for the following sections shown in Fig. Ex. 2.20.

Question 21: Determine the torsional constants (C_x, C_y) and moments of inertia (I_{xx}, I_{yy}) of the following sections shown in Fig. Ex. 2.21.

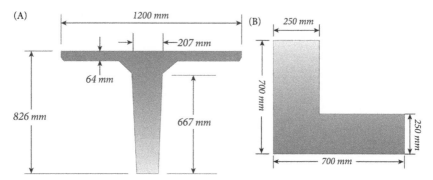

Figure Ex. 2.20 (A) A plain concrete T-section. (B) A plain concrete L-section.

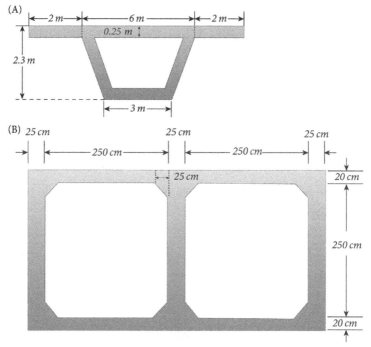

Figure Ex. 2.21 (A) A plain concrete culvert section. (B) A plain concrete box-girder section.

Question 22: Determine the geometric center and the moment of inertia about the centroidal axis of the above dam section shown in Fig. Ex. 2.22. All dimensions are in meters.

Question 23: Determine the shear center of the following sections shown in Fig. Ex. 2.23.

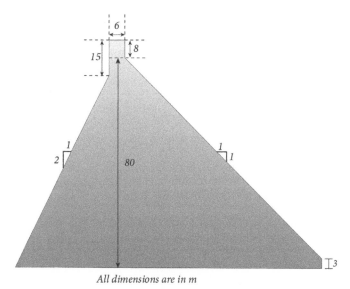

Figure Ex. 2.22 The cross-section of a dam.

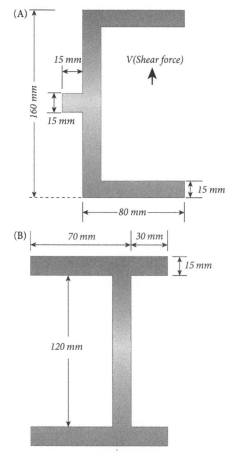

Figure Ex. 2.23 (A) An unsymmetrical channel section. (B) An unsymmetrical I-section.

SYMBOLS AND NOTATION

A	Area of the cross-section
Δ	Longitudinal deformation due to axial load along the members axis
v	Deformation due to end moment and end shear along the vertical direction
α	Rotation at the end due to end moment and end shear along the vertical direction
θ	Twist angle along the member due to torque at the end
σ_{xx}	Normal stress along the x-axis
σ_{yy}	Normal stress along the y-axis
σ_{zz}	Normal stress along the z-axis
υ	Poisson's ratio
τ_{xy}	Shearing stress in the x-plane in the y-direction
τ_{yz}	Shearing stress in the y-plane in the z-direction
τ_{xy}	Shearing stress in the z-plane in the x-direction
ε_x	Axial strain in the x-direction
ε_y	Axial strain in the y-direction
ε_z	Axial strain in the z-direction
ρ	Stress factor (depends on the section)
γ_{xy}	Shearing strain in the x-plane in the y-direction
γ_{yz}	Shearing strain in the y-plane in the z-direction
γ_{zy}	Shearing strain in the z-plane in the x-direction
y	Distance from the origin to the centroid of the shape along the y-direction
x	Distance from the origin to the centroid of the shape along the x-direction
τ_T	Lateral torsional buckling constant
ϕ	Change in the angle of twist
ϕ_p	Angle between the principal axis uu and the positive x-axis (principal angle)
GC	Geometric center
E	Modulus of elasticity of the material of the cross-section
F	Applied force
f_b	Stresses at the bottom most fiber of a cross-section
f_t	Stresses at the top most fiber of a cross-section
f_{cr}	Cracking strength of concrete
G	Shear modulus
I_x	Moment of inertia about the x-axis
I_y	Moment of inertia about the y-axis
I_{22}	Moment of inertia about 2-axis
I_{33}	Moment of inertia about 3-axis
I_u	Principal moment of inertia about uu
I_v	Principal moment of inertia about vv
J	Torsional constant
K	Spring constant
kd	Depth of neutral axis
L	Length of the member
M_x	Bending moment about the x-axis
M_y	Bending Moment about the y-axis
M_{cr}	Cracking moment of a concrete section
NA	Neutral axis

P	External force
PC	Plastic centroid
Q	Shear plane
q	Shear stress due to torsion
r	Radius of gyration
r_x	Rotation about the x-axis
r_y	Rotation about the y-axis
r_z	Rotation about the longitudinal or z-axis
SA_x	Shear area along x-axis
SA_y	Shear area along y-axis
SC	Shear center
S_x	Section modulus in the x-direction
S_y	Section modulus in the y-direction
S_2	Section modulus calculated in direction along 2
S_3	Section modulus calculated in direction along 3
SA_3	Shear area along 2-axis
S_{cr}	Shear area along 3-axis
T	Applied torque
T_z	Moment about the z-axis or torsion
U	Deformation
u_x	Movement along the x-axis
u_y	Movement along the y-axis
u_z	Movement along the member axis or the z-axis
V_x	Shear force in x-direction
V_y	Shear force in y-direction
W_{zz} or C_w	Warping constant
w_z	Out-of-plane movement (distortion) of the cross-section's points along the longitudinal axis
Z_2	Plastic section modulus about 2-axis
Z_3	Plastic section modulus about 3-axis

REFERENCES

American Institute of Steel Construction (AISC 360) (2005). *Steel construction manual* (13th ed.). New Delhi: Tata McGraw-Hill.

Ashby, M.F., & Jones, D.R. (2012). Engineering materials 1—An introduction to properties, applications and design (4th ed.). ISBN: 978-0-08-096665-6.

Beer, F. P., Johnston, E. R., & Dewolf, J. T. (2004). *Mechanics of materials* (3rd ed.). New Dehli, India: Tata McGraw-Hill Publishing Company Limited.

Gere, J. M., & Goodno, B. J. (2012). *Mechanics of materials* (8th ed.). USA: Cengage Learning.

MacGregor, J. G., Wight, J. K., Teng, S., & Irawan, P. (1997). *Reinforced concrete: Mechanics and design* (Vol. 3). Upper Saddle River, NJ: Prentice-Hall.

Mehta, P. K. (1986). *Concrete. Structure, properties and materials.* Upper Saddle River, NJ: Prentice-Hall.

Nawy Edward, G. (2005). *Reinforced concrete: A fundamental approach* (5th ed.). New Jersey: Prentice Hall.

Neville, A. M. (1995). *Properties of concrete*. New York, NY: John Wiley & Sons. 1996. ISBN: 0-582-23070-5.

Neville, A. M. (2006). *Concrete: Neville's insights and issues*. London: Thomas Telford.

Nilson, A. H., Darwin, D., & Dolan, C. W. (2005). *Design of concrete structures*. India: McGraw-Hill Education. ISBN: 0070598541, 9780070598546.

FURTHER READING

Jones, D. R., & Ashby, M. F. (2005). *Engineering materials 2: An introduction to microstructures, processing and design*. Boston, MA: Butterworth-Heinemann.

Kosmatka, S. H., & Panarese, W. C. (2002). *Design and control of concrete mixtures* (Vol. 5420). Skokie, IL: Portland Cement Association.

Leet, K. (1982). *Reinforced concrete design*. McGraw-Hill Higher Education (online). ISBN-10: 0070370249, ISBN-13: 978-0070370241.

Park, R., & Paulay, T. (1975). *Reinforced concrete structures*. New York, NY: John Wiley & Sons.

Sinha, S. N. (2014). *Reinforced concrete design*. India: Tata McGraw-Hill Education. ISBN-10: 9351342476, ISBN-13: 978-9351342472.

Wang, C. K., & Salmon, C. G. (1979). *Reinforced concrete design*. New York, NY: Harper & Row, Publishers. Incorporated. ISBN: 0700225145.

Winter, G. H., & Nilson, A. H. (1986). *Design of concrete structures*. New York, NY: McGraw-Hill Inc. ISBN 10: 0070465614, ISBN 13: 9780070465619.

CHAPTER THREE

Axial–Flexual Response of Cross-Sections

How are stresses produced in different parts of the cross-sections due to external actions? How can the stress resultants be computed from section geometry and material models?

CROSS-SECTION RESPONSE

Overview

The response of the cross-section pertains to the determination of the stress and strain state for a particular set of actions or deformations of the member, or computation of the stress resultants and the capacity for a particular failure criterion. In this chapter, we will focus on the axial–flexural response, primarily concerning the axial load and bending moments, more specifically, axial and bending deformations of cross-sections of beam-column-type members. A unified approach will be developed for various types of sections and materials, using some of the concepts and formulations presented in Chapter 2, Understanding Cross-Sections, for defining the cross-sections and computing the properties.

Actions, Stresses, Stress Resultants, and Capacity

In general, the stresses can be viewed as being produced either as a result of strains from deformations, or from the actions directly as shown in Fig. 3.1. The computation of these stresses can be carried out from the actions by using the appropriate cross-sections properties. They can be summed together to obtain stress resultants which will be in equilibrium with the applied actions. In a general continuum, six stress components are produced corresponding to six strain components (as already discussed in chapter: Understanding Cross-Sections), for the cross-sections of

Structural Cross Sections
DOI: http://dx.doi.org/10.1016/B978-0-12-804443-8.00003-8
137

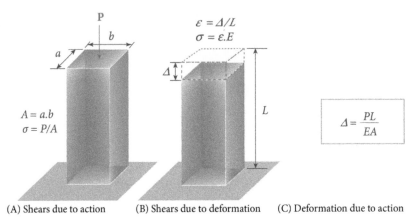

(A) Shears due to action   (B) Shears due to deformation   (C) Deformation due to action

Figure 3.1 (A)Stress can be computed as direct result of action, irrespective of the material or, (B) due to deformation, using material properties. (C) Both methods can be related by the action—deformation relationship for certain cases.

beam-column-type members, only two stresses are of interest from the design point of view:

- the normal, or the axial stress, along the local axis 1,
- the in-plane shear stress, on the local plane 2—3.

Although the stress and strain are related and linked to each other by the material stiffness matrix as discussed in Chapter 2, Understanding Cross-Sections, there are instances when it is possible for the stress and strain to be independent of each other. For example, if a rod of a certain cross-section and material is freely suspended or is placed on a frictionless plane, and then subjected to a temperature change, the rod will expand or contract producing an axial strain. However, no corresponding axial stress will be produced. On the other hand, if the ends of rod are restrained from movement, then almost a constant stress will be developed in the cross-section without the corresponding strain. In this case, the reactions produced at the end restraints or supports and the stress in cross-section will be related through the cross-sectional properties. Consider another example where a concrete column is subjected to axial load. The elastic stress and corresponding strain will be directly related to the load. However, if this same load is sustained for a long time, say 1 year, the stress will remain the same but the strain will increase due to creep phenomenon.

We have also seen in Chapter 2, Understanding Cross-Sections, that internal stresses can be directly determined from external actions using the assumptions of linear strain variation and linear stress distribution.

For such cases, where stresses are determined directly from actions, a check on the level of stress and its distribution against the "allowable" stress in the material can give some indication of the available reserve strength. This process is shown in Fig. 3.2. The allowable stress is generally assumed to be some fraction of the specified or characteristic strength of the material. This approach is widely used in the ASD (Allowable Stress Design) or WSD (Working Strength Design) procedures for timber and steel sections as well as for prestressed concrete. This was also used in the past for reinforced concrete sections as well.

However, for most reinforced concrete, prestressed concrete and composite sections, these conditions are not satisfied, especially near failure (or ultimate) state. In some cases, the stress distribution is dependent upon cross-section geometry, material properties as well as the type and magnitude of applied actions, and cannot be determined easily or explicitly. For these sections, a reverse process is used to determine the stresses based on predefined failure criteria, and corresponding stress resultants and then compared with applied actions that are considered with an adequate margin of safety, as shown in Fig. 3.3.

For reinforced concrete, the WSD approach is considered by most current design practices and codes to be inadequate and obsolete for proper incorporation and determination of the factor of safety (FOS), because it gives no indication of the actual capacity and performance of cross-sections beyond the allowable stress range, especially near

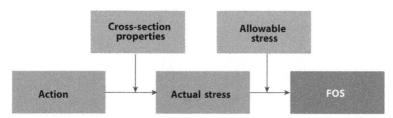

Figure 3.2 Determination of actual stresses due to applied actions as the basis for cross-section design.

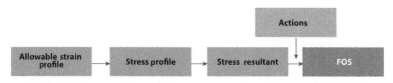

Figure 3.3 Determination of internal stress resultants as the starting point for cross-section design.

and beyond failure conditions in case of overload. The stress–control methods also rely on assumptions such as linear elastic material relationships, which may not be true for concrete and steel at certain stress levels. The alternative approach is to determine the stress resultants independently by using a predefined failure criterion, either by limiting the maximum strain value (hence limiting the stresses and deformations indirectly) or by limiting the maximum stress value directly. From this failure criterion, the stress distribution is determined across the cross-sections by using the appropriate material models. The summation of these stresses gives the "Internal Stress Resultant" or the "Intrinsic Capacity" of the cross-sections. These stress resultants are then compared with the applied actions to obtain the available FOS or margin of safety. If this margin is too large or too small, the section parameters (or the material properties) are revised and the process is repeated until a suitable or reasonable safety or performance level is obtained. Therefore the determination of internal stress resultant is crucial to determine the margin of safety and hence to meet the key design objective.

The axial strain alone gives rise to axial stresses and to the axial stress resultant, which is a summation of the stress over the area, and the bending moment stress resultant, which is a product of the axial stress resultant and its distance from a particular reference axis. Most of the discussion in this chapter will focus on "compact" and "plastic" sections, in which the buckling of cross-section elements under compressive stress does not occur before yielding or failure of the material.

EXTERNAL ACTIONS AND INTERNAL STRESSES
Combined Axial Stress—The Basic Equation

The axial stresses are produced by a combination of (see Fig. 3.4):

- The axial strain induced due to axial deformation, produced due to axial load.
- The axial strain induced due to curvature deformation, produced due to bending moment about any axis.
- The axial strain induced due to out-of-plane distortion of the cross-section due to warping, produced due to torsion.
- Axial strain produced due to restrained expansion or contraction of members.

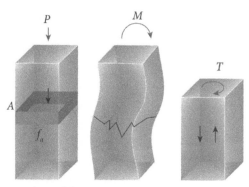

Figure 3.4 Actions causing axial stresses.

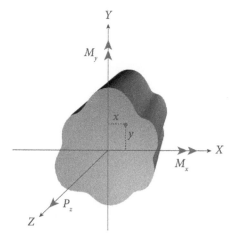

Figure 3.5 Actions causing elastic axial stress at any point on the cross-section.

The properties needed to determine axial stress due to all corresponding actions therefore are:

- the cross-sectional area, A,
- the moment of inertia about the axis of bending, or about the two principal axes, I,
- the warping constant, W.

If warping can be neglected, or if there is no significant torsion, the elastic axial stress at any point located at x-, y-coordinates on the cross-section, as shown in Fig. 3.5, can be easily determined from the combined stress equation as follows:

$$f_a = \frac{P_z}{A} + \frac{M_x y}{I_{xx}} + \frac{M_y x}{I_{yy}} \qquad (3.1a)$$

The combined stress equation, in the above form only makes sense for homogenous sections, for linear strain variation across the cross-section, and for linear stress—strain relationship. So it is directly applicable to compact or plastic steel sections, wood sections and for uncracked concrete sections, all within the linear stress—strain range. The equation can also be used for sections with more than one material, if they are transformed to an equivalent section with respect to some base material and all other materials have been converted to the base material using the appropriate value of modular ratio (the ratio of their elastic moduli). It means that if the cross-section is made up of several shapes of different materials, then the stress at a point in a particular shape, determined by the combined stress equation, should also be multiplied by the modular ratio of the material at that point, as shown in Eq. (3.1b).

$$f_{a_i} = n_i \left(\frac{P_z}{A} + \frac{M_x y}{I_{xx}} + \frac{M_y x}{I_{yy}} \right) \tag{3.1b}$$

where n_i is the modular ratio for any material "i" with respect to a base material with elastic modulus E_{base}.

$$n_i = \frac{E_i}{E_{base}} \tag{3.2}$$

This is especially important for reinforced concrete sections, prestressed concrete sections, and for composite sections when using their equivalent transformed properties. However, even in that case, the assumption of linear stress variation is needed. Fig. 3.6 shows a 3D plot of the axial stress variation in a composite section made up of concrete and steel.

Fig. 3.7 shows the axial stress distributions in some of the common cross-sections, both homogenous and heterogeneous.

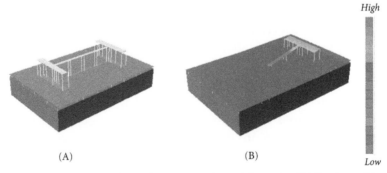

Figure 3.6 Axial stresses depend on material modular ratio. (A) Section with axial load. (B) Section with moment.

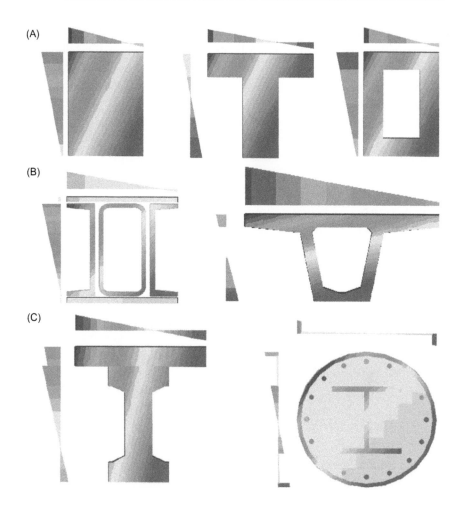

Figure 3.7 Application of combined stress equation to determine axial stress distribution of various types of cross-sections (CSI Section Builder). (A) The axial stress at any point on the cross-section for a combination of axial load and biaxial moment can be determined by using the combined stress equation. (B) The combined stress equation can be used to determine stress distribution for prestressed concrete sections, as well as for built-up steel sections. (C) In a composite section, the axial stress in each component of different material is computed on the basis of its modular ratio with respect to a base material. Here two examples are shown. Examples of a concrete section, made of concrete portions of different strength (and modulus of elasticity a steel tube), filled with reinforced concrete and an I-section embedded in concrete are shown.

The Usefulness and Applicability of the Combined Stress Equation

The combined stress equations (3.1) are generally used in the following context:

- To compute the elastic axial stresses in steel and other metal sections, for un-factored loads.
- For computing stressing in homogenous sections, such as masonry, timber, etc. before onset of cracking or failure in tension.
- To determine the tensile and compressive stresses in prestressed beams and girders. The prestressing force is represented by the axial action P_z, and the eccentricity of the force is represented by bending moments. These bending moments are added to the applied moments due to loads. In this case, the most important considerations are the stages of section built up, prestressing application and load application. Another consideration is the difference in concrete strength of various parts of the section. For example, for a particular prestressed section in a bridge girder, at least eight different cases need to be considered for evaluating the maximum and minimum stresses.
- Determination of the cracking stresses in reinforced concrete sections after determination of the neutral axis and transformed properties.
- Determination of the maximum stress in steel sections subjected to axial load and moment. However, a modified form of this equation is used for the actual design of steel sections using the concept of combined stress ratio, described next.

The Combined Stress Ratio for Axial Stress

The concept of combined stress ratio is an extension of the combined stress equations (3.1). For a homogenous perfectly elastic isotropic material, if the total combined stress is less than a specified or allowable value, then the section is considered to be "safe" for the specified combination of actions. However, in practice, the combined stress equation cannot be used directly to determine the adequacy of a section because, for most materials, the allowable or limiting stress values for axial load and bending moments about the x- and y-axes are not the same. Therefore the combined stress equation is modified in the form of stress ratios, rather than actual stress values, as follows.

When all allowable stresses are the same,

$$r = \frac{f}{F} \leq 1.0 \qquad (3.3a)$$

When different allowable stresses are specified for each action,

$$r = \frac{f_a}{F_a} + \frac{f_{bx}}{F_{bx}} + \frac{f_{by}}{F_{by}} \leq 1.0 \qquad (3.3b)$$

This equation simply means that for a cross-section to be acceptable, the sum of ratios between the actual stress and the limiting stress for each stress component should be less than or equal to 1.0. It is important to note that the concept of combined stress ratio can be used both for allowable stress design as well as for ultimate or load and resistance factor design. The combined stress ratio concept is mostly used in the design of steel sections.

Interaction of Stresses Due to Axial Load and Moment

As discussed in the previous sections, the axial stresses primarily depend on the simultaneous presence of axial load, bending moment about the x-axis, and bending moment about the y-axis. This dependency allows to define a "failure surface" for any limiting value of stress at a particular point or any location on the cross-section. For example, if the maximum stress anywhere on the section is limited to a certain value, then infinite combinations of axial load (P), bending moment about x-axis (M_x), and bending moment about y-axis (M_y) can be obtained that will produce this stress. A plot of these infinite combinations will define the failure surface for the specified criteria. Such surfaces based on internal stress resultants are discussed later in this chapter. Similar interaction plots (3D representation of all possible P, M_x, and M_y points) can be obtained for other stress conditions as well. To obtain such a surface, all we need to do is to vary each action one by one with some finite interval while keeping the other two constant until limiting stress value is obtained. We can plot all such points to obtain the desired failure surface. A typical interaction surface for a reinforced column section is shown in Fig. 3.8. This topic will be discussed in greater detail in the subsequent sections.

Fig. 3.9 shows an axial load versus moment (in one direction) interaction diagram for an isotropic homogeneous material. As mentioned earlier, it shows all the possible combinations of axial load and moment which result in development of failure stresses in material. Any point (e.g., E) representing the actions (factored axial load and applied moment) inside the diagram will correspond to a safe combination of P and M

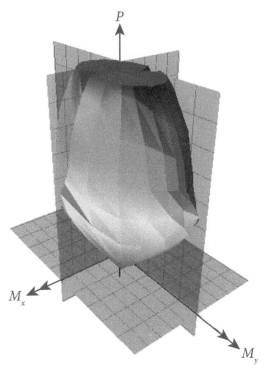

Figure 3.8 A Typical interaction surface between axial load and moments about two orthogonal axes for a given axial strain limit (CSiCOL).

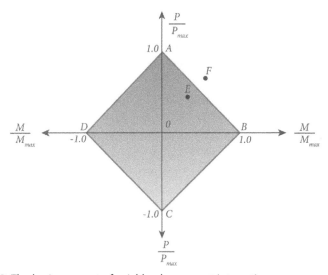

Figure 3.9 The basic concept of axial load−moment interaction.

while a point outside the diagram (e.g., F) will imply the failure of section under those actions. The shape of these interaction diagrams vary widely for a variety of practical materials. For reinforced concrete sections, the shape of $P-M$ interaction depends on cross-sectional shape, behavior, and strengths of individual constituent materials (steel and plain concrete) and reinforcement configuration in section. As an example, Fig. 3.10A shows the $P-M$ interaction (also sometimes referred to as failure curve) for a material with no tensile capacity. The interaction is only shown in compression side of the diagram for moments in both clockwise and counterclockwise directions. A common example of such behavior is plain concrete, where sometimes it is not important or required to consider its tensile strength. Similarly, Fig. 3.10B shows the interaction for a material having a tensile strength equal to one-half of its compressive strength. The diagram shifts towards tension side of axial load depending upon exhibited capacity in tension. At points B and D, failure mechanism changes from compression-controlled to tension-controlled and therefore, often referred to as balanced points. The development of such $P-M$ interaction curves and the factors affecting their shape will be discussed in detail in later sections.

Principal Stresses and the Mohr's Circle
The Basic Concept
In a general continuum or 3D solid, six stress components are present. However, in a beam-column-type member cross-section, not all of them are equally relevant. It has been illustrated in various elementary strength of materials text books (Den Hartog, 2014; Singer & Pytel, 1980; Timoshenko, 1930, 1953; Volterra & Gaines, 1971) that if an element from a 3D solid is cut at 45 degrees, the two pairs of shear stresses (on vertical and horizontal faces) combine in such a manner that their action is same as that of two pairs of normal stresses (one tensile and one compressive) acting on that 45 degrees face and of value equal to that of the shear stresses. Therefore the six stresses acting on a 3D element can be combined into a pair of inclined compressive stresses and a pair of inclined tensile stresses that act at right angles to each other. These are known as principal stresses, which arise from the interaction of axial and shear stresses. The development of principal stresses as a result of interaction between stresses, in a beam element subjected to combined axial force, shear force, and moment, is shown in Fig. 3.11 (see Fig. 3.12).

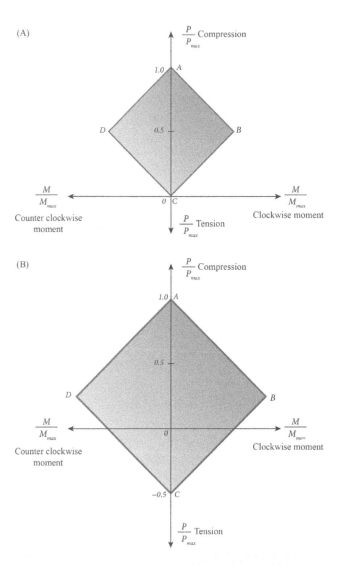

Figure 3.10 Variation in shape of axial load—moment interaction diagram for a material with different tensile strengths, compared to compressive (based on Wight & MacGregor, 2012). (A) A material with no tensile strength (B) A material with tensile strength equal to one-half of its compressive strength.

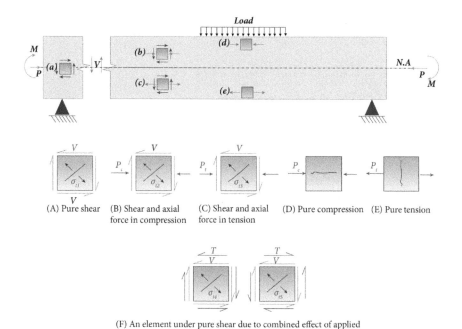

(A) Pure shear   (B) Shear and axial   (C) Shear and axial   (D) Pure compression   (E) Pure tension
force in compression   force in tension

(F) An element under pure shear due to combined effect of applied
shear and shear stresses developed due to torsion

Figure 3.11 Principal stresses in a beam ($\sigma_{t1} \neq \sigma_{t2} \neq \sigma_{t3} \neq \sigma_{t4} \neq \sigma_{t5}$).

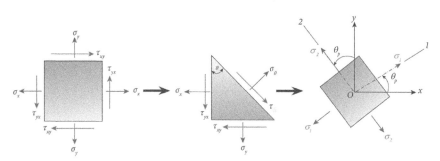

Figure 3.12 Transformation of direct and shear stresses in a given $x-y$-coordinate system in to principal stresses in $1-2$ principal coordinate system.

The principal maximum stress and a principal minimum stress can be obtained using a well-known 2D graphical representation known as Mohr's circle. This technique (see Fig. 3.13) is extremely useful as it enables to visualize the relationships between the normal and shear stresses acting on various inclined planes at a point in a stressed body. Mohr's circle can also be used to determine the angle of principal stresses, maximum shear stresses, and stresses on inclined planes.

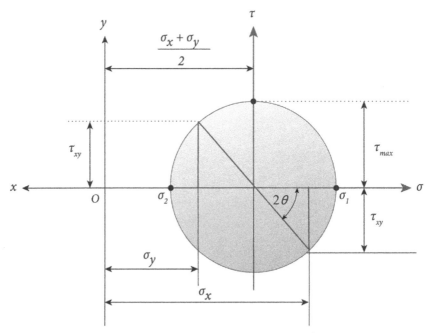

Figure 3.13 The Mohr's circle is a very simple tool for determining the value and direction of the principle stresses from axial stress and the shear stress.

To establish Mohr's circle, we first start with the stress transformation equations for plane stress at a given location,

$$\sigma_{x1} = \frac{\sigma_x + \sigma_y}{2} + \frac{\sigma_x - \sigma_y}{2} \cos 2\theta + \tau_{xy} \sin 2\theta \qquad (3.4)$$

$$\tau_{x1y1} = \frac{\sigma_x + \sigma_y}{2} \sin 2\theta + \tau_{xy} \cos 2\theta \qquad (3.5)$$

If we vary θ from 0 to 360 degrees, we will get all possible values of σ_{x1} and τ_{x1y1} (the transformed values) for a given stress state. Combining the above equations, we have,

$$\left(\sigma_{x1} - \frac{\sigma_x + \sigma_y}{2}\right)^2 + \tau_{x1y1}^2 = \left(\frac{\sigma_x - \sigma_y}{2}\right)^2 + \tau_{xy}^2 \qquad (3.6)$$

This is the equation of a circle, plotted on a graph where the abscissa is the normal stress and the ordinate is the shear stress. This is easier to see if we interpret σ_x and σ_y as being the two principal stresses, and τ_{xy} as being the maximum shear stress. Then we can define the average stress, σ_{avg}, and a "radius" R (which is just equal to the maximum shear stress),

$$\sigma_{avg} = \left(\frac{\sigma_x + \sigma_y}{2}\right) \quad \text{and} \quad R = \sqrt{\left(\frac{\sigma_x + \sigma_y}{2}\right)^2 + \tau_{xy}^2} \qquad (3.7)$$

The circle equation above now takes on a more familiar form,

$$(\sigma_{x1} - \sigma_{avg})^2 + \tau_{x1y1}^2 = R^2 \qquad (3.8)$$

The circle is centered at the average stress value, and has a radius R equal to the maximum shear stress τ_{max}, as shown in Fig. 3.13.

The general equations for three stress components using this principle are:

$$\sigma_1 = \frac{\sigma_x + \sigma_y}{2} + \sqrt{\left(\frac{\sigma_x - \sigma_y}{2}\right)^2 + \tau_{xy}^2} \qquad (3.9a)$$

$$\sigma_2 = \frac{\sigma_x + \sigma_y}{2} - \sqrt{\left(\frac{\sigma_x - \sigma_y}{2}\right)^2 + \tau_{xy}^2} \qquad (3.9b)$$

$$\tau_{max} = \frac{1}{2}(\sigma_1 - \sigma_2) = \sqrt{\left(\frac{\sigma_x - \sigma_y}{2}\right)^2 + \tau_{xy}^2} \qquad (3.9c)$$

The rotations and directions of these stresses are shown in Fig. 3.12. In columns and beams, generally only a single axial stress is of interest and the axial stress in other direction is neglected. The combination of a single axial stress, say σ_x with in-plane shear τ_{xy} can be represented by simplified equations as:

$$\sigma_1 = \frac{\sigma_x}{2} + \sqrt{\left(\frac{\sigma_x}{2}\right)^2 + \tau_{xy}^2} \qquad (3.10a)$$

$$\sigma_2 = \frac{\sigma_x}{2} - \sqrt{\left(\frac{\sigma_x}{2}\right)^2 + \tau_{xy}^2} \qquad (3.10b)$$

$$\tau_{max} = \frac{1}{2}(\sigma_1 - \sigma_2) = \sqrt{\left(\frac{\sigma_x}{2}\right)^2 + \tau_{xy}^2} \qquad (3.10c)$$

Significance of Principal Stresses in RC Beam Design

In reinforced concrete beams, it is interesting to note that the inclinations and magnitudes of principle stresses vary considerably across the cross-section depth at any location and along the beam length, depending on the presence of bending moment, shear force, axial force, and torsion. At neutral axis location, they are always inclined at 45 degrees to the longitudinal axis as axial stresses are zero. Since the principle tension stresses resulting from either shear alone or from the combined action of shear and bending can cause potential cracking, it should be properly taken care of in design of concrete members. This principle tension (or diagonal tension) governs the "shear stress" capacity of concrete section and the principle compression determines the direction and magnitude of compression struts in the strut-and-tie model for shear and torsion design. Fig. 3.14 shows the distribution of axial stresses, shear stresses, maximum principal stresses and their directions, along the length and depth of a beam element subjected to vertical loads.

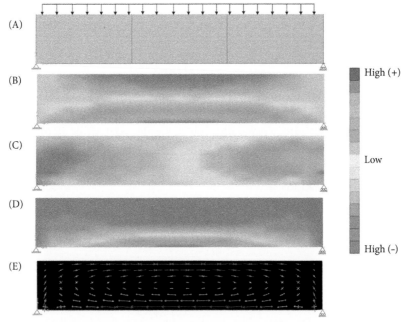

Figure 3.14 Axial, shear, and principle stress distribution in a beam. (A) Beam subjected to vertical loads, (B) axial stress, (C) shear stress, (D) maximum principal stress, and (E) principle stress directions.

The arrows in (E) are showing the change in principal directions along the length and depth of beam. It can be seen that their inclination at half-way along the depth approach to approximately 45 degrees with respect to longitudinal axis.

While the discussion of stresses produced by actions is useful to understand the relation between actions, stress, and strain interaction, this is not of much practical use for designing the cross-sections beyond the linear elastic response range, especially for reinforced, composite and prestressed concrete. For this, our understanding of internal stress resultants, independent of external actions is important.

AXIAL—FLEXURAL STRESS RESULTANTS
The Diversity of the Problem and the Need for Unified Approach

Generally the design of structural members can be divided and categorized in many ways based on type of member, type of material, type of applied action, type of cross-section shape, design code, and design method. Often each of these types and their combinations are treated separately and differently for the purpose of determining design strength of members, which is largely derived from the strength and capacity of its cross-section. Table 3.1 shows the wide spectrum and diversity of concrete cross-section design problem.

As discussed in Chapter 1, Structures and Structural Design and Chapter 2, Understanding Cross-Sections, the structural member cross-section can be classified into many categories, and often each category is traditionally treated separately with different theoretical treatment or design procedures. However, the boundaries between these classifications are often arbitrary and transitional. For example, un-reinforced section can be a special case of reinforced section with "zero" reinforcement. Similarly a reinforced concrete section with round bars and composite section with steel shapes is essentially similar as-far-as the strength/ capacity is concerned. A rectangular section or a tee/flanged section is a special case of a general polygon shape, and singly or doubly reinforced sections are special cases of arbitrary reinforcement layout. It would therefore, be advantageous to develop an integrated and unified treatment of all classes of cross-sections. This will not only provide better

Table 3.1 The Diversity of Concrete Cross-Section Design Problem

Types of Materials and Their Combinations	Types of Actions	Location of Reinforcement	Stress–Strain Curve	Cross-Section Shape	Design Approach and Method
Un-reinforced concrete	Uniaxial bending, M_x or M_y only	Singly reinforced	Simplified rectangular block	Rectangular	Allowable stress design (*ASD*)
Reinforced concrete	Uniaxial bending and axial force, M_x or M_y and P	Doubly reinforced	Semiparabolic and full parabolic curves	Circular	Ultimate strength design (*USD*)
Partially prestressed concrete	Biaxial bending, M_x and M_y	Arbitrarily reinforced	Curves for confined and unconfined concrete	Flanged	Load resistance factor design (*LRFD*)
Fully prestressed concrete	Biaxial bending and axial force, M_x and M_y and P		Linear elastic	General polygonal	Performance-based design (*PBD*)
Fiber reinforced concrete			Bilinear elasto-plastic	Multicell sections	
Steel–concrete composite			Elastic, post elastic, strain hardening		

understanding of the intrinsic behavior but also yield in numerical formulations suitable for implementation into general-purpose computer programs.

With some critical thinking, it is possible to unify various classifications of the cross-sections, to develop a unified set of governing equations for determining the axial–flexural response. Such unified approach is specifically useful for application to computer-based solutions and implementations.

The Unification of Cross-Section Materials

Let us consider a concrete section with some normal reinforcement, some prestressing strands, and a steel section placed inside. When this section is subjected to biaxial moment and axial load, we have a very general case of partially prestressed, composite beam-column.

Let us also consider that the strain versus depth relationship for each material is known. Similarly, let us assume that the stress–strain relationship for each material is also available. This relationship can be represented by a set of discrete points, or a continuous function, and will enable us to handle arbitrary stress–strain curves of concrete, as well as steel. Developing the stress resultant equation for this case will enable us to handle practically any case of cross-section analysis and design. This will cover concrete sections that are completely un-reinforced to fully prestressed to composite sections, either made up of concrete and steel, or from concrete of different strength or different stress–strain characteristics.

Also, as far as the axial–flexural capacity is concerned, the prestressing strand is just another reinforcement type with higher (or different) yield strength. In fact, we shall see later that any type of material can be considered, in the evaluation of axial–flexural capacity. The prestressing force is just another axial load. Therefore we will consider a partially prestressed column as our basic cross-section.

The Unification of Cross-Section Shapes and Configurations

Almost every cross-section can be defined by using multiple shapes and points and almost all shapes can be represented by polygons. The general polygons can be further subdivided into simple quadrilaterals, and ultimately into triangles. Therefore we can develop the equations for a general cross-section made up of several polygons and points having different properties. This will cover practically any cross-section, and we do not need to handle rectangular, tee or circular sections as specific cases.

Points representing rebars and prestressing strands with arbitrary location, area, and stress—strain curve will enable us to handle all cases of reinforcement, from singly reinforced to arbitrarily reinforced and prestressed sections.

The Unification of Line-Type Structural Members

How do we differentiate between a beam and a column? The differentiation cannot be based on orientation alone. In general, a fame member is subjected to axial load and two moments. Therefore it is obvious that the axial—flexural theory should be developed for the general beam-column. When using the formulation for beams, the axial load may be neglected (or better it may always be included even small). So it is quite apparent that there is no real need to differentiate between beam and column cross-sections, or between uniaxial and biaxial loads to determine stress resultants. In fact, the general stress resultants, as we shall see later, are independent of the applied actions, and are an intrinsic property of the cross-section. The general case of a section undergoing bending about an arbitrary axis will cover biaxial loaded columns, as well as beams. Therefore, for our development, we will consider a cross-section undergoing axial and biaxial—flexural deformations (see Fig. 3.15).

The Unification of Design Approaches and Design Codes

In general, three basic approaches are prevalent for cross-section design in different design codes:
- The Working Strength Design (WSD)
- The Ultimate Strength Design (USD)
- The Performance-Based Design (PBD)

The key difference between the WSD and USD, as far as the determination of axial—flexural stress resultants is concerned, is the way the stress distribution is determined. In WSD, the stress profile is specified directly, is often linear, and is a fraction of the material strength (called the allowable stress or the working stress). In USD, the strain profile is specified directly, from which the stress profile is determined based on the material stress—strain relationship. In PBD, the expected material properties, often without the use of arbitrary capacity reduction factors for various levels of risk, are used to determine cross-section deformations, strength, and seismic response. However, a general set of equations based on equilibrium conditions will satisfy the requirements of the WSD, USD, and consequently PBD.

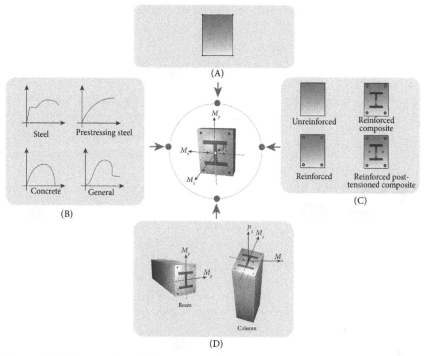

Figure 3.15 A general partially prestressed, composite beam-column cross-section, subjected to bending about arbitrary axis *m*—*n*. (A) General polygon shape (B) Unified materials (C) Unified composition (D) Unified design type.

We have seen from the above discussion that it is possible and convenient to deal with the cross-section flexural capacity problem in a general and unified manner. We will now look at the development of a set of general equations to determine the axial—flexural stress resultants.

THE GENERAL STRESS RESULTANT EQUATIONS

The Basic Assumptions, Their Necessity, and Validity

The development and presentation of flexural theory for concrete beams and columns is based on the following basic assumptions (MacGregor et al., 1997; Nawy et al., 2005; Nilson et al., 2010).

1. Plane sections remain plane after deformation.
2. Concrete does not take any tension.
3. There is perfect bond between various parts of the section.

4. The maximum strain in concrete is limited to some value, say 0.003 or 0.0035, etc.

5. Internal stress resultants and external actions are in equilibrium.

However, the only necessary assumption is the last one that relates to the equilibrium between the internal stress resultants and external actions for sections at rest. All other assumptions have been made to simplify the development of the governing equations and the corresponding design procedures. Here we will review the validity and necessity of these assumptions and then develop the governing equations based on the assumption of equilibrium alone.

1. It is also not necessary to assume that plane sections remain plane after deformation. A nonlinear strain distribution across the section width and depth may exist. This consideration, in no way, violates the basic equilibrium condition. Let us consider that all components of the cross-section are capable of deformation, independently. It means that the connection between various parts may not be perfect, and there may be a shear lag or lack of perfect bond among them. All we need in this case is a strain distribution function across the section for each component of the cross-section. However, in most cases, the assumption of perfect bond and hence a single strain function for all parts of the section considerably simplifies the computational aspects, but fails to account for problems involving bond slip, and in case of deep beams.

2. Another consideration is that the stress in each component can be determined in accordance with the stress—strain relationship defined for that component, with or without local buckling failure. This means that the entire section does not have to be plastic. It is possible that some parts of the section are made from separate shapes with their stress—strain curves defined in such a way that it fails even before reaching its yield strain. The above assumption only signifies that section capacity will be computed by using stress—strain relationship defined for each section component. So the assumption of zero tensile stress in concrete becomes redundant if the concrete stress—strain curve includes the tension part as well.

The above two considerations are enough to develop a general axial—flexural theory applicable to a wide range of cross-sections (hot-rolled steel, built-up and composite sections), design methods, and design code. They do not rely on any of the assumptions used in conventional flexural theory for reinforced concrete sections.

The Basic Stress Resultant Equations—Simple Formulation

The major difference between various design codes to determine the cross-section strength comes from using a different approach for incorporating FOS and different definition of the material stress—strain model. We have already addressed the issue of different material models. Fig. 3.16 shows three different approaches to determine the moment capacity of a reinforced concrete beam. Approach (a) is the most basic approach with an assumed linear elastic stress profile. The resultant compressive and tensile force can easily be determined in this case which can be multiplied with internal moment arm to determine resultant moment capacity. Approach (b) is used in USD where nonlinear stress profile with a specified maximum strain is used. However, as discussed in Chapter 2, Understanding Cross-Sections, this actual nonlinear stress profile is simplified in to an idealized stress block, which makes the determination of resultant compressive force and neutral axis depth convenient. Approach (c) is the determination of complete moment—curvature response of section which has a lot of applications in various latest PBD procedures. Moment—curvature response of sections will be discussed in detail in Chapter 5, Response and Design for Shear and Torsion.

As an example, let us consider a simple case of a simply supported singly reinforced concrete beam. The conventional design using ultimate strength approach requires the approximation of actual compressive stress distribution in concrete (which may have the form of a rising parabola) in to a rectangular stress block. This equivalent rectangular stress block

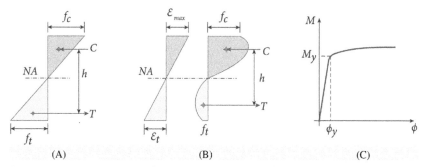

Figure 3.16 Determination of moment capacity using various design approaches: (A) using allowable stresses directly as required in Working Stress Design (WSD); (B) using a specified maximum strain as required in Ultimate Strength Design (USD); and (C) using complete moment—curvature curve as required in various performance-based design procedures.

(often referred to as Whitney's block) is a convenient way to determine internal compressive force. ACI 318 building code defines the depth of stress block as $a = \beta_1 c$, where c is the actual neutral axis depth and β_1 is a coefficient such that the area of equivalent rectangular stress block is approximately the same as that of the parabolic compressive block. This will result in same value of compressive force in both cases. Moreover, ACI 318 code recommends a value of 0.85 times the maximum 28-day compressive strength of concrete (f_c') as the average compressive stress (width of rectangular stress block). A maximum allowable strain value of 0.003 is adopted as safe limiting strain.

Now applying the basic assumptions of flexural theory, we can easily derive the moment capacity of this simple case. The equilibrium condition requires that the maximum tensile force in steel rebars $(T = A_s f_y)$ must be balanced by the total compressive force in concrete (i.e., the volume of equivalent rectangular stress block in Fig. 3.17). Therefore

$$C = T \tag{3.11}$$

$$0.85 f_c' b a = A_s f_y \tag{3.12}$$

$$a = \frac{A_s f_y}{0.85 f_c' b} \tag{3.13}$$

If the area of reinforcement (A_s) normalized by gross area of section (bd) is denoted by reinforcement ratio $(\rho = A_s / bd)$, the equation for a can be rewritten as,

$$a = \frac{\rho d f_y}{0.85 f_c'} \tag{3.14}$$

The internal resistance of section (nominal moment capacity) can be determined by applying moment equilibrium about either the position of resultant C or T.

$$M_n = A_s f_y \, jd \tag{3.15}$$

$$M_n = 0.85 f_c' b a \, jd \tag{3.16}$$

where jd is the distance between resultant compressions and tensile forces. For the simplified rectangular stress block,

$$jd = d - \frac{a}{2} \tag{3.17}$$

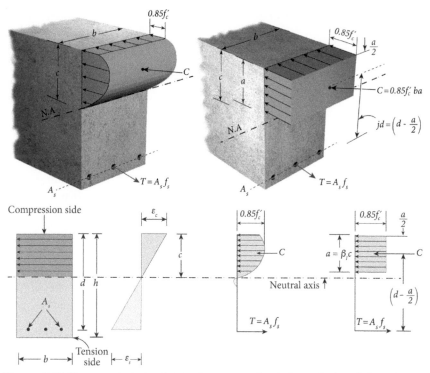

Figure 3.17 Stress and strain distribution across depth in a singly reinforced concrete beam. (From left to right) Beam cross-section; strain distribution; actual stress block; and assumed equivalent rectangular stress block.

Therefore the nominal moment is,

$$M_n = A_s f_y \left(d - \frac{a}{2} \right) \tag{3.18}$$

$$M_n = 0.85 f_c' b a \left(d - \frac{a}{2} \right) \tag{3.19}$$

ACI code recommends a strength reduction factor (ϕ) of 0.9 to be applied to nominal moment strength. The depth of neutral axis can be determined by equating the ratios of sides of similar triangles (see Fig. 3.17), as follows:

$$\frac{c}{d} = \frac{\epsilon_c}{\epsilon_c + \epsilon_s} \tag{3.20}$$

$$c = \left(\frac{0.003}{0.003 + f_y/E} \right) d \tag{3.21}$$

ACI code recommends the following values for stress block depth factor (β_1):

$$\beta_1 = \left\{ \begin{array}{c} 0.85, for\ 17 < f_c' < 28\ \text{MPa} \\[2mm] 0.85 - 0.05 \left(\dfrac{f_c' - 28}{7} \right), for\ f_c' > 28\ \text{MPa} \end{array} \right\} \qquad (3.22)$$

In line with the above discussion, let us first consider a cross-section (as shown in Fig. 3.18) made up of a single shape of a single material under flexural rotation about an inclined axis due to biaxial bending, and a known location of neutral axis (NA). Let us also assume that we have defined a strain distribution across the section, and the stress is defined in terms of strain. We therefore, have strain ε and stress σ as a function of x and y.

A summation of this stress across this general, two-dimensional cross-section area will give us the required axial–flexural stress resultants for the particular strain distribution as:

$$P_z = \iint\limits_{x\ y} \sigma(x, y)\ dx\ dy \qquad (3.23a)$$

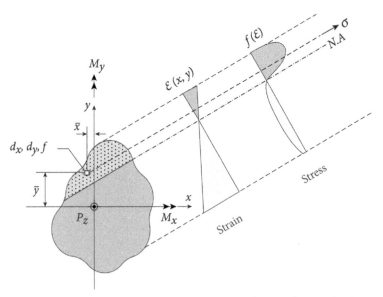

Figure 3.18 Strain and stress on a cross-section made up of a single shape and material, subjected to bending about an arbitrary axis.

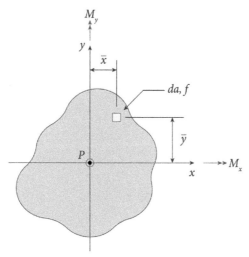

Figure 3.19 Stress resultants for a 2D stress field continuum.

$$M_x = \int_x \int_y \sigma(x, y) \, dx \, dy \cdot \bar{y} \qquad (3.23b)$$

$$M_y = \int_x \int_y \sigma(x, y) \, dx \, dy \cdot \bar{x} \qquad (3.23c)$$

It can be seen that these stress resultants are a function of stresses that in turn are a function of strains. The strain itself is a function of location on the cross-section. Therefore these equations are a generalized expression to find the force, generated by a certain stress, in a plane. This plane may be steel, concrete, timber, or any other arbitrary material with an assumed or known stress—strain relationship (see Fig. 3.19).

These general equations are valid for any shape, composition, material or strain distribution on the continuum. It is basically dependent on the final stress distribution, including the effect of stresses induced due to applied load deformation, prestress, prestrain, time-dependent stress, and time-dependent strain. It is, however, not practical to determine this stress distribution, and to evaluate these integrals directly for a general cross-section with arbitrary stress distribution. Several procedures for evaluating these integrals have been developed. Some are based on Green's Theorem, which evaluate the area integrals by integrating along the boundary (Weisstein, E. W. *Green's theorem*, Wolfram Web Resource),

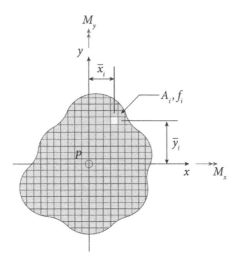

Figure 3.20 Stress resultants for a 2D stress field using stress fibers.

while others are based on converting the areas into line strips. Most of these procedures or models are based on certain assumptions and are specific to certain type of cross-sections and materials. If the cross-section is discretized into small fibers, as shown in Fig. 3.20, then the above equations can be converted into a simple summation form as follows:

$$P = \sum_{i=1}^{n} \sigma_i(x, y)\, A_i \tag{3.24a}$$

$$M_x = \sum_{i=1}^{n} \sigma_i(x, y) A_i \bar{y}_i \tag{3.24b}$$

$$M_y = \sum_{i=1}^{n} \sigma_i(x, y) A_i \bar{x}_i \tag{3.24c}$$

This form of equations is suitable for computer-aided determination of stress resultants. To use the above equations, the cross-section must first be divided into very small fibers (or mesh elements) and then the stress must be determined at the center of each fiber. The accuracy of the stress resultants determined in this manner would depend on the size of mesh and on the shape, size, and composition of the cross-section. This approach offers unique solutions to determine the axial–flexural stress resultants and the complete capacity surface of general cross-sections made up of different materials. This simple formulation addresses the case of cross-sections subjected to a particular level of induced strain, such as from external loads and moments only. The two-step process consists of,

1. dividing a general cross-section into appropriate fibers or mesh elements;
a. determining stress in each fiber considering all load and material characteristics.

Now let us add some points on the section, each with a specified area, and a stress function related to the strain at that point, in the same way as for the shape above. Then the total stress resultants from two components will become as shown in Eq. (3.25a), the second half of the equation expresses the force generated in a point. This point may be an ordinary reinforcing bar or a prestressed strand. The force generated in these points is simply a summation of the force generated in the individual point. It may be noted that different points will have different contribution to the total force. This will depend on the points, locations, and stress—strain relation, defined for it.

$$P_z = \iint\limits_{x\ y} \sigma(x, y)\, dx\, dy + \sum_{i=1}^{n} A_{Pi} \sigma_{Pi}(x, y) \tag{3.25a}$$

$$M_x = \iint\limits_{x\ y} \sigma(x, y)\, dx\, dy \cdot \overline{y} + \sum_{i=1}^{n} A_{Pi}\, \sigma_{Pi}(x, y)\ y_{Pi} \tag{3.25b}$$

$$M_y = \iint\limits_{x\ y} \sigma(x, y)\, dx\, dy \cdot \overline{x} + \sum_{i=1}^{n} A_{Pi}\, \sigma_{Pi}(x, y)\ x_{Pi} \tag{3.25c}$$

The above concept can be extended to determine the stress resultants for cross-sections made up from several shapes and points of different materials. This can cover plain concrete, reinforced concrete, prestressed concrete, and composite concrete sections (see Fig. 3.21). This method to determine the stress resultants is also the basic principle behind the method referred to as the "fiber model" of the cross-section analysis, in which each fiber or part of the section is treated independently to determine the total stress resultant, and these fibers extend to nearby sections for determining strain profiles.

$$P_z = \iint\limits_{x\ y} \sigma_C(x, y)\, dx\, dy + \sum_{i=1}^{n} A_{Pi}\, \sigma_{Pi}(x, y) + \sum_{i=1}^{n} A_{Si}\, \sigma_{Si}(x, y) \tag{3.26a}$$

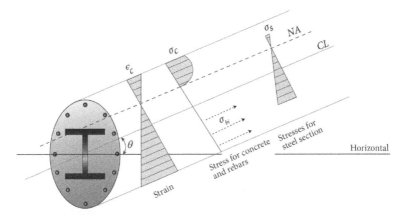

Figure 3.21 The stress resultants for a general cross-section made from several materials can be computed for a particular neutral axis angle direction and depth by first using assuming a strain profile, then computing stresses in each material at various location and then summing them.

$$M_x = \iint_{x\ y} \sigma_C(x, y)\ dx\ dy \cdot \bar{y} + \sum_{i=1}^{n} A_{Pi}\ \sigma_{Pi}(x, y)\ y_{Pi} + \sum_{i=1}^{n} A_{Si}\ \sigma_{Si}(x, y)\ y_{Pi}$$

(3.26b)

$$M_y = \iint_{x\ y} \sigma_C(x, y)\ dx\ dy \cdot \bar{x} + \sum_{i=1}^{n} A_{Pi}\ \sigma_{Pi}(x, y)\ x_{Pi} + \sum_{i=1}^{n} A_{Si}\ \sigma_{Si}(x, y)\ x_{Pi}$$

(3.26c)

Since we have developed the general equations for determining the stress resultants, we will explore some ways to use them for practical analysis and design.

Integrating Design Codes

The above equations represent the theoretical stress resultant values and need to be modified by appropriate capacity reduction factors before using these for design purposes (as shown in Eqs. (3.22)−(3.24)). The capacity reduction factors are generally specified by the design codes. Two types of capacity reduction factors are in use at the cross-section level. The first is applied to the total value of the stress resultant, such as the "ϕ" factors used in the ACI and many other design codes. The second is applied to each material stress separately, such as the "γ_m"

factors in the British Standards and other limit state design procedures. The above equations can therefore be modified to include both types of factors. The modified equations thus become universal and can be adapted for several design codes. For example, while using for ACI codes, all γ_m factors can be set to 1, whereas use the appropriate ϕ factors for axial and bending components depending on the strain value. For use in BS code, the ϕ factors will be 1 while appropriate values of γ_m will be used for concrete and steel components.

$$N_z = \phi_1 \left[\frac{1}{\gamma_1} \int_x \int_y f(x, y) \, dx \, dy + \frac{1}{\gamma_2} \sum_{i=1}^{n} A_i f_i(x, y) \right] \quad (3.27a)$$

$$M_x = \phi_2 \left[\frac{1}{\gamma_1} \int_x \int_y f(x, y) \, dx \, dy . \bar{y} + \frac{1}{\gamma_2} \sum_{i=1}^{n} A_i f_i(x, y) \, y_i \right] \quad (3.27b)$$

$$M_y = \phi_3 \left[\frac{1}{\gamma_1} \int_x \int_y f(x, y) \, dx \, dy . \bar{x} + + \frac{1}{\gamma_2} \sum_{i=1}^{n} A_i f_i(x, y) \, x_i \right] \quad (3.27c)$$

These equations, although are fairly general but still do not consider all the nonlinearities involved in the reinforced concrete, prestressed concrete, and composite concrete. For example, the following effects are not considered:

- prestrain on the cross-section components due to loading history,
- effect of shrinkage strain,
- effect of temperature strain,
- effect of steel relaxation,
- effect of prestresses,
- effect of nonlinear strain distribution,
- effect of viscoelastic strain, such as creep.

Therefore an extended formulation of the stress resultant equations is developed to handle the above effects.

The Generalized Cross-Section and Materials

Now consider a completely general cross-section consisting of a combination of shapes of different materials, including concrete and steel. The cross-section may also contain holes, reinforcements, and prestressing

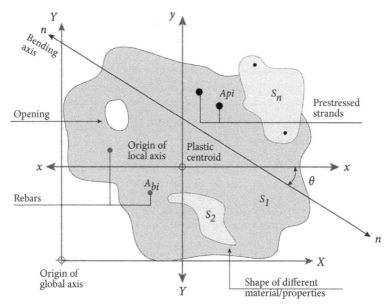

Figure 3.22 A generalized representation of cross-section having several shapes and materials.

strands at any arbitrary location. The shapes in the section may partially or fully overlap each other or may be completely inside or outside each other's boundaries. Let us also consider that all parts of the section have not been built at the same time and that the cross-section has been loaded incrementally. This means that at any given time and loading state the strains in adjacent materials at the same location may not be the same. A generalized representation of such a cross-section is shown in Fig. 3.22.

Such complex cross-sections can be represented by a combination of two basic entities, the polygons and the points. The polygons are used to represent solid shapes and holes, and the points represent the conventional reinforcing bars and prestressed strands. Curved shapes and boundaries are modeled by several straight edges. If required, even the large reinforcing bars or prestressing strands can be represented by n-sided polygons to achieve greater accuracy. Each different material or same material with different properties or loading history is represented by a separate entity. For the convenience of computations, the global $X-Y$-coordinate system is defined such that the entire section (all shapes, holes, reinforcement, and prestressed strands) lies in the first quadrant. The plastic centroid of the section is chosen as the origin of the local xy axes, and located at xp and yp.

The materials used in a cross-section are defined separately. For each material, the following five relationships can be defined:

- the basic stress–strain relationship, used to obtain basic stress for a given strain;
- the time-dependent stress modification relationship, used to model change in concrete strength and modulus of elasticity with time;
- the stress modification relationship, used to model prestress conditions, such as residual stresses or other nonstrain-dependent stresses;
- the time-dependent strain relationship, used to model creep, shrinkage, and relaxation;
- the strain modification relationship, used to model prestrain, slippage, local buckling, nonlinear strain distribution, cross-section warping, and strain due to temperature gradient.

Once the materials are defined, they are assigned to corresponding entities on the cross-section. It is important to note that each of these relationships can be assigned to different entities in the section independently. In theory, therefore, any combination of basic stress–strain relationships, time-dependent stress variations, time-dependent strain variation, and prestrain conditions can be used for the same entity within the cross-section.

EXTENDED FORMULATION OF STRESS RESULTANT EQUATIONS

Determination of Stress Field

The stress at any point on the cross-section can be divided into two parts. The part of stress that is a function of the strain at that location (f_ε), and the part of the stress that is independent of the strain (f_o). Consider that the stress field is defined on a plane using the local normalized coordinate system u–v, as:

$$f(u, v) = f_o(u, v) + f_\varepsilon(\varepsilon(u, v)) \tag{3.28a}$$

The stress field on the cross-section is also dependent on time, considering the effect of creep, shrinkage, relaxation, and incremental loading. The total stress field function then becomes:

$$f(u, v, t) = f_{ot}(u, v, t) + f_{\varepsilon t}(\varepsilon(u, v, t)) \tag{3.28b}$$

It is assumed here that the first part of the stress function is directly specified for each material and can simply be added to the stress determined from the strain distribution. In most cases the initial or direct stress component is zero and is not considered in the determination of capacity stress resultants. The strain-dependent stress can be determined purely from the stress—strain relationship, once the strain distribution at any given time t, has been determined. In light of the above equation, the possible cases that may occur are described in Table 3.2.

In the first four cases, described in Table 3.2, the stresses are not dependent on the strain. In the first two cases, the stress is also time independent, which indicates that changes in strain and time will have no effect on such stresses. So, only changes in the stress will alter the stress profile, while changes in strain will have no effect on the stress distribution. The stress field function in the third case is time dependent and independent of strain. This is encountered in phenomena such as

Table 3.2 Possible Cases of Stress Induced in a Cross-Section and Its Dependence on Strain, Stress—Strain Relationship, and Time

Case	Stress	Stress—Strain Relationship	Strain	Examples
1	Strain independent Time independent	Irrelevant	Does not change with time	Residual stresses
2	Strain independent Time independent	Irrelevant	Changes with time	Basic unrestrained creep, shrinkage, and temperature variations
3	Strain independent Time dependent	Irrelevant	Does not change with time	Basic relaxation
4	Strain independent Time dependent	Irrelevant	Changes with time	No phenomenon known to authors
5	Strain dependent	Does not change with time	Does not change with time	Instantaneous elastic deformation
6	Strain dependent	Does not change with time	Changes with time	Certain models of creep and shrinkage
7	Strain dependent	Changes with time	Does not change with time	Aging of concrete
8	Strain dependent	Changes with time	Changes with time	Certain models of creep, shrinkage, and temperature change

the relaxation of steel. In the fourth case, the stress and strain may vary independently of each other with time and there does not seem to be any example of such case. In last four cases, the stresses are strain dependent, which in turn may or may not be time dependent. Changes in the strain modify the stress accordingly. The stress–strain relationship may also change with time, in such case the new stress values at a given time instant for the modified strains will have to be computed from the modified stress–strain relationship.

In accordance with the above discussion and the cases, the extended stress equation may be rewritten as:

$$f = f_o(u, v) + f_{o\varepsilon}(\varepsilon_o(u, v)) + f_{ot}(u, v, t) + f_\varepsilon(\varepsilon(u, v)) + f_{\varepsilon t}(\varepsilon(u, v, t)) \quad (3.29)$$

While the strain equation may be written as:

$$\varepsilon = \varepsilon_o(u, v) + \varepsilon(u, v) + \varepsilon_t(u, v, t) \quad (3.30)$$

The first three terms of stress equation correspond to the strain-independent stresses and may be induced during the making of the cross-section or by some time-dependent phenomenon. The stress profile for these three terms is generated even before the orientation of the reference strain plane is fixed, as they are not dependent of the stains. The last two stress terms are strain dependent and can only be computed once the strain is known. A summary of notations used in above equations is explained as follows:

$f(u, v)$	=	A general time-independent stress field (function of u and v)
$f_o(u, v)$	=	Initial strain-independent, time-independent component of stress field on the cross-section
$f_t(u, v, t)$	=	A general time-dependent stress field (function of u and v)
$f_\varepsilon(\varepsilon(u, v))$	=	Strain-dependent, time-independent component of stress field
$f_{ot}(u, v, t)$	=	Time-dependent, strain-independent stresses as a function of u, v, and t
$f_{\varepsilon t}(\varepsilon(u, v, t))$	=	Strain- and time-dependent stresses as a function of u, v, and t
$f_{o\varepsilon}(\varepsilon_o(u, v))$	=	Strain-independent stresses induced while building up of the cross-section as a function of u and v
$\varepsilon_o(u, v)$	=	Strain induced while the building up of section as a function of u and v
$\varepsilon(u, v)$	=	Time-independent strain as a function of u and v
$\varepsilon_t(u, v, t)$	=	Time-dependent strain as a function of u, v, and t

Generation of Stress Profile

The overall process for stress profile generation to be used for determining the stress resultants can be summarized in the following steps:

1. Determine stresses that do not depend on the strain. A practical example of such a case may be the residual stresses, which are not caused by any strains.
2. Determine stresses that do depend on the strain where strain is time independent. This is the most common case encountered and may be referred to as Instantaneous Elastic Deformation.
3. Determine stresses that depend on time but are independent of strain variations. These may be encountered in cases such as relaxation of steel where, for the same strain level the stresses in steel are reduced over a period of time.
4. Determine stresses that are dependent on strain and time. These may include stresses such as creep and relaxation.

The first step is often not relevant to concrete sections and can be ignored. The last two steps are time dependent and may only be carried out if there are time-dependent stresses in the section. It is possible that the stress—strain relationship of the material may change over time. In such instances the new stress values are computed using the modified stress—strain relationship for the given strain level.

The stress profile generated will depend on the orientation of the reference strain plane, time elapsed, as well as on the type of stress being considered. Initially the stress profile generated due to a particular strain profile is determined. This strain profile will depend on the stress offsets calculated from the reference strain plane. These will include both time-dependent and time-independent stains. Strain-independent stresses are then incorporated on this stress profile, if present. The resulting stress profile is used (integrated) to find the stress resultants that give a single point on the capacity surface for specific orientation of the strain profile at a specific time. For each change in the orientation of the strain profile or the strain-independent stresses or both, the stress profile is recomputed giving another set of stress resultants. The process of determining the stress profiles due to the change in the orientation of the reference strain plane and time are given in Tables 3.3 and 3.4.

Table 3.3 Stress Profile Calculation Procedure With Varying Strain and Time

Event	Case (See Table 3.2)	Process to Determine Stress Profile	Result
Reference strain profile is changed at a given time instant	1–4	Stresses are independent of strain. No change in stress profile occurs.	Stress Profile
	5 and 6	New strain distribution alters the stress profile without changing the original stress–strain relationship (properties) of the material. New stresses are read from the original stress–strain curve to obtain new strains.	
	7 and 8	New strain distribution changes the strain profile, which is calculated using the modified stress–strain relationship (properties) of the material.	
Reference strain profile is fixed, time changes from t_o to t	1, 2, and 5	Strain may or may not vary with time. As the stress or strain is not time dependent, the stress profile is not altered.	
	3 and 4	Stress changes with time, independent of the strain and stress–strain relationship. For Case 4, strain is also time dependent but as stresses are not dependent on strains the stress profile is not changed.	
	6	Strain varies with time changing the stress profile while the stress–strain relationship remains unaltered, so the stresses are read from the same stress–strain relationship for the new strain values.	
	7	Strain remains constant but the stress–strain relationship is modified with time. Stress is read for the same strain on the modified stress–strain curve.	
	8	Strain and material properties (stress–strain relationship) both change with time. Stress is read for the new strain on the modified stress–strain curve.	

Table 3.4 Incorporating Various Phenomenon in the Calculation of Stress Profile

S. No.	Phenomenon	Corresponding Case in Table 3.2	Incorporated Through
1	Creep	2, 6, and 8	Reference strain profile and creep model
2	Shrinkage	2, 6, and 8	Reference strain profile and shrinkage model
3	Relaxation	3	Reference strain profile and relaxation model
4	Temperature variation	2, 6, and 8	Reference strain profile and temperature model
5	Aging of concrete	7	Reference strain profile and aging model
6	Elastic deformation	5	Reference strain profile
7	Nonlinear strain profile	–	Reference strain profile
8	Nonlinear stress profile	–	Reference strain profile
9	Confinement in concrete	–	Reference strain profile and confinement model

Determination of Strain Distribution

It can be seen from the previous discussion that the value of any particular stress resultant component corresponds to a particular stress distribution, which in turn corresponds to a particular strain distribution on the cross-section. The stress resultant represents the cross-section capacity vector if the strain distribution is derived from a failure criterion or is based on a limiting strain. It is obvious that infinite strain distribution profiles are possible, which satisfy the failure criterion and represent corresponding infinite points on infinite capacity surfaces. A compatible and consistent strain distribution, however, will yield all points on a single capacity surface. A particular stain distribution will yield a single point on the capacity surface and will correspond to a particular equilibrium condition. For computational purposes, the general strain distributions can be reduced to a finite number of compatible strain profiles leading to a finite number of points that define the interaction surface.

All the position and time-dependent strain functions for various parts of the cross-section, and for different materials, are defined in terms of magnitude and direction as offsets from this plane. So to obtain a particular strain distribution, the reference strain plane is located on the cross-section, and then at each point on the cross-section, the basic strain is determined from the position of the reference strain plane. This basic strain is then modified by using various strain offsets specified for the material at that location. It may be noted here that the strain at different points of a cross-section, defined as offsets from the reference strain plane are compatible with each other for that orientation of the plane as shown in Fig. 3.23. The strains may change with the orientation of the reference plane. The modified strain profile is used to compute the strain-dependent stress profiles using the procedure defined in Table 3.3. The reference strain plane and the strain and stress profiles are defined using normalized local axis (see Fig. 3.24).

For the simple reinforced concrete sections deforming under the plane section assumption with no time-dependent strain variations, the reference plane itself becomes the final strain profile, as all strain offsets are zero. In this case, the pivot point is located at the first concrete compression fiber on the cross-section, and the reference strain plane is at an offset of 0.003 or some other arbitrary value, from the cross-section. In this case, the intersection of reference strain plane with the cross-section plane becomes the neutral axis.

To define any particular strain distribution, the concept of a Reference Strain Plane is introduced here, which may be defined as a plane at some known or assumed location and orientation with respect to the unde-formed plane of the cross-section. The location and orientation of this plane is defined by using a pivot point, a pivot line, and the rotation about the pivot line. The pivot line can be located above or below the cross-section at a distance equal to the reference, or limiting strain, at the pivot point on the cross-section defined in the failure criterion; positive location (above) represents compressive strain. This pivot line is located parallel to the axis of cross-section curvature; thus the reference strain plane is free to rotate about this pivot line matching with the curvature of the cross-section. This is shown in Fig. 3.23.

COMPUTING THE BIAXIAL-FEXURAL STRESS RESULTANTS

The general equations for axial—flexural stress resultants applicable to cross-sections made from multiple shapes and materials have been developed earlier in the chapter. Now, we will develop general procedures for solution of these equations, and look at some of the

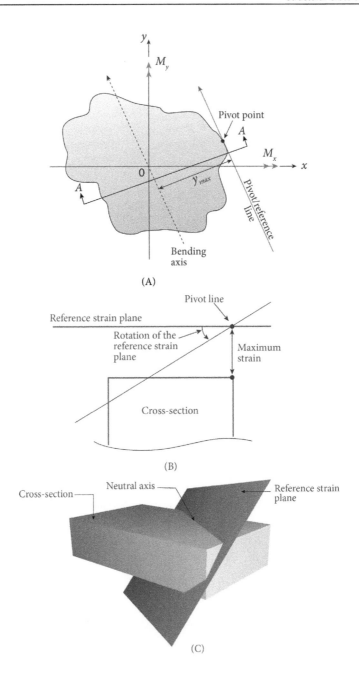

Figure 3.23 The concept of reference strain plane. (A) The location of pivot line and pivot point, (B) the strain plane in 2D, and (C) 3D representation of strain plane.

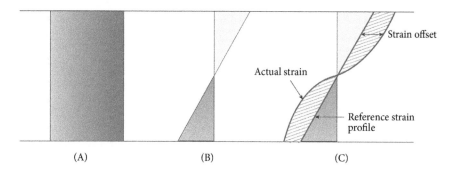

Figure 3.24 Reference strain and actual strain profile. (A) Cross-section (B) Reference strain profile (C) Actual strain with offsets from reference strain profile.

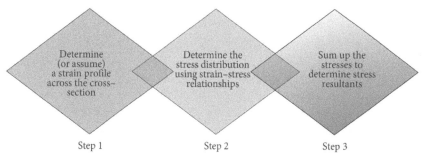

Figure 3.25 A simple three-step process of determining stress resultants.

computational aspects involved in the process. Although the equations are quite general, we will primarily focus on their application to concrete, reinforced concrete, and composite sections. There are two approaches that can be used to determine the stress resultants in a typical beam-column cross-section. A simple and straightforward procedure based on known strain profile can be divided into three steps (see Fig. 3.25).

The overall procedure for computing stress resultants for any general cross-section using the proposed model can be outlined as:

1. define the material models in terms of basic stress—strain functions and time-dependent stress and strain functions;
2. convert these functions to discretized surfaces or curves in their respective local axes;
3. model the geometry of the cross-section using polygon shapes and points;
4. assign the material models and functions to various polygons and points;

5. discretize the cross-section into primary, nonoverlapping polygons and independent point entities and determine the governing material models (using primary meshing);

6. locate the reference strain plane based on the current failure criterion;

7. compute the basic stress profiles for all materials using the reference strain profile and initial conditions;

8. modify the stress profiles for each material based on appropriate material functions;

9. for each material stress profile, carry out secondary and tertiary meshing of the polygons and compute the corresponding stress resultant for the resulting triangles and points;

10. modify the stress resultants using the appropriate material-specific and strain-dependent capacity reduction factors;

11. compute the total stress resultants for all material stress profiles.

If required, Steps 6—10 can be repeated for other locations of the reference strain plane (see Fig. 3.26).

Each of these steps requires significant amount of work when dealing with generalized cross-sections, materials, and deformation. We will now address these problems one by one and expand upon the concepts already discussed for the unified approach to the determination of axial—flexural response.

Determining the Strain Profile

So far we have talked a lot about the importance of the strain profile in the solution of the general stress resultant equations and for determining the capacity interaction surfaces. We will now take a closer look at the strain profiles, the concept of the neutral axis, the relationship between the strain profile and the neutral axis, and some simplified methods for determining the strain profile.

What Strain Profile to Use?

The big question is how to determine the strain profile (or strain distribution), or what strain distribution can be assumed in a particular situation. Ideally the strain distribution should be obtained directly from the applied actions, using the strain—deformation relationships. Unfortunately the direct strain—deformation relationships are available for very limited shapes and material characteristics and often do not consider factors like cracking in reinforced concrete, shear lag in steel section,

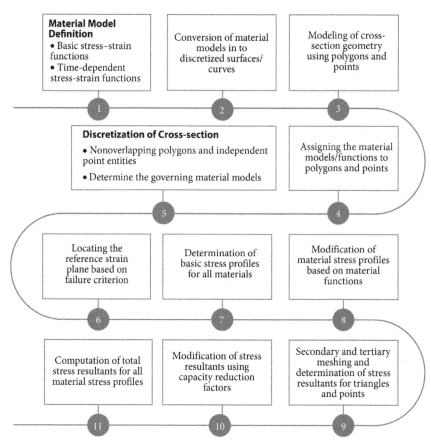

Figure 3.26 Step-by-step procedure to determine stress resultants for a general cross-section.

the buckling of cross-sectional elements, combination of heterogeneous materials, and composites of various materials.

The usual practice in structural mechanics is therefore, to assume a "compatible strain distribution." The "compatibility" considers the type of deformation and the type of cross-section materials. For a general cross-section under a combination of deformations, a general axial strain distribution will be a function of x and y. For example, in steel sections, especially slender sections, the strain varies along the depth as well as the width. This variation results from the shear lag, local buckling, and rapid variation in cross-section stiffness.

However, for reinforced and prestressed sections of shallow beams and columns, the assumption of plane sections can be made in the flexural

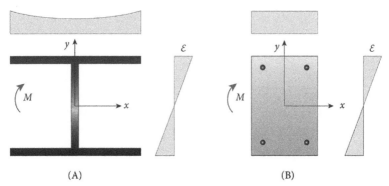

Figure 3.27 The strain variation across cross-section depth and width: (A) steel section and (B) reinforced concrete section.

zone and the strain becomes a function of y only. In case of biaxial bending, the y-distance is measured perpendicular to the neutral axis. In this case, the strain distribution can be determined with respect to the neutral axis. Before we discuss the practical application of this assumption, we will further explore the concept of neutral axis. The concept of unidirectional and bidirectional strain variation is shown in Fig. 3.27.

The Concept of Neutral Axis

An important concept related to the practical determination of strain distribution is the Neutral Axis. What is the real meaning and significance of this term, used so often and so extensively in structural mechanics, analysis, and design? In essence, the neutral axis is a line produced by the intersection of the undeformed cross-section plane and the deformed surface as shown in Fig. 3.23C. This produces a surface of zero strain oriented along the longitudinal axis of the beam-type member.

The zero-strain surface may or may not be within the bounds of the cross-section. For example, for a member undergoing axial deformation only, the zero-strain surface does not exist, except at infinity. The concept of the neutral axis can also be extended to a plate-type member and in some cases even to solid-type members. As long as a surface (which may not be a plane) can be identified having zero strain under a particular deformation, then any cross-section that intersects the surface can indicate the location of the neutral axis. Sometimes there may be more than one surface of zero strain within a member. For example, for a deep beam subjected to heavy concentrated loads, the region near the load or support is in a complex state of deformations. In this case, there could

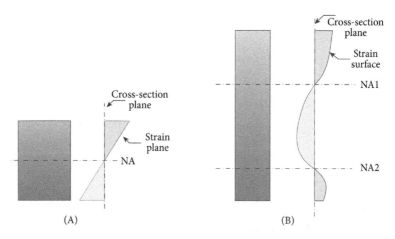

Figure 3.28 The concept of neutral axis: (A) linear strain distribution (shallow beam) and (B) nonlinear strain distribution (deep beam).

be several surfaces of zero strain producing a strain profile shown in Fig. 3.28, with two or more corresponding "neutral axes." So, strictly speaking, the conventional concept of neutral axis is only relevant to shallow beam-type members, under flexural deformations.

Where Is the Neutral Axis?

For an axial—flexural member, the orientation and depth of the neutral axis depends on the applied actions. The direction of the neutral axis would generally be perpendicular to the direction of the eccentricity vector (deformed from the moment and axial loads). This however is strictly true only for fully symmetrical sections, such as circles or to some extent square sections, for general sections the direction is related to the direction of the resultant eccentricity, but the exact relationship is generally not known. The depth of the neutral axis is related to the level of axial load in addition to the moments. Again for a general unsymmetrical, heterogeneous section, there is no closed-form solution available to determine the depth of the neutral axis.

Most computer programs handling general cross-sections determine the direction and depth of neutral axis using a combination of iterative and interpolative techniques. In general, the direction of the neutral axis gives the resultant moment in the same direction as the applied moment is the correct direction for these actions. The neutral axis depth is determined when the axial stress resultant is equal to applied load.

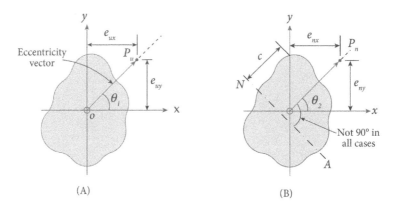

Figure 3.29 The location of neutral axis. Orientation and depth determined simultaneously to match $\theta_1 = \theta_2$, $P_u = P_n$ and corresponding eccentricities. (A) Eccentricity vector for applied action (B) Eccentricity vector for internal stress resultants.

The problem is, that for different neutral axis depths for the same neutral axis angle, the direction of resultant moment changes for unsymmetrical sections. Therefore the direction and depth need to be determined simultaneously and not independently. Another problem is that the direction and depth of neutral axis will change depending on the load combination being considered. It is therefore, more convenient to generate the entire interaction surface for various increments of neutral axis angle and depth and then locate the desired location by searching and interpolation (see Fig. 3.29).

The Practical Strain Distribution

It is quite obvious from the above discussion and the stress—strain relationship described later, that the usual basis of using a linear strain distribution, with the assumption of plane sections, and a limited strain of 0.003 (or similar value) with a rectangular strain—stress relationship for concrete, is a highly simplified representation of a fairly complex phenomena. However, this simplification gives a good estimation of load-deformation behavior for practical design application for usual cases.

Using the assumption of a plane section and linear strain distribution, the strain profile becomes a straight line. If the neutral axis depth is known, the strain line can be pivoted on the neutral axis, and the slope of this line can be varied to generate infinite strain profiles. However, in most cases, the maximum strain or the failure strain is specified. In such cases, the maximum strain is used as the pivotal value and the slope of the strain line (which is also the inverse of the curvature) can be varied to generate several strain profiles. Alternatively the maximum strain value is fixed, and the neutral axis is varied to change the slope of the strain line and generate the strain profiles. This last approach is the most common method used in manual as well as computer-aided generation of interaction surfaces and curves. For the purpose of solving the stress resultant equations, the general strain profile can be defined for each material in the cross-section in the form of an array of strain values corresponding to a normalized or unit depth.

Determination of Stress From Strain

Once the strain profile has been determined or assumed, the corresponding stress profile in the material can be determined from the stress—strain relationship defined for the particular material. A detailed discussion of the determination of stress—strain relationships for various materials is beyond the scope of this book and can be found in various reference books (Chen & Saleeb, 1994; Chen, 2007; Samali, Attard, & Song, 2013). However, we have briefly discussed the stress—strain relationships for concrete and steel in Chapter 2, Understanding Cross-Sections. The stress—strain relationship for concrete is generally complex and highly nonlinear for which, several model curves have been proposed. They vary from complex parabolic curves to simple rectangular stress blocks. As discussed in Chapter 2, Understanding Cross-Sections, the stress—strain curve for steel is also fairly complex when considered in its entirety. It may have at least three regions: the elastic, the plastic, and the strain hardening. However, in most cases, a simplified bilinear elasto-plastic curve is used in computations. A linear elastic relationship is used until yield strain is reached, after which, a completely plastic relationship is used where stress remains constant for infinite strain.

Discretization of Cross-Section and Stress Field

Discretization of the cross-section converts a planer continuum to fibers or a mesh element of finite size and position. Typically the accuracy of discretized solutions, such as finite element method depends on the mesh characteristics (unless higher order elements are used), and the accuracy often improves with mesh refinement. For a general cross-section, represented by a number of complex polygons and points, the points need not be discretized, and each is taken as an individual fiber.

The basic procedure to mesh the cross-sections for the purpose of computing geometric properties is introduced in Chapter 2, Understanding Cross-Sections. However, for the purpose of determining the stress resultants, some additional considerations and meshing may need to be done to incorporate the variation of stress over the cross-section. The procedure may be summarized as follows.

Each of the complex polygons is discretized in three levels:

1. *Primary meshing based on cross-section geometry*: All complex polygons are divided into simple, convex polygons and overlapping regions are resolved in such a way that polygons of only one material occupy the overlapping regions. This meshing is performed only once at the start of the computation process, and normally results in a few mesh elements.

2. *Secondary meshing based on the stress variation along the bending direction*: This meshing is performed for each polygon generated by the primary meshing, as and when needed for a particular stress variation along the bending direction. The secondary mesh generates simple triangular elements. The number of triangles generated for each polygon at this level depends on the number of points used to represent the stress profile along the bending direction. For a rectangular stress block, only two triangles per polygon may be sufficient to integrate the stress field accurately.

3. *Tertiary meshing based on the stress variation normal to the direction of bending*: This meshing is only carried out if the stress is not constant across the width of the section at a particular location along the direction of bending. If required, this meshing subdivides the triangles obtained in secondary meshing based on the number of points used to model the stress profile normal to the bending direction, and produces finer triangles. In most cases this meshing is not needed because the transverse stress variation is often not considered in conventional mechanics (see Fig. 3.30).

After the final mesh has been generated, the stress variation on each triangular element is considered linear in both directions and can be integrated exactly as shown below. It can be seen from the above methodology that the meshing or discretization of the cross-section for the purpose of stress integration depends both on the cross-section and variations in the stress field. The fineness of the mesh is automatically

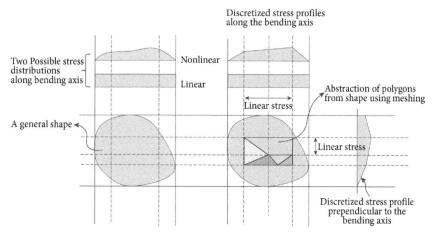

Figure 3.30 The process of discretization of stress field based on cross-section geometry and the stress variation.

controlled by the complexity of the cross-section and the nature of variation in the stress field. This approach is faster and more accurate than the conventional integration methods, which are based on uniform or predetermined discretization of the cross-section considering its geometry alone. Moreover, the size of the basic polygon mesh has no effect on the accuracy of the method as the final discretization is performed considering the nonlinearity of the stress–strain field in addition to the shape and size of the polygon.

Computation of Final Stress Resultants

The computation of the final stress resultants is carried out by considering the points (rebars) and shapes (polygons). The determination of stress resultants for the rebar or point areas is relatively simple. The stress in each bar is computed from the stress profile and is multiplied by the bar area to determine the force. The axial stress resultant is simply the summation of all bar forces. The resultant moment about the x-axis can be determined by multiplying the bar force by the y-distance from the bar to the reference axis. The resultant moment about the y-axis can be obtained in a similar way as summarized in the equations below.

$$P_{z1} = \sum_{i=1}^{n} A_i f_i \tag{3.31a}$$

$$M_{x1} = \sum_{i=1}^{n} A_i f_i y_i \tag{3.31b}$$

$$M_{y1} = \sum_{i=1}^{n} A_i f_i x_i \tag{3.31c}$$

However, it is not a trivial task to determine the stress resultants for a general polygonal shape under a general stress profile as shown in the discussion above. The axial stress resultant is actually the volume of the stress profile, defined as a surface stretched over the shape. The moment stress resultant is the product of the volume of the stress profile and the distance of its centroid from the respective axis. This leads to the integration of the stress function over the shape area. It is quite difficult to formulate closed-form equations for calculating the volume and centroid of this general stress volume over the general cross-section shapes. However, if the cross-section shapes have been meshed into quads and finally into triangles, we can convert the integration problem to a summation problem, similar to the one used for the bars. The stress at each vertex of the triangle can be determined from the stress profile function, and then, all we need is a solution for calculating the volume and centroid of a triangle with known stresses at three vertices, as shown in Fig. 3.31. Once the centroid and volume is known, the stress resultants can be computed as:

$$P_{z2} = \sum_{j=1}^{m} V_j \tag{3.32a}$$

$$M_{x2} = \sum_{j=1}^{m} V_j y_j \tag{3.32b}$$

$$M_{y2} = \sum_{j=1}^{m} V_j x_j \tag{3.32c}$$

And the total resultants are:

$$P_z = P_{z1} + P_{z2} \tag{3.33a}$$

$$M_x = M_{x1} + M_{x2} \tag{3.33b}$$

$$M_y = M_{y1} + M_{y2} \tag{3.33c}$$

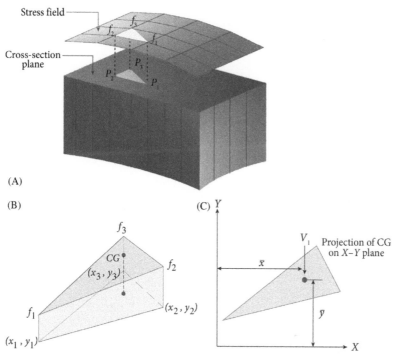

Figure 3.31 The determination of volume and centroid of the stress prism. (A) The triangle in the meshed stress surface; (B) the volume and centroid of stress prism; and (C) the projection of centroid and volume on the x–y-plane.

The determination of the volume and especially the centroid of a triangle with three stress values is not as simple as it seems. The geometric centroid of the triangle is not the centroid of the volume. One way to solve this problem is to divide the triangular volume into three general quads, determine the volume and centroid of each quad, and then compute the total volume and volume centroid using transformation rules, as shown in the following procedure (see Fig. 3.32).

The volume and center of gravity of the prism are computed by dividing the prism into three tetrahedrons (a, b, c). Center of gravity and volume of one tetrahedron (see Fig. 3.31) is given as:

$$x_c = \frac{\left(x_i + x_j + x_k + x_l\right)}{4} \tag{3.34a}$$

$$y_c = \frac{\left(y_i + y_j + y_k + y_l\right)}{4} \tag{3.34b}$$

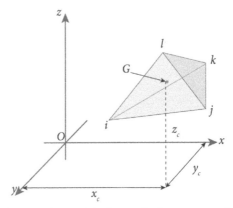

Figure 3.32 The volume and centroid of a tetrahedron as part of the stress prism.

$$z_c = \frac{(z_i + z_j + z_k + z_l)}{4} \tag{3.34c}$$

$$V = \frac{1}{6} \begin{vmatrix} 1 & x_i & y_i & z_i \\ 1 & x_j & y_j & z_j \\ 1 & x_k & y_k & z_k \\ 1 & x_l & y_l & z_l \end{vmatrix} \tag{3.35}$$

The total volume and center of gravity are computed as:

$$V_i = V_a + V_b + V_c \tag{3.36}$$

$$\bar{x}_i = \frac{V_a \bar{x}_a + V_b \bar{x}_b + V_c \bar{x}_c}{V} \tag{3.37a}$$

$$\bar{y}_i = \frac{V_a \bar{y}_a + V_b \bar{y}_b + V_c \bar{y}_c}{V} \tag{3.37b}$$

A sample program code to determine some of the properties from stress profile is given in appendix of this book.

The Use of General Stress Resultant Equations

The general stress resultant equations, although quite general in scope, cannot be used readily for cross-section design especially for reinforced and composite concrete. The equations provide the stress resultant value of a completely defined cross-section for a particular definition of strain profile and stress—strain relationship. In practice, the cross-section parameters are not fully known in advance. The amount and distribution of reinforcement is generally not known, and needs to be determined for a particular set of design actions. Even if the cross-section parameters were known in advance, the strain profile compatible with the applied actions cannot be determined directly from the applied actions. Therefore the equations do not provide a direct solution for the cross-section design problem. The equations can however be used to generate capacity surfaces and curves that can then be used indirectly for section capacity investigation and design. Here we will discuss the origin, meaning, and use of various capacity surfaces and curves.

THE CAPACITY INTERACTION SURFACE

As can be seen in the previous development, there are three stress resultants, all of which can be determined for a particular strain profile on the cross-section. So if the strain profile is varied or changed, for each variation, we will get different values of these stress resultants. These three stress resultants can be plotted in a 3D space and will generate a continuous surface for all possible variations of the strain profile on a particular cross-section. This is generally termed as the stress resultant "Interaction Surface." If the strain profiles used to generate the surface are derived from material failure conditions, then this surface becomes the "Capacity Surface" or the "Failure Surface." Any combination of applied actions P, M_x, and M_y, that is inside the volume enclosed by this surface is considered safe, whereas any combination that results in a point that is outside this surface is considered unsafe. Fig. 3.33 shows typical cross-section and its ultimate capacity interaction surface.

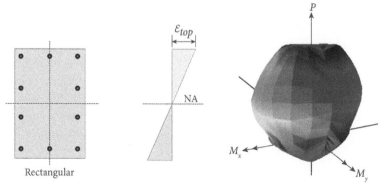

Figure 3.33 A rectangular cross-section and its ultimate axial—flexural capacity interaction surface for a specific ultimate strain, ε_{top}.

Generation of the Interaction Surface

Simplified Procedure

To generate the full interaction surface, the neutral axis is rotated about the plastic centroid by discrete intervals, and for each angle, the depth of neutral axis is varied form the one extreme end of the cross-section to the other in discrete intervals. A discrete surface is generated using stress resultant at each variation. The generated surface may then be smoothened by using the b-splines and Bézier curves. This smoothening is useful in removing some of the inaccuracies that may be generated due to numerical round off, and due to the discrete steps used in the neutral axis orientation, depth of neutral axis, and the integration steps. It also allows for generation of fewer points than would otherwise be required to obtain a smooth surface, improving the performance of the programs.

Extended Procedure

A similar approach is used to generate the interaction surfaces for the extended computational model as discussed earlier. The strain reference plane, as described earlier, is pivoted about the line perpendicular to the eccentricity vector. As the strain distribution over the section is not available the reference plane is rotated from 0 to 90 degrees about the pivot line. For each orientation of the plane, the strain profile is computed and modified to account for the effects of prestressing, relaxation, creep, shrinkage, and warping. Strain values due to these phenomena are added to the assumed strain profile at the given location over the cross-section. The stress—strain relationships for all the materials present in the section are known. Using the modified strain profile and material properties, the stress profile is calculated. Strain-independent stresses, if present, are added to the obtained stress profile. Stress resultants are computed using this final stress profile. Similarly, stress resultants for each rotation of the strain reference plane about its pivot from 0 to 90 degrees for each interval of the pivot line from 0 to 360 degrees are computed. Each rotation of the strain reference plane for a given orientation of the pivot line gives a set of values for P, M_x, and M_y. These computed values of stress resultant (P, M_x, M_y) are plotted in 3D space and joined by a smooth curved surface to obtain the interaction surface of the section. The shape and values of the interaction surface plots are also dependent on the choice of capacity reduction factors, as shown earlier (see Fig. 3.34).

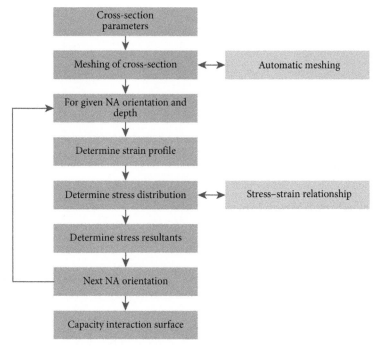

Figure 3.34 Generation of the interaction surface.

Visualization and Interpretation

This interaction surface or capacity surface is generally plotted in such a way that the axial stress resultant P_z is along z-axis, M_x is along x-axis, and M_y is along y-axis. Some typical interaction surfaces are shown in Fig. 3.35. The shape of the interaction surface primarily depends on the cross-section parameters, like the geometry, material properties, amount and location of reinforcement, etc. This surface is an intrinsic property of the cross-section alone and does not depend upon the applied actions. Each point on the surface, however, can be viewed as the equilibrium condition for a set of applied actions that will produce the particular strain profile compatible with the computation of stress resultants at that point. In fact, this condition can be indirectly used to determine the strain profile that is compatible with the applied actions.

As the interaction surface exists in three-dimensional space, it cannot be plotted on a two-dimensional paper space directly. It is, therefore, not practical to use the interaction surfaces for manual design. The interaction surface can however be converted to two-dimensional

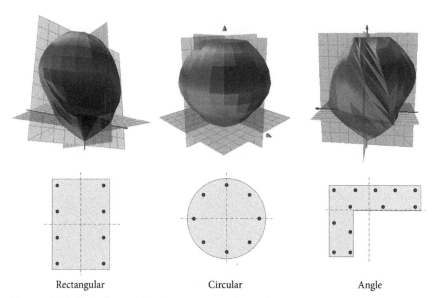

Rectangular   Circular   Angle

Figure 3.35 Typical capacity interaction surfaces for various cross-section shapes (CSiCOL).

curves by appropriate "slicing" of the surface. The two most common types of curves derived from the interaction surface are discussed here (see Fig. 3.36).

The Moment—Moment Interaction Curve

For a particular value of the axial stress resultant (which may correspond to an applied action), the capacity surface can be sliced on the M_x-M_y-plane, to obtain the plot between moments M_x and moment M_y. This plot between the moment capacity about the x- and y-axes provides several useful insights into the behavior of the cross-sections. First, it shows how the moment capacity varies around the cross-section at a particular load level. For symmetrical sections with symmetrical reinforcement the shape of $M-M$ curve remains symmetric and similar at different load levels. For unsymmetrical sections, the shape can change significantly at various load levels. This means that the efficiency of the cross-section varies depending on the level of axial load and the direction of the resultant moment. In general, for biaxial moments of nearly equal values, the most efficient shape is the circular shape, followed by the octagonal, hexagonal, and square. If the resultant moment direction is clearly known for all load combinations, then the section shape and reinforcement

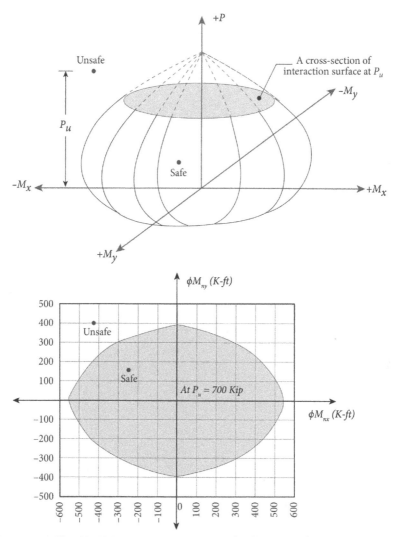

Figure 3.36 The M–M interaction curve is a plot between the moment M_x and moment M_y at a specified level of axial load (based on CSiCOL).

pattern can be selected to yield highly efficient sections using the M–M plots. A special case of the M–M plot obtained at P = 0 provides the biaxial moment capacity curve for a beam section. For a column section, this curve is a representation of capacity change for neutral axis angle variation from 0 to 360 degrees. This is shown in Figs. 3.37–3.40 for a circular, rectangular, angular and T-sections of reinforced concrete.

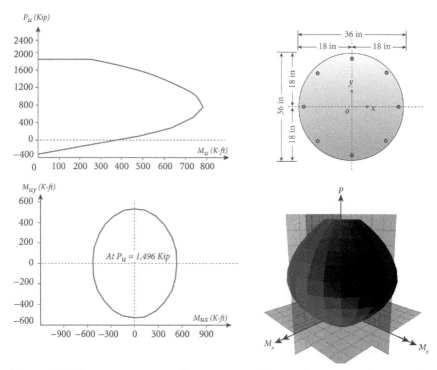

Figure 3.37 The most symmetrical interaction surfaces and curves are for a circular RC section (CSiCOL).

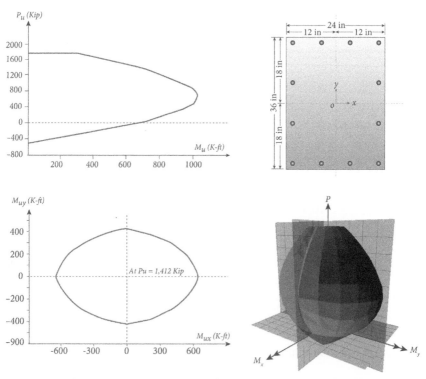

Figure 3.38 The $P-M$ and $M-M$ curves for a reinforced concrete rectangular RC section (CSiCOL).

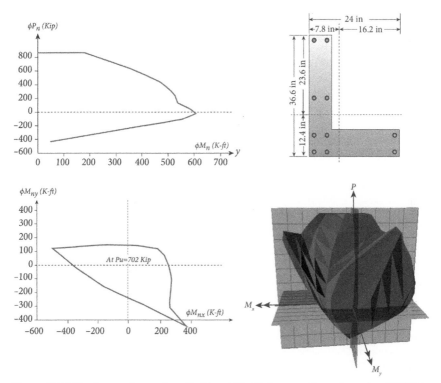

Figure 3.39 The interaction surface, $P-M$ and $M-M$ diagrams for an L section (CSiCOL).

Generally, these curves are assumed to be symmetrical and follow some second-order polynomial function. The load contour and other approximate methods for calculating biaxial capacity are also based on such assumptions (Chen, 1997; Hassoun, 1985; Varghese, 2010). These assumptions are perfectly applicable to circular section with symmetrical rebar layout, approximately correct for rectangular sections and are completely inadequate for flanged and unsymmetrical sections.

The Load–Moment (PM) Interaction Curves

If the capacity surface is sliced vertically along any angle about the origin, we obtain a plot between resultant moment and axial stress resultant, often termed as the $P-M$ interaction curve. This is a very common and useful tool for design and investigation of columns. Special $P-M$ curves can be obtained by slicing the capacity surface along the x- and y-axes. These plots are then used for the special case of $P-M_x$ or $P-M_y$ interaction when bending is clearly about one of the principle axis. When the

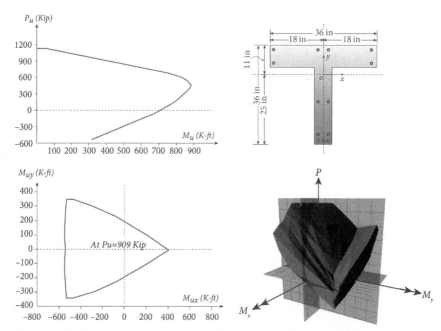

Figure 3.40 The capacity interaction surface, *P–M* and *M–M* interaction diagrams for a T-section (CSiCOL). (It is important to note that the small change in neutral axis angle for a T-section will significantly change the axial load capacity due to the flanged nature of the section, where each rotation of the neutral axis brings different shape and parts of the section under compression. For this reason, if *M–M* diagrams are plotted at different values of axial load (from low to high values), we can observe a complete reversal of the diagram shape.)

bending is on an inclined axis, it is difficult to establish a relationship between the direction of applied moment resultant and the direction of capacity resultant moment. For a column section, this curve is a representation of the capacity change for the neutral axis variation from one edge to other extreme edge of the cross-section.

Generally, only half of these curves are considered, assuming that the cross-section and the rebar layout is symmetrical. However, for unsymmetrical sections or rebar layout, both sides of the curve should be plotted and considered, depending upon the direction of the moment, or the angle of neutral axis should be considered from 0 to 360 degrees. Each of the above curves can be generated for various values of yield strength of steel f_y, compressive strength of concrete f_c, and rebar size. These multiple curves can be used for selecting the appropriate value of any desired parameter. For example, a particular section and rebars can be selected and curves can be generated for different values of concrete strength. The appropriate concrete strength can then be selected for particular loading conditions. This can be very useful in high-rise buildings where column size and rebars can be preselected but the concrete strength can be varied for different floors. Similarly, curves can be generated for given concrete strength, rebar grade, and arrangement, but for different diameters of rebars. Such curves can be useful for selecting the suitable size of bars, thus maintaining a particular layout. The graphical presentation of the interaction curves not only provides a direct means for the design of various combinations of applied loads for a given section, but also gives an insight of understanding of the cross-section behavior and rebar arrangement.

Capacity Reduction Factors

As mentioned earlier, different design codes use different types of capacity reduction factors to modify the theoretical stress resultants (or capacity) determined using the above approach. This is shown in Eq. (3.15) where the capacity reduction factors are incorporated and need to be included in the computation process. Here we will compare the effect of these capacity reduction factors on the load—moment curves of two different shapes. Fig. 3.41 shows the change in $P-M$ interaction diagrams for capacity reduction factors used from six different design guidelines (AC1 318-02 to 11, ACI 318-99, CSA A23.3 94-04, BS 8110-97, Euro Code 2, and IS 456-2000).

It can be seen that the shape of the $P-M$ curves is significantly modified, especially in the low axial load range based on the choice of capacity reduction factor. It is important to note that ACI 318-11 prescribes different capacity reduction factors for different cross-sectional shapes and the distribution of reinforcement by relating the factor to the level of strain, rather than fixing it for all sections or particular load or material type. On the other hand, the constant capacity reduction factors prescribed by other codes simply results in a uniform scaling of different portions of the curve.

Effect of Material Strengths and Section Depth on $P-M$ Interaction of RC Sections

As mentioned earlier, various factors affect the shape of $P-M$ interaction curve for a reinforced concrete section. These factors include the

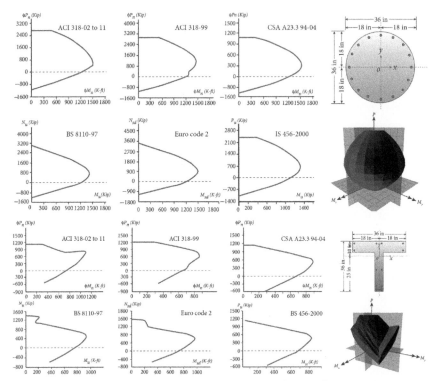

Figure 3.41 The effect of capacity reduction factors from different codes on the shape of *PM* interaction diagrams (CSiCOL).

behavior and mechanical properties of individual constituent materials (steel and plain concrete), section shape and reinforcement configuration. Fig. 3.42 shows the $P-M$ interaction curve for four rectangular sections with varying depth-to-width ratios from 1 to 4. The amount of longitudinal reinforcement was kept the same. A uniform increase in section capacity against compression-controlled failure can be observed, while the tensile capacity (as governed by tensile strength of concrete and steel) remains the same.

Similarly, Fig. 3.43 shows the effect of concrete compressive strength on $P-M$ curve of a rectangular RC section. All other parameters were kept constant except concrete strength. Again the section capacity on compression side of curve is increased. Fig. 3.44 shows the effect of increasing the yield strength of longitudinal rebars in the same rectangular section. Although the increase in tension capacity is relatively higher than compressive, the overall section capacity is enhanced with shifting of

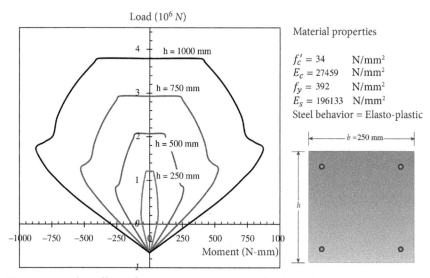

Figure 3.42 The effect of an increase in section depth (and gross area) on $P-M$ interaction diagram.

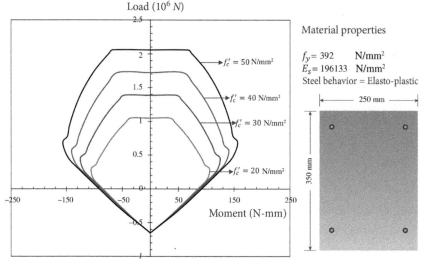

Figure 3.43 The effect of an increase in compressive strength of concrete on $P-M$ interaction diagram.

balanced point to lower levels of axial load. The distribution of longitudinal bars in the RC sections can also significantly affect the shape of $P-M$ interaction diagram. Fig. 3.45 schematically shows this effect where the interaction of moment and axial force in both directions (clockwise and

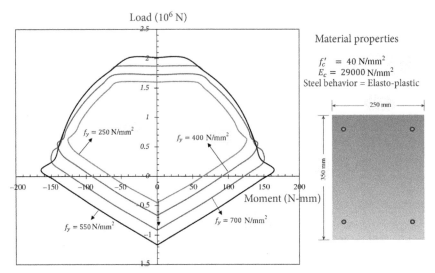

Figure 3.44 The effect of an increase in yield strength of steel on $P-M$ interaction diagram.

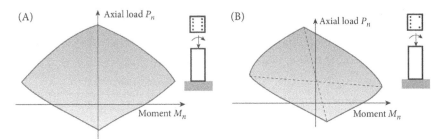

Figure 3.45 The effect of (A) symmetrical or (B) unsymmetrical distribution of longitudinal reinforcement on $P-M$ interaction diagram (based on Wight & MacGregor, 2012).

counterclockwise) is not same. The complete interaction diagram is skewed with a shift in balanced failure points.

BIAXIAL–FLEXURAL CAPACITY
The Definition of Biaxial Bending

The notion of biaxial bending is in fact a simplification of the flexural behavior for the purpose of theoretical convenience. The actual bending in a flexural member is about a single axis, defined earlier as the Neutral

Axis, which is in line with the major principle axis. If this neutral axis happens to be parallel to one of the coordinate axes, then we call this "uniaxial" bending. If the neutral axis (or the principle axis) is inclined with respect to the coordinate axis, then bending can be resolved into two components parallel to each coordinate axis, hence the term Biaxial Bending. A cross-section may be subjected to "biaxial bending" with respect to the coordinate axis due to any of the following reasons:

- The applied axial load is located away from the plastic centroid, and is not on any of the coordinate axes, passing through the centroid. The applied moment vector or the applied eccentricity vector is not parallel to any of the coordinate axes.
- The basic cross-section geometry, or reinforcement layout, or material distribution is not symmetrical with respect to any of coordinate axes.
- The principle axes are not parallel to the coordinate axes.

All cases of biaxial bending can be converted to simple bending, and all cases of simple bending can be converted to cases of biaxial bending purely by changing the reference axis with respect to the bending vector. The bending vector can be defined in many ways as explained next. Fig. 3.46 and Table 3.5 show various cases of uniaxial and biaxial bending.

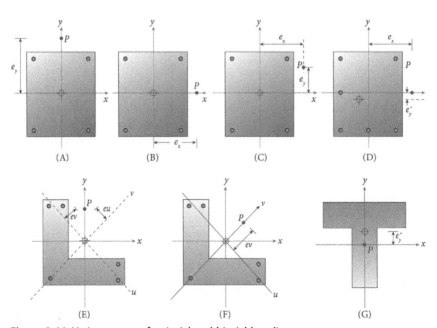

Figure 3.46 Various cases of uniaxial and biaxial bending.

Table 3.5 Various Cases of Uniaxial and Biaxial Bending

Case	Description	Bending Moment M_x	M_y	Bending Type	Remarks
A	Symmetrical section Symmetrical rebars	$P.e_y$	—	Uniaxial	Loading and section symmetrical
B	Symmetrical section Symmetrical rebars	—	$P.e_x$	Uniaxial	Loading and section symmetrical
C	Symmetrical section Symmetrical rebars	$P.e_y$	$P.e_x$	Biaxial	Unsymmetrical loading and symmetrical section
D	Symmetrical section Unsymmetrical rebars	Pe_y'	$P.e_x$	Biaxial	Symmetrical loading and unsymmetrical section
E	Unsymmetrical section and rebars	$P.e_y$ $P.e_v$	— $P.e_u$	Biaxial about uv Uniaxial about xy	Loading symmetry depends on choice of axis
F	Unsymmetrical section and rebars	$P.e_v$	—	Uniaxial about uv Biaxial about xy	Loading symmetry depends on choice of axis
G	Unsymmetrical section	Pe_y'	—	Uniaxial even for $e = 0$	Loading symmetrical about section centroid, not concentric with plastic centroid

The Applied Eccentricity Vector

This is the equivalent eccentricity or hypothetical location of the applied load with respect to the plastic centroid. This eccentricity includes the actual eccentricity of the load with respect to the geometric centroid of the cross-section plus the eccentricity due to applied moment resultant, plus the eccentricity of the geometric centroid from the plastic centroid. It is of interest to note that the physical location of the load is often specified with respect to the geometric centroid, which may not be the same as the plastic centroid. The final applied eccentricity, when connected to the plastic centroid, provides the eccentricity vector. The angle of this vector with respect to the x-axis gives the direction of eccentricity (see Fig. 3.47).

The Applied Moment Vector

The total applied eccentricity vector can be used to determine the equivalent resultant moment as:

$$M_t = P_t e_t \tag{3.38}$$

The eccentricity can be resolved along the coordinate axis to determine the total or equivalent applied moment components about the coordinate axis, and the axial load can be shifted back to the plastic

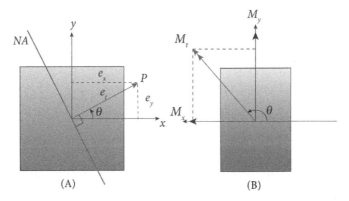

Figure 3.47 (A) The applied eccentricity and (B) moment vectors.

centroid. If the total moments are known, then the equivalent eccentricity can be computed as:

$$e_x = e_t \cos(\theta_e) \quad \text{and} \quad e_y = e_t \sin(\theta_e) \qquad (3.39)$$

$$M_x = P_t e_y \quad \text{and} \quad M_y = P_t e_x \qquad (3.40)$$

It is interesting to note that the positive eccentricity e_y corresponds to M_x moment in the negative sense (counterclockwise) whereas a positive e_x corresponds to clockwise M_y. This disparity sometimes creates confusion in the computation and interpretation of stress resultants. Fig. 3.47 shows the eccentricity vectors and moment using the right-hand rule coordinates.

The Resultant Moment Vector

The internal stress resultants are almost always determined with respect to the coordinate axis through the plastic centroid. The internal moments about the two axes can be used to determine the resultant moment as:

$$M_n = \sqrt{(M_{nx})^2 + (M_{ny})^2} \qquad (3.41)$$

and its direction with respect to the x-axis as:

$$\theta_n = \tan^{-1}\left(\frac{M_{ny}}{M_{nx}}\right) \qquad (3.42)$$

The resultant moments can also be used to compute equivalent eccentricity, perpendicular to the moment direction.

Computing Cross-Section Capacity Ratio

The concept of capacity ratio is introduced earlier in this chapter, where the ratio of the actual stress to the allowable stress gives the capacity ratio (or utilization ratio, or demand-to-capacity D/C ratio). However, for a column where axial load and moments act simultaneously, a conventional definition of capacity ratio is not applicable directly. The axial–flexural capacity of a general beam–column cross-section can be represented by a position vector C from the origin with a specific magnitude and direction. If the vector is represented in spherical coordinate space, then this point is denoted by $C(r, \theta, \alpha)$. A locus of all values of θ and α with a particular value of r will define a closed surface around the origin, constituting a volume. In this case, the vector r is a function of (θ, α). This surface represents the capacity of the cross-section for specified criteria used to generate the points (see Fig. 3.48).

If the capacity is represented in Cartesian coordinates then it can be defined as $C(x, y, z)$. The vector OC can be resolved in a number of ways for the purpose of computation and for interpretation. If the vertical component along the z-axis represents the axial load capacity vector, then the horizontal component on the xy-plane represents the vector of moment capacity along the axis of bending. The moment vector can be further resolved into two components, M_x and M_y, along each of the principle axes. In this form, the capacity surface may be called as the interaction surface as it represents the interaction between axial load and two principal moments. Any applied load point that lies inside the volume bounded by the interaction surface is "safe," whereas a point outside is "unsafe." The generation of capacity interaction surfaces and

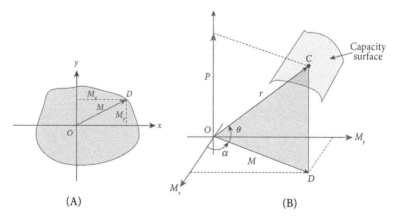

Figure 3.48 The definition of capacity or load vectors. (A) Cartesian coordinates (B) Spherical coordinates.

its visualization has been discussed earlier in this chapter. We will now use the definition of the capacity vector and the actual load vectors to compute the capacity ratio, a concept similar to combined stress ratio but more general and intuitive.

The capacity ratio is in principle defined as the ratio between the applied actions to the corresponding capacity of the member or the cross-section as the case may be. The capacity ratio is easy to define and conceive when dealing with one independent action and the corresponding capacity. For example, in the case of a beam, the applied moment can be compared to the moment capacity to compute the capacity ratio. In the case of a column cross-section, however, the definition of the capacity and the capacity ratio is not easy to define. In fact, several capacities and capacity ratios can be defined and computed. The first is the axial load capacity ratio; this is defined as the ratio between the axial load and the axial load capacity in the absence of any moments. This would be given as:

$$R_x = \frac{P_u}{\phi P_{no}} \tag{3.43}$$

However, if the column is also subjected to simultaneous moments, as often is the case, then the definition of the capacity ratio becomes a bit ambiguous. Of course, in that case we could define the so-called true capacity ratio which is the ratio of the length of applied action vector and the capacity vector along the same direction. The length of the applied action vector is given as:

$$|A| = \sqrt{P_u^2 + M_{ux}^2 + M_{uy}^2} \tag{3.44}$$

whereas, the length of the capacity vector is given by

$$|C| = \sqrt{\phi P_n^2 + \phi M_{nx}^2 + \phi M_{ny}^2} \tag{3.45}$$

and,

$$R = \frac{|A|}{|C|} \tag{3.46}$$

This is shown in Fig. 3.49, and ϕ is used to represent all relevant capacity reduction factors, in accordance with the respective codes.

It is important to emphasize that the value of P_n, M_{nx}, and M_{ny} must be obtained from the capacity interaction surface by extending the applied action vector until it intersects the capacity surface.

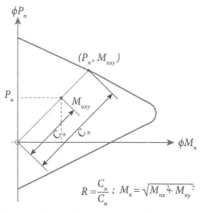

Figure 3.49 The capacity ratio based on true capacity vector.

Other capacity ratios can be defined, e.g., the moment capacity ratio for a given axial load. In this case, the capacity ratio is given by the ratio of the length of moment vector to the moment capacity vector.

$$R = \frac{\sqrt{M_{ux}^2 + M_{uy}^2}}{\sqrt{\phi M_{nx}^2 + \phi M_{ny}^2}} \tag{3.47}$$

or simplified as

$$R = \frac{M_{uxy}}{\phi M_{nxy}} \tag{3.48}$$

The moment capacity is determined by cutting the capacity interaction surface at the given axial load to get a moment–moment interaction curve. This is shown in Fig. 3.50.

Yet another way to define the capacity ratio is to fix both the axial load and one of the moments and then compute the ratio between the remaining moment and the corresponding moment capacity. Sometimes the sum of the individual ratios between the applied axial load and axial load capacity at zero moments, the applied moment M_x and maximum moment capacity about x-axis for zero axial load and zero moment about y-axis, and the ratio between the applied moment about y-axis and the corresponding capacity. This ratio is given by:

$$R = \frac{P_u}{\phi P_{no}} + \frac{M_{ux}}{\phi M_{nxo}} + \frac{M_{uy}}{\phi M_{nyo}} \tag{3.49}$$

This is similar to the combined stress ratio used for the design of steel column and beams. It is obvious from the above discussion that there is

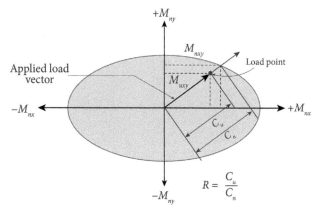

Vector moment capacities at P_u

Figure 3.50 The capacity ratio based on moment ratios at given P_n.

no single value of the capacity ratio in the case of column cross-sections subjected to axial load and biaxial moment. Each approach will give a different value of the capacity ratio, depending upon the relative valve of the applied actions.

For a steel column section, the concept of stress ratio can be used directly as follows:

$$R = \frac{f_a}{F_a} + \frac{f_{bx}}{F_{bx}} + \frac{f_{by}}{F_{by}} \leq 1.0 \tag{3.50}$$

where f_a, f_{bx}, and f_{by} are the actual stresses due to axial load, bending moment about x-axis, and bending moment about y-axis, respectively, while F_a, F_{bx}, and F_{by} are the limiting values of corresponding stresses.

CODE-BASED DESIGN FOR FLEXURE

Various design codes, standards, and guidelines provide simplified procedures for designing and proportioning the structural members against various applied actions. Code-based procedures are simplified, convenient, and widely used by practicing engineers. This section in intended to provide an overview of code-based design procedure for flexural actions using three most widely used codes, ACI 318 (2014), BS 8110 (1997), and EC 2 (2004) as example. Fig. 3.51 shows the typical

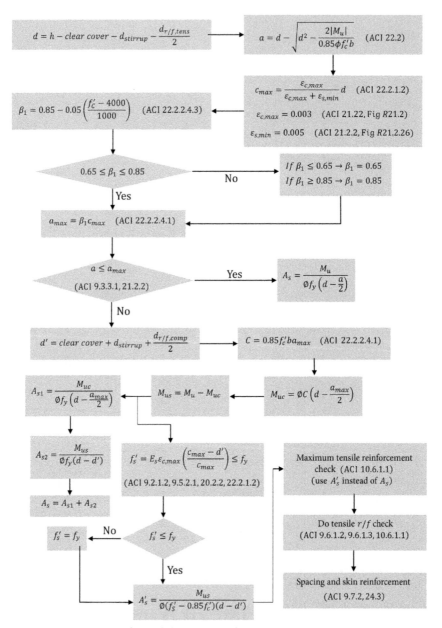

Figure 3.51 ACI 318-14 flexural design procedure.

flexural design procedure according to ACI 318-14. The required inputs include the section dimensions, rebar sizes, material capacities and elastic moduli, and ultimate shear forces. ACI 318-14 can be referred further for detailed explanation of all notations and parameters. It is important to note that these design procedures are not applicable to members in high seismic intensity area and are applicable to constituent flexural members of ordinary moment resisting frames.

On similar lines, Figs. 3.52A,B and 3.53A,B show the procedure for the flexural design of beams as prescribed in BS 8110 (1997) and EC 2 (2004), respectively. These references can be referred further for detailed explanation of all notations and parameters.

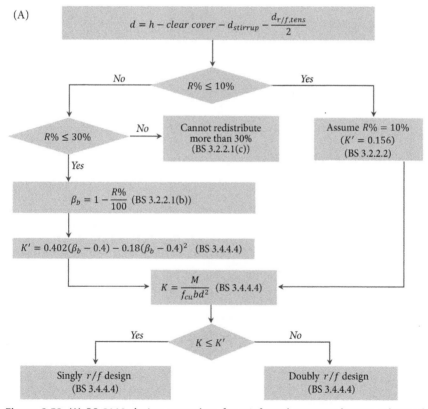

Figure 3.52 (A) BS 8110 design procedure for reinforced concrete beams subjected to flexural actions. (B) Flexural design procedures for singly- and doubly-reinforced beams according to BS 8110.

Singly reinforced beam design

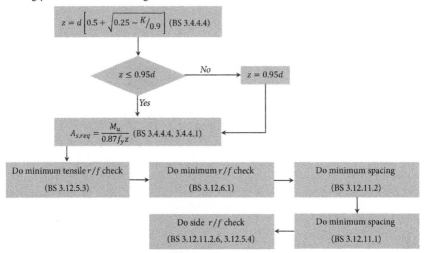

Doubly reinforced beam design

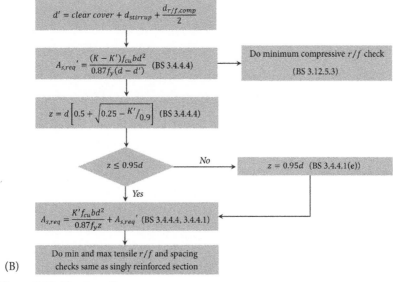

(B)

Figure 3.52 (Continued).

(A)
Section information

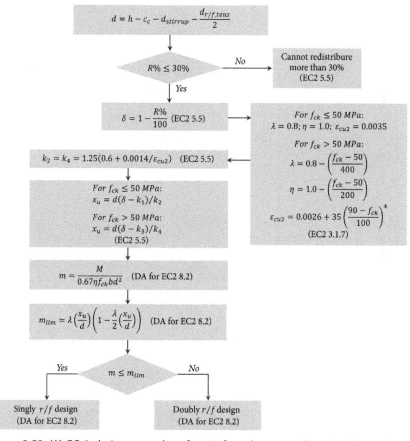

Figure 3.53 (A) EC 2 design procedure for reinforced concrete beams subjected to flexural actions. (B) Flexural design procedures for singly- and doubly-reinforced beams according to EC 2.

(B) Singly reinforced design   Doubly reinforced design

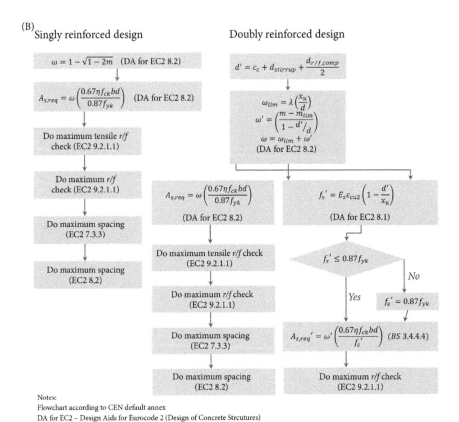

Notes:
Flowchart according to CEN default annex
DA for EC2 – Design Aids for Eurocode 2 (Design of Concrete Strcutures)

Figure 3.53 (Continued).

SOLVED EXAMPLES

Direct and Shear Stress on a Rotated Element

Question 1: An element shown in Fig. Ex. 3.1 is taken from a loaded beam member. Calculate the direct and shear stresses if this element is rotated with a clockwise angle of 30 degrees with respect to x-axis.

Given

$\sigma_x = 80 \text{ N/mm}^2$

$\sigma_y = 30 \text{ N/mm}^2$

$\tau_{xy} = 25 \text{ N/mm}^2$

$\theta = 30°$

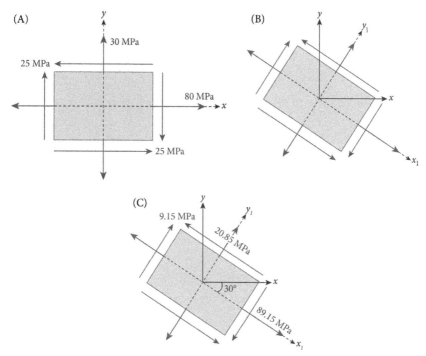

Figure Ex. 3.1 (A) Plane stresses on element taken from a loaded beam. (B) Element inclined at an angle $\theta = 30°$. (C) Stresses in an element inclined at an angle $\theta = 30°$.

Solution

The orientation of element is at clockwise angle of 30 degrees as shown in figure, where the axis x_1 is at an angle $\theta = 30°$

$$2\theta = -60°$$

$$\sin 2\theta = \sin(-60°) = -0.866$$

$$\cos 2\theta = \cos(-60°) = 0.5$$

Now, from stress transformation equations, we can calculate the stresses as follows:

$$\sigma_{x1} = \frac{\sigma_x + \sigma_y}{2} + \frac{\sigma_x - \sigma_y}{2} \cos 2\theta + \tau_{xy}\sin 2\theta$$

$$\sigma_{x1} = \frac{80 + 30}{2} + \frac{80 - 30}{2} \cos(-60°) + (-25)\sin(-60°)$$

$$\sigma_{x1} = 55 + 12.5 + 21.65 = 89.15 \text{ N/mm}^2$$

Similarly for σ_{x2}

$$\sigma_{x2} = \frac{\sigma_x + \sigma_y}{2} - \frac{\sigma_x - \sigma_y}{2}\cos 2\theta - \tau_{xy}\sin 2\theta$$

$$\sigma_{x2} = \frac{80 + 30}{2} - \frac{80 - 30}{2}\cos(-60°) - (-25)\sin(-60°)$$

$$\sigma_{x2} = 55 - 12.5 - 21.65 = 20.85 \text{ N/mm}^2$$

And shear stress for the given orientation τ_{x1y1} is given by

$$\tau_{x1y1} = -\frac{\sigma_x - \sigma_y}{2}\sin 2\theta + \tau_{xy}\cos 2\theta$$

$$\tau_{x1y1} = -\frac{80 - 30}{2}\sin(-60°) + (-25) \times \cos(-60°)$$

$$\tau_{x1y1} = 21.65 - 12.5 = 9.15 \text{ N/mm}^2$$

Result

Direct and shear stresses acting on an element when rotated with clockwise angle of $\theta = 30°$ are:

$$\sigma_{x1} = 89.15 \text{ N/mm}^2$$

$$\sigma_{y1} = 20.85 \text{ N/mm}^2$$

$$\tau_{x1y1} = 9.15 \text{ N/mm}^2$$

Principal Axes and Principal Stresses

Question 2: Determine the principle axis for the stress element in Problem 1 and calculate the principal stress and maximum shear stress. Verify that the shear stress in principal direction is zero (see Fig. Ex. 3.2).

Given

$\sigma_x = 80 \text{ N/mm}^2$
$\sigma_y = 30 \text{ N/mm}^2$
$\tau_{xy} = 25 \text{ N/mm}^2$

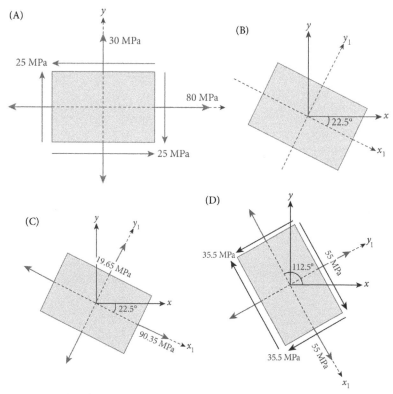

Figure Ex. 3.2 (A) Element in a plane stress. (B) Element inclined to principal axes $x_1 - y_1$. (C) Principal stresses. (D) Maximum shear stresses.

Solution
Principal Orientation and Axes:

$$\tan 2\theta = \frac{2\tau_{xy}}{\sigma_x - \sigma_y}$$

$$\tan 2\theta = \frac{2 \times (-25)}{80 - 30} = -1$$

Therefore

$$2\theta = 45°$$

$$2\theta = 135° \rightarrow \theta = 67.5°$$

$$2\theta = 315° \rightarrow \theta = 157.5°$$

Principle Stresses:
For $\theta = 67.5°$

$$\cos 2\theta = \cos(135°) = -0.707$$

$$\sin 2\theta = \sin(135°) = 0.707$$

$$\sigma_{x1} = \frac{\sigma_x + \sigma_y}{2} + \frac{\sigma_x - \sigma_y}{2}\cos 2\theta + \tau_{xy}\sin 2\theta$$

$$\sigma_{x1} = \frac{80 + 30}{2} + \frac{80 - 30}{2}\cos(135°) + (-25)\sin(135°)$$

$$\sigma_{x1} = 55 - 17.67 - 17.67 = 19.65 \text{ N/mm}^2$$

For $\theta = 157.5°$

$$\cos 2\theta = \cos(315°) = 0.707$$

$$\sin 2\theta = \sin(315°) = -0.707$$

$$\sigma_{x1} = \frac{\sigma_x + \sigma_y}{2} + \frac{\sigma_x - \sigma_y}{2}\cos 2\theta + \tau_{xy}\sin 2\theta$$

$$\sigma_{x1} = \frac{80 + 30}{2} + \frac{80 - 30}{2}\cos(315°) + (-25)\sin(315°)$$

$$\sigma_{x1} = 55 + 17.67 + 17.67 = 90.35 \text{ N/mm}^2$$

Therefore
$\sigma_{x1} = 19.65 \text{ N/mm}^2$ at $\theta = 67.5°$
$\sigma_{x1} = 90.35 \text{ N/mm}^2$ at $\theta = 157.5°$

Alternatively, σ_1 and σ_2 can be calculated as

$$\sigma_{1,2} = \frac{\sigma_x + \sigma_y}{2} \pm \sqrt{\left(\frac{\sigma_x - \sigma_y}{2}\right)^2 + \left(\tau_{xy}\right)^2}$$

$$\sigma_{1,2} = \frac{80 + 30}{2} \pm \sqrt{\left(\frac{80 - 30}{2}\right)^2 + (25)^2}$$

$$\sigma_{1,2} = 55 \pm \sqrt{(25)^2 + (25)^2}$$

$$\sigma_{1,2} = 55 \pm 35.35 \text{ N/mm}^2$$

Therefore

$\sigma_1 = 90.35 \text{ N/mm}^2$

$\sigma_2 = 19.65 \text{ N/mm}^2$

Shear stress at $\theta = 67.5°$ is given by:

$$\tau_{x1y1} = -\frac{\sigma_x - \sigma_y}{2}\sin 2\theta + \tau_{xy}\cos 2\theta$$

$$\tau_{x1y1} = -\frac{80 - 30}{2}\sin(135°) + (-25) \times \cos(135°)$$

$$\tau_{x1y1} = 17.675 - 17.675 = 0 \text{ N/mm}^2$$

Similarly, for $\theta = 157.5°$ shear stress is zero. Hence, it is verified that shear stress is zero in principal directions.

Maximum Shear Stress

$$\tau_{max} = \sqrt{\left(\frac{\sigma_x - \sigma_y}{2}\right)^2 + \left(\tau_{xy}\right)^2}$$

$$\tau_{max} = \sqrt{\left(\frac{80 - 30}{2}\right)^2 + (-25)^2}$$

$$\tau_{max} = 35.35 \text{ N/mm}^2$$

The orientation of maximum shear stress is:

$$\theta_{s1} = \theta_1 - 45° = 157.5° - 45° = 112.5°$$

And the normal stresses acting on the planes of maximum shear stresses are:

$$\sigma_{av} = \frac{\sigma_x - \sigma_y}{2} = \frac{80 + 30}{2} = 55 \text{ N/mm}^2$$

Result

Principal stresses and maximum shear of the stress element are:

$\sigma_{x1} = 19.65 \text{ N/mm}^2$ at $\theta = 67.5°$

$\sigma_{x1} = 90.35 \text{ N/mm}^2$ at $\theta = 157.5°$

$\sigma_1 = 90.35 \text{ N/mm}^2$

$\sigma_2 = 19.65 \text{ N/mm}^2$

$\tau_{max} = 35.35 \text{ N/mm}^2$

$\theta_{s1} = 112.5°$

$\sigma_{av} = 55 \text{ N/mm}^2$

Neutral Axis of the Sections

Question 3: Determine the neutral axis of the sections given below (see Fig. Ex. 3.3).

Solution
An Inverted T-Section

Let $X-Y$ be the reference axis across the bottom of the cross-section to calculate neutral axis. The origin O of the xy-coordinates is placed at the centroid of the cross-sectional area, and therefore the x-axis becomes the neutral axis of the cross-section.

The given section is divided into two rectangles having area A_1 and A_2. Let y_1 and y_2 be the distance from $X-Y$-axis to the centroids of the area A_1 and A_2, respectively.

$$A_1 = 300 \times 100 = 30,000 \text{ mm}^2$$

$$y_1 = \frac{100}{2} = 50 \text{ mm}$$

$$A_2 = 200 \times 100 = 20,000 \text{ mm}^2$$

$$y_2 = 100 + \frac{200}{2} = 200 \text{ mm}$$

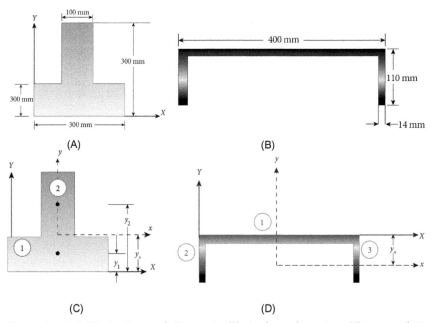

Figure Ex. 3.3 (A) An inverted T-section. (B) A channel section. (C) Inverted T-sections with centroidal axis at y_n distance from the base. (D) Inverted T-sections with centroidal axis at y_n distance from X-axis.

Distance of centroidal axis from reference axis is calculated as:

$$y_n = \frac{\sum y_i A_i}{\sum A_i}$$

$$y_n = \frac{30,000 \times 50 + 20,000 \times 200}{30,000 + 20,000} = 110 \text{ mm}$$

Thus the position of neutral axis is 110 mm from the bottom edge of the section.

Channel Section

Let $X-Y$ be the reference axis across the bottom of the cross-section to calculate neutral axis. The origin O of the xy-coordinates is placed at the centroid of the cross-sectional area, and therefore the x-axis becomes the neutral axis of the cross-section.

The given section is divided into two rectangles having area A_1, A_2, and A_3. Let y_1, y_2, and y_3 be the distance from $X-Y$-axis to the centroids of the area A_1, A_2, and A_3, respectively.

$$A_1 = 400 - 2 \times 14 = 5208 \text{ mm}^2$$

$$y_1 = \frac{14}{2} = 7 \text{ mm}$$

$$A_2 = 110 \times 14 = 1540 \text{ mm}^2$$

$$y_2 = \frac{110}{2} = 55 \text{ mm}$$

$$A_3 = A_2 = 1540 \text{ mm}^2$$

$$y_3 = y_2 = 55 \text{ mm}$$

Distance of centroidal axis from reference axis is calculated as:

$$y_n = \frac{\sum y_i A_i}{\sum A_i}$$

$$y_n = \frac{5208 \times 07 + 2 \times 1540 \times 55}{5208 + 2 \times 1540} = 24.84 \text{ mm}$$

Thus the position of neutral axis is 24.84 mm from the top edge of the section.

Result

1. The position of neutral axis for inverted T-section is 110 mm from the bottom edge of the section.
2. The position of neutral axis for channel section is 24.84 mm from the top edge of the section.

Question 4: Determine location of neutral axis for a 250 mm × 300 mm rectangular RC section for (a) balanced conditions, and (b) for the two given reinforcement details as shown in Fig. Ex. 3.4. $f_c' = 25$ MPa and $f_y = 400$ MPa. Determine whether the given cross sections are under-reinforced or over-reinforced sections.

Solution (a)

Here for the given section the neutral axis for balanced section is determined by assuming that both concrete and steel reaches their ultimate

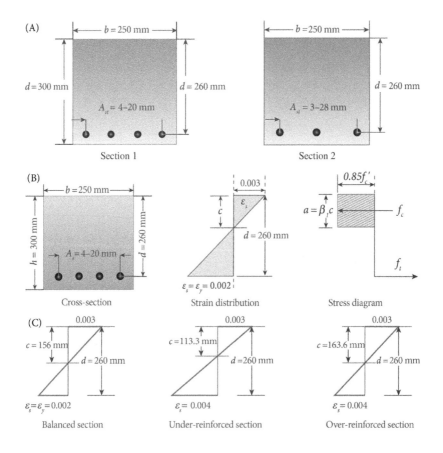

Figure Ex. 3.4 (A) Singly-reinforced rectangular sections. (B) Strain and stress distribution diagram of single reinforced section. (C) Strain distribution for different stages.

strain at the same time. Therefore ultimate strain in concrete is 0.003 and ultimate strain in steel $\varepsilon_s = \varepsilon_y$

$$\varepsilon_y = \frac{f_y}{E_s} = \frac{400}{2 \times 10^5} = 0.002$$

The strain distribution and stress diagram is shown in Fig. Ex. 3.4B.

For Balanced Section

From the similar triangles in strain diagram, we can write that:

$$C = \frac{0.003}{(0.003 + \varepsilon_s)} d = 156 \text{ mm}$$

Therefore neutral axis lies at the depth of 156 mm from the top edge of the section for balanced section.

For Section 1

We know that neutral axis divides the compression and tension area in the section, so for the equilibrium of the force acting on the section.

Compression force = tension force

$$F_c = F_t$$

$$0.85 f_c' ab = f_y A_{st}$$

$$A_{st} = 4\pi \frac{d^2}{4} = 4\pi \frac{20^2}{4} = 1256.64 \text{ mm}^2$$

$$C = \frac{f_y A_{st}}{\beta_1 \left(0.85 f_c'\right) b}$$

$$C = \frac{400 \times 1256.64}{0.85 \times 0.85 \times 25 \times 250} = 111.31 \text{ mm}$$

Thus the neutral axis for the given reinforcement is 111.31 mm and is less than neutral axis depth for balanced sections. So the given section is under reinforced section.

For Section 2

$$F_c = F_t$$

$$0.85 f_c' ab = f_y A_{st}$$

$$A_{st} = 3\pi \frac{d^2}{4} = 3\pi \frac{28^2}{4} = 1847.26 \text{ mm}^2$$

$$C = \frac{f_y A_{st}}{\beta_1 \left(0.85 f_c'\right) b}$$

$$C = \frac{400 \times 1847.26}{0.85 \times 0.85 \times 25 \times 250} = 163.63 \text{ mm}$$

Thus the neutral axis for the given reinforcement is 163.63 mm and is greater than neutral axis depth for balanced sections. So the given section is over reinforced section.

Result

1. Neutral axis for balanced sections lies at a depth of 156 mm from the top edge of the section.
2. Neutral axis for Section 1 lies at a depth of 113.31 mm and the section is under-reinforced section.
3. Neutral axis for Section 2 lies at a depth of 163.63 mm and the section is over-reinforced section.

P–M Interaction Curve

Question 5: Determine the concentric axial load capacity for the rectangular reinforced concrete column section, as shown in Fig. Ex. 3.5.

Given

$f_c' = 25 \text{ N/mm}^2$
$f_y = 400 \text{ N/mm}^2$
$E_s = 2 \times 10^5 \text{ N/mm}^2$
$b = 400 \text{ mm}$
$D = 400 \text{ mm}$
$d = 340 \text{ mm}$
$A_{s1} = 4 - 28 \text{ mm bars}$
$A_{s2} = 4 - 28 \text{ mm bars}$

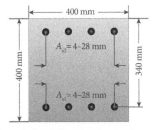

Figure Ex. 3.5 A rectangular reinforced concrete section.

Solution
Gross section area of concrete $(A_g) = 400 \times 400 = 160,000$ mm^2

$$\text{Area of steel } (A_{s1}) = 4\pi \frac{d^2}{4} = 4\pi \frac{28^2}{4} = 2463.1 \text{ mm}^2$$

$$\text{Area of steel } (A_{s2}) = 4\pi \frac{d^2}{4} = 4\pi \frac{28^2}{4} = 2463.1 \text{ mm}^2$$

Area of steel $(A_s) = 2463.1 + 2463.1 = 4926.02$ mm^2

$$r = \frac{A_{st}}{A_g} = \frac{4926.02}{160,000}$$

$$r = 0.031 = 3.1\%$$

$$\varepsilon_y = \frac{f_y}{E_s} = \frac{400}{200,000} = 0.002$$

Calculation of the Concentric Axial Load Capacity

$$P_o = 0.85 f_c' \left(A_g - A_{st} \right) + f_y A_{st}$$

$$P_o = (0.85 \times 25) \times (160,000 - 4926.02) + (400 \times 4926.02)$$

$$P_o = 3,295,322.08 + 1,970,408 = 5,265,730.08 \text{ N} = 5265.73 \text{ kN}$$

Assuming the capacity reduction factor, $\varphi = 0.7$, thus:

$$\varphi P_o = 0.7 \times 5265.73 = 3686.01 \text{ kN}$$

Result
The concentric axial load capacity of given reinforced concrete section is 3686.01 kN.

Question 6: For the RC section given in Question 5, determine the point on its $P-M$ interaction diagram corresponding to balanced failure.

Solution
To determine a complete interaction diagram, a number of strain distributions must be considered and the corresponding values of P_n, M_n, φP_n, and φM_n are calculated. The determination of these values for a strain profile corresponding to balanced failure is given below.

Computing φP_n and φM_n for Balanced Failure $\left(\varepsilon_{s1} = -\varepsilon_y \right)$

The first step is to determine neutral axis depth (C) and the strains in the reinforcement. The column cross-section and the strain distribution

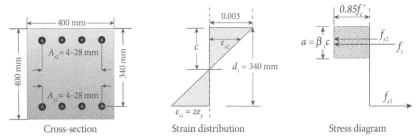

Cross-section   Strain distribution   Stress diagram

Figure Ex. 3.6 Strain and stress distributions for a rectangular reinforced concrete section.

corresponding to $\varepsilon_{s1} = \varepsilon_y(z = -1)$ are shown in Fig. Ex. 3.6. The strain in the bottom layer of steel is $-1\varepsilon_y = -0.002$. From similar triangles, the depth of NA is calculated as:

$$C = \frac{0.003}{0.003 + 0.002} d_1 = 204 \text{ mm}$$

Using similar triangles, the strain in compression reinforcement is:

$$\varepsilon_{s2} = \frac{C - d_1}{C} \times 0.003$$

$$\varepsilon_{s2} = \frac{204 - 60}{204} \times 0.003 = 0.00212$$

And

$$\varepsilon_{s1} = -0.002$$

The stress in reinforcement layer 2:

$$f_{s2} = \varepsilon_{s2} \times E_s = 0.00212 \times 200,000 = 424 \text{ N/mm}^2$$

$$-f_y \leq f_{s2} \leq f_y$$

Therefore $f_{s2} = 400 \text{ N/mm}^2$.

Now, calculating the depth of the equivalent rectangular stress block

$$a = \beta_1 c = 0.85 \times 204 = 173.4 \text{ mm}$$

This is less than h; therefore this value can be used. Now we can calculate the forces in the concrete and steel. Force in concrete from stress block:

$$F_c = 0.85 f_c' ab = 0.85 \times 25 \times 173.4 \times \frac{400}{1000} = 1473.9 \text{ kN}$$

The distance $d_1 = 340$ mm to reinforcement layer 1 exceeds $a = 173.4$ mm. Hence, this layer of steel lies outside the compression block and does not displace concrete include in the area when computing C_c. Thus

$$F_{s1} = f_{s1} A_{s1} = -400 \times 2463.1 = -985.24 \text{ kN}$$

Reinforcement layer 2 lies in the compression zone, since $a = 173.4$ mm exceeds $d_2 = 60$ mm. Hence, we must allow for the stress in the concrete displaced by the steel when we compute F_{s2}.

$$F_{s2} = \left(f_{cs2} - 0.85 f_c' \right) A_{s2}$$

$$F_{s2} = (400 - 0.85 \times 25) \times 2463.1 = 932.9 \text{ kN}$$

The nominal load capacity, P_n is found by summing the axial force components

$$P_n = C_c + \sum F_{si} = 1473.9 - 985.24 + 932.9 = 1421.56 \text{ kN}$$

Since $\varepsilon_{s1} = -\varepsilon_y$, this is balanced failure condition and $P_n = P_b$.

From Fig. Ex. 3.6, the moment of C_c, F_{s1}, and F_{s2} about the centroid of the section is

$$M_n = C_c \left(\frac{h}{2} - \frac{a}{2} \right) + F_{s1} \left(\frac{h}{2} - d_1 \right) + F_{s2} \left(\frac{h}{2} - d_2 \right)$$

$$M_n = 1473.9 \times \left(\frac{400}{2} - \frac{173.9}{2} \right) - 985.24 \times \left(\frac{400}{2} - 340 \right)$$

$$+ 932.9 \left(\frac{400}{2} - 60 \right)$$

$$M_n = 166,624.36 + 137,933.6 = 435,163.9 \text{ kN-m} = 435.163 \text{ kN-m}$$

Therefore $M_n = M_b = 435.163$ kN-m.

Finally, we can calculate φ, φP_n, and φM_n according to ACI Sec. B.9.3.2. The strain ε_t in the layer of reinforcement farthest from the compression face is $\varepsilon_{s1} = -0.002 = -\varepsilon_y$. Thus

Assuming the capacity reduction factor, $\varphi = 0.7$

$$\varphi P_n = 0.7 \times 1421.56 = 995.1 \text{ kN}$$

$$\varphi M_n = 0.7 \times 435.163 = 304.61 \text{ kN-m}$$

Result

For balance failure, the axial load capacity for the given RC section is 995.1 kN and the corresponding moment capacity is 304.61 kN−m.

Question 7: For the RC section given in Question 5, determine the corresponding point on its $P-M$ interaction diagram when strain in bottom layer of reinforcement reaches to (a) two and (b) four times their yield strain capacity.

Solution

a. Computing φP_n and φM_n for $z = -2$

To illustrate the calculation of φ for cases falling between the compression-controlled limit and the tension-controlled limit, the computation will be repeated for the strain distribution corresponding to $z = -2$ (with $\varepsilon_{s1} = -2\varepsilon_y$) as shown in Fig. Ex. 3.6.

The strain in the bottom layer of steel is $-2\varepsilon_y = -0.004$. From similar triangles, the depth of NA is calculated as:

$$C = \frac{0.003}{0.003 + 0.004} d_1 = 145.71 \text{ mm}$$

Using similar triangles, the strain in compression reinforcement is:

$$\varepsilon_{s2} = \frac{C - d_1}{C} \times 0.003$$

$$\varepsilon_{s2} = \frac{145.71 - 60}{145.71} \times 0.003 = 0.00176$$

And

$$\varepsilon_{s1} = -0.002$$

The stress in reinforcement layer 2:

$$f_{s2} = \varepsilon_{s2} \times E_s = 0.00176 \times 200,000 = 352 \text{ N/mm}^2$$

Since

$$-f_y \leq f_{s2} \leq f_y$$

Therefore $f_{s2} = 352 \text{ N/mm}^2$.

Now, calculating the depth of the equivalent rectangular stress block

$$a = \beta_1 c = 0.85 \times 145.71 = 123.85 \text{ mm}$$

Force in concrete from stress block:

$$F_c = 0.85 f_c' \, ab = 0.85 \times 25 \times 123.85 \times \frac{400}{1000} = 1052.72 \text{ kN}$$

$$F_{s1} = f_{s1} A_{s1} = -400 \times 2463.1 = -985.24 \text{ kN}$$

Reinforcement layer 2 lies in the compression zone, since $a = 123.85$ mm exceeds $d_2 = 60$ mm. Hence, we must allow for the stress in the concrete displaced by the steel when we compute F_{s2}.

$$F_{s2} = \left(f_{cs2} - 0.85 f_c' \right) A_{s2}$$

$$F_{s2} = (352 - 0.85 \times 25) \times 2463.1 = 814.67 \text{ kN}$$

The nominal load capacity, P_n is found by summing the axial force components

$$P_n = C_c + \sum F_{si} = 1052.72 - 985.24 + 814.67 = 884.45 \text{ kN}$$

From Fig. Ex. 3.6, the moment of $C_c, F_{s1},$ and F_{s2} about the centroid of the section is

$$M_n = C_c \left(\frac{h}{2} - \frac{a}{2} \right) + F_{s1} \left(\frac{h}{2} - d_1 \right) + F_{s2} \left(\frac{h}{2} - d_2 \right)$$

$$M_n = 1052.72 \times \left(\frac{400}{2} - \frac{145.71}{2} \right) - 985.24 \times \left(\frac{400}{2} - 340 \right)$$

$$+ \, 884.45 \times \left(\frac{400}{2} - 60 \right)$$

$$M_n = 133,848.08 + 137,933.6 + 114,375.8 = 386,157.48 \text{ kN} - \text{mm}$$

$$= 386.157 \text{ kN} - \text{m}$$

Finally, we can calculate $\varphi, \varphi P_n,$ and φM_n according to ACI Sec. B.9.3.2. The strain ε_t in the layer of reinforcement farthest from the compression face is $\varepsilon_{s1} = -0.004 = -2\varepsilon_y$. Thus

$$\varphi = 0.56 - 68\varepsilon_t = 0.56 - 68 \times (-0.004) = 0.832$$

$$\varphi P_n = 0.832 \times 884.45 = 735.86 \text{ kN}$$

$$\varphi M_n = 0.832 \times 386.157 = 321.60 \text{ kN} - \text{m}$$

b. Compute φP_n and φM_n for $z = -4$

Using the same Fig. Ex. 3.6, repeating the same procedure, for $Z = -4, \varepsilon_{s1} = -0.008$, neutral axis depth is calculated as:

$$C = \frac{0.003}{0.003 + 0.008} d_1 = 92.73 \text{ mm}$$

Using similar triangles, the strain in compression reinforcement is:

$$\varepsilon_{s2} = \frac{C - d_1}{C} \times 0.003$$

$$\varepsilon_{s2} = \frac{92.73 - 60}{92.73} \times 0.003 = 0.00106$$

And

$$\varepsilon_{s1} = -0.002$$

The stress in reinforcement layer 2:

$$f_{s2} = \varepsilon_{s2} \times E_s = 0.00106 \times 200,000 = 212 \text{ N/mm}^2$$

Since

$$-f_y \leq f_{s2} \leq f_y$$

Therefore $f_{s2} = 212 \text{ N/mm}^2$.

Now, calculating the depth of the equivalent rectangular stress block

$$a = \beta_1 c = 0.85 \times 92.73 = 78.82 \text{ mm}$$

Force in concrete from stress block:

$$F_c = 0.85 f_c' ab = 0.85 \times 25 \times 78.82 \times \frac{400}{1000} = 669.97 \text{ kN}$$

$$F_{s1} = f_{s1} A_{s1} = -400 \times 2463.1 = -985.24 \text{ kN}$$

Reinforcement layer 2 lies in the compression zone, since $a = 78.82$ mm exceeds $d_2 = 60$ mm. Hence, we must allow for the stress in the concrete displaced by the steel when we compute F_{s2}.

$$F_{s2} = \left(f_{cs2} - 0.85 f_c' \right) A_{s2}$$

$$F_{s2} = (212 - 0.85 \times 25) \times 2463.1 = 469.84 \text{ kN}$$

The nominal load capacity, P_n is found by summing the axial force components

$$P_n = C_c + \sum F_{si} = 670 - 985.24 + 469.84 = 154.6 \text{ kN}$$

From Fig. Ex. 3.6, the moment of C_c, F_{s1}, and F_{s2} about the centroid of the section is

$$M_n = C_c\left(\frac{h}{2} - \frac{a}{2}\right) + F_{s1}\left(\frac{h}{2} - d_1\right) + F_{s2}\left(\frac{h}{2} - d_2\right)$$

$$M_n = 670 \times \left(\frac{400}{2} - \frac{78.82}{2}\right) - 985.24 \times \left(\frac{400}{2} - 340\right)$$

$$+ \ 469.84 \times \left(\frac{400}{2} - 60\right)$$

$$M_n = 107,803 + 137,933.6 + 65,777.6 = 311,514.2 \text{ kN}-\text{mm} = 311.51 \text{ kN}-\text{m}$$

Finally, we can calculate φ, φP_n, and φM_n according to ACI Sec. B.9.3.2. The strain ε_t in the layer of reinforcement farthest from the compression face is $\varepsilon_{s1} = -0.008 = -4\varepsilon_y$. Thus

$$\varphi = 0.9$$

$$\varphi P_n = 0.9 \times 154.6 = 139.14 \text{ kN}$$

$$\varphi M_n = 0.9 \times 311.38 = 280.35 \text{ kN}-\text{m}$$

Result

The axial load capacity of the given RC section, corresponding to bottom steel strain of two times and four times of yield strain, are 735.95 kN and 139.14 kN is 995.1 kN. The corresponding moment capacities are 321.60 kN–m and 280.35 kN–m.

Question 8: For the RC section given in Question 5, determine the axial tensile capacity. Also plot the *PM* interaction diagram using results from Questions 5–7.

Solution

Calculation of the Capacity in Axial Tension

The strength under such a loading is equal to the yield strength of the reinforcement in tension as given by,

$$P_n = \sum_{i=1}^{n} f_y A_{si} = -400 \times 2463.1 \times 2 = 1,970,480 \text{ N} = 1970.48 \text{ kN}$$

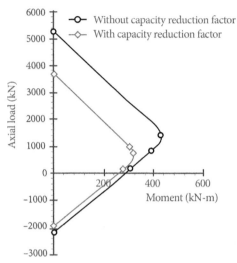

Figure Ex. 3.7 *P–M* interaction diagram.

$\varphi = 0.9$, therefore

$$\varphi P_{nt} = 0.9 \times 2210.48 = 1773.432 \text{ kN}$$

Result

The results from Questions 5–8 are summarized in table below and the *PM* interaction curve is shown in Fig. Ex. 3.7.

Coordinates of *PM* Interaction Curve

Z	P_n(kN)	M_n(kN–m)	φ	φP_n(kN)	φM_n(kN–m)
0	5013.73	0	0.7	3509.611	0
−1	1421.56	435.163	0.7	995.1	304.61
−2	884.45	386.18	0.832	735.86	321.30
−4	154.6	311.51	0.9	139.14	280.35
	1970.48	0	0.9	− 1773.432	0

Question 9: An RC cross-section is shown in Fig. Ex. 3.8. The compressive strength of concrete is $f_c' = 30$ MPa. Yield strength of steel is 420 MPa. Generate *PM* curve for the cross-section provided $E_c = 30,000$ MPa, $E_s = 2 \times 10^5$ MPa, $\beta_1 = 0.85$ MPa, and $\varepsilon_y = 0.00207$.

Given

$$f_y = 420 \text{ MPa}$$
$$\varepsilon_y = 0.00207$$

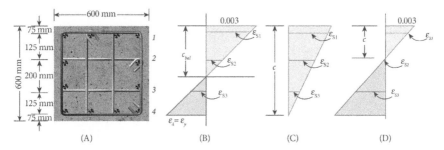

Figure Ex. 3.8 An RC cross-section with various strain points corresponding to points on *P–M* interaction curve. (A) A reinforce concrete column cross-section (B) Balanced point (C) Point above balanced (D) Point below balanced.

$$f'_c = 30 \text{ MPa}$$
$$E_c = 30,000 \text{ MPa}$$
$$\beta_1 = 0.85$$
$$E_s = 2 \times 10^5 \text{ MPa}$$
$$A_s = 645 \text{ mm}^2 (\text{one bar})$$

Solution
Steel Areas and Depth

Layer No. 1   Area = (A_{s1}) = 2580 mm²   Depth = d_1 = 75 mm
Layer No. 2   Area = (A_{s2}) = 1290 mm²   Depth = d_2 = 200 mm
Layer No. 3   Area = (A_{s3}) = 1290 mm²   Depth = d_3 = 400 mm
Layer No. 4   Area = (A_{s4}) = 2580 mm²   Depth = d_4 = 525 mm

Axial Compression Load (P_o)

$$C_c = 0.85 f'_c bh = 0.85 \times 30 \times 600 \times 600 = 9,180,000 \text{ N} = 9180 \text{ kN}$$

$$C_{si} = A_{si}(f_y - 0.85 f'_c)$$

$$C_{s1} = A_{s1}\left(f_y - 0.85 f'_c\right) = 2580(420 - 30 \times 0.85) = 1,017,810 \text{ N} = 1017.81 \text{ kN}$$

$$C_{s2} = A_{s2}\left(f_y - 0.85 f'_c\right) = 1290(420 - 30 \times 0.85) = 508,905 \text{ N} = 508.905 \text{ kN}$$

$$C_{s3} = A_{s3}\left(f_y - 0.85 f'_c\right) = 1290(420 - 30 \times 0.85) = 508,905 \text{ N} = 508.905 \text{ kN}$$

$$C_{s4} = A_{s4}\left(f_y - 0.85 f'_c\right) = 2580(420 - 30 \times 0.85) = 1,017,810 \text{ N} = 1017.81 \text{ kN}$$

$$P_o = 12,233,430 \text{ N}$$

Axial Tensile Load (P_t)

$$T_{si} = A_{si}f_y$$

$$T_{s1} = A_{s1}f_y = 2580 \times 420 = 1,083,600 \text{ N} = 1083.6 \text{ kN}$$

$$T_{s2} = A_{s2}f_y = 1290 \times 420 = 541,800 \text{ N} = 541.8 \text{ kN}$$

$$T_{s3} = A_{s3}f_y = 1290 \times 420 = 541,800 \text{ N} = 541.8 \text{ kN}$$

$$T_{s4} = A_{s4}f_y = 2580 \times 420 = 1,083,600 \text{ N} = 1083.6 \text{ kN}$$

$$P_t = 3,250,800 \text{ N}$$

For Balanced Point

Fig. Ex. 3.8B refers,

$$C = \frac{0.003}{0.003 + \varepsilon_y} (d)$$

$$C = \frac{0.003}{0.003 + \varepsilon_y} (525) = 310.65 \text{ mm}$$

$$\varepsilon_{si} = \frac{C - d_i}{C} (0.003)$$

$$\varepsilon_{s1} = \frac{C - d_1}{C} (0.003) = \frac{310.65 - 75}{310.65} \times (0.003) = 2.275 \times 10^{-3} > \varepsilon_y$$

$$\varepsilon_{s2} = \frac{C - d_2}{C} (0.003) = \frac{310.65 - 200}{310.65} \times (0.003) = 1.068 \times 10^{-3}$$

$$\varepsilon_{s3} = \frac{C - d_3}{C} (0.003) = \frac{310.65 - 400}{310.65} \times (0.003) = -8.62 \times 10^{-4}$$

Now

$$f_{s1} = 420 \text{ MPa}$$

$$f_{s2} = \varepsilon_{s2}E_s = 1.068 \times 10^{-3} \times 2 \times 10^5 = 213.6 \text{ MPa}$$

$$f_{s3} = \varepsilon_{s3}E_s = -8.62 \times 10^{-4} \times 2 \times 10^5 = -172.4 \text{ MPa}$$

$$f_{s4} = -420 \text{ MPa}$$

And

$$C_c = 0.85\beta_1 f_c' bC = 0.85 \times 30 \times 600 \times 0.85 \times 310.65$$
$$= 4,040,003 \text{ N} = 4040.003 \text{ kN}$$

$$C_{s1} = A_{s1}\left(f_{s1} - 0.85 f_c'\right) = 2580 \times (420 - 0.85 \times 30) = 1,017,810 \text{ N}$$
$$= 1017.810 \text{ kN}$$

$$C_{s2} = A_{s2}\left(f_{s2} - 0.85 f_c'\right) = 1290 \times (213.6 - 0.85 \times 30) = 242,649 \text{ N}$$
$$= 242.649 \text{ kN}$$

$$T_{s3} = A_{s3} \times f_{s3} = 1290 \times (-172.4) = -222,396 \text{ N} = -222.396 \text{ kN}$$

$$T_{s4} = A_{s4} \times f_{s4} = 2580 \times (-420) = -1,083,600 \text{ N} = -1083.600 \text{ kN}$$

$$P_{bal} = 3,994,466.25 \text{ N} = 3994.466 \text{ kN}$$

$$M_{bal} = C_c\left(\frac{h}{2} - \frac{\beta_1 C}{2}\right) + \sum_{i=1}^{4} C_{si}\left(\frac{h}{2} - d_i\right) + \sum_{i=1}^{4} T_{si}\left(d_i - \frac{h}{2}\right)$$

$$= 1,197,936,246 \text{ N}-\text{mm}$$

$$\phi_{bal} = \frac{0.003}{C} = \frac{0.003}{310.6} = 9.65 \times 10^{-6} \, (1/\text{mm})$$

For Points Above Balanced

Fig. Ex. 3.8C refers,

$$C = d_4 = 525 \text{ mm}$$

$$\varepsilon_{si} = \frac{C - d_i}{C}(0.003)$$

$$\varepsilon_{s1} = \frac{C - d_1}{C}(0.003) = \frac{525 - 75}{525} \times (0.003) = 2.571 \times 10^{-3} > \varepsilon_y$$

$$\varepsilon_{s2} = \frac{C - d_2}{C}(0.003) = \frac{525 - 200}{525} \times (0.003) = 1.857 \times 10^{-3}$$

$$\varepsilon_{s3} = \frac{C - d_3}{C}(0.003) = \frac{525 - 400}{525} \times (0.003) = 7.14 \times 10^{-4}$$

Now

$$f_{s1} = 420 \text{ MPa}$$

$$f_{s2} = \varepsilon_{s2} E_s = 1.857 \times 10^{-3} \times 2 \times 10^5 = 371.4 \text{ MPa}$$

$$f_{s3} = \varepsilon_{s3} E_s = 7.14 \times 10^{-4} \times 2 \times 10^5 = 142.8 \text{ MPa}$$

And

$$C_c = 0.85\beta_1 f_c' bC = 0.85 \times 30 \times 600 \times 0.85 \times 525 = 6,827,625 \text{ N}$$
$$= 6827.625 \text{ kN}$$

$$C_{s1} = A_{s1}\left(f_{s1} - 0.85 f_c'\right) = 2580 \times (420 - 0.85 \times 30) = 1,017,810 \text{ N}$$
$$= 1017.810 \text{ kN}$$

$$C_{s2} = A_{s2}\left(f_{s2} - 0.85 f_c'\right) = 1290 \times (371.4 - 0.85 \times 30) = 449,565 \text{ N}$$
$$= 449.565 \text{ kN}$$

$$C_{s3} = A_{s3}\left(f_{s3} - 0.85 f_c'\right) = 1290 \times (142.8 - 0.85 \times 30) = 151,317 \text{ N}$$
$$= 151.317 \text{ kN}$$

$$P = 8,446,317 \text{ N} = 8446.317 \text{ kN}$$

$$M = C_c\left(\frac{h}{2} - \frac{\beta_1 C}{2}\right) + \sum_{i=1}^{4} C_{si}\left(\frac{h}{2} - d_i\right) + \sum_{i=1}^{4} T_{si}\left(d_i - \frac{h}{2}\right)$$

$$= 783,705,721.9 \text{ N--mm}$$

$$\phi = \frac{0.003}{C} = \frac{0.003}{525} = 5.714 \times 10^{-6}\,(1/\text{mm})$$

Points Below Balanced

Fig. Ex. 3.8D refers,

$$\text{Set } \varepsilon_s = 2\varepsilon_y = 2 \times 0.00207 = 4.14 \times 10^{-3}$$

$$C = \left(\frac{0.003}{0.003 + \varepsilon_s}\right)d = \left(\frac{0.003}{0.003 + 4.14 \times 10^{-3}}\right) \times 525 = 220.588 \text{ mm}$$

$$\varepsilon_{si} = \frac{C - d_i}{C}(0.003)$$

$$\varepsilon_{s1} = \frac{C - d_1}{C}(0.003) = \frac{220.588 - 75}{220.588} \times (0.003) = 1.979 \times 10^{-3}$$

$$\varepsilon_{s2} = \frac{C - d_2}{C}(0.003) = \frac{220.588 - 200}{220.588} \times (0.003) = 2.798 \times 10^{-4}$$

$$\varepsilon_{s3} = \frac{C - d_3}{C}(0.003) = \frac{220.588 - 400}{220.588} \times (0.003) = -2.44 \times 10^{-4}$$

Now

$$f_{s1} = \varepsilon_{s1}E_1 = 0.00197 \times 2 \times 10^5 = 394 \text{ MPa}$$

$$f_{s2} = \varepsilon_{s2}E_s = 2.798 \times 10^{-4} \times 2 \times 10^5 = 55.96 \text{ MPa}$$

$$f_{s3} = -420 \text{ MPa}$$

$$f_{s4} = -420 \text{ MPa}$$

$$C_c = 0.85\beta_1 f_c' bC = 0.85 \times 0.85 \times 30 \times 600 \times 0.85 \times 220.588$$
$$= 2,868,642.9 \text{ N} = 2868.6429 \text{ kN}$$

$$C_{s1} = A_{s1}\left(f_{s1} - 0.85 f_c'\right) = 2580 \times (394 - 0.85 \times 30) = 950,730 \text{ N} = 950.73 \text{ kN}$$

$$C_{s2} = A_{s2}\left(f_{s2} - 0.85 f_c'\right) = 1290 \times (55.96 - 0.85 \times 30) = 39,293.4 \text{ N}$$
$$= 39.293 \text{ kN}$$

$$T_{s3} = A_{s3}f_y = 1290 \times (-420) = -541,800 \text{ N} = -541.8 \text{ kN}$$

$$T_{s4} = A_{s4}f_y = 2580 \times (-420) = -1,083,600 \text{ N} = -1083.6 \text{ kN}$$

$$P = 2,233,059.9 \text{ N} = 2233.0599 \text{ kN}$$

$$M = C_c\left(\frac{h}{2} - \frac{\beta_1 C}{2}\right) + \sum_{i=1}^{4} C_{si}\left(\frac{h}{2} - d_i\right) + \sum_{i=1}^{4} T_{si}\left(d_i - \frac{h}{2}\right)$$

$$= 1,107,480,588 \text{ N–mm}$$

$$\phi = \frac{0.003}{C} = \frac{0.003}{220.588} = 13.6 \times 10^{-6} (1/\text{mm})$$

Result
The final $P–M$ curve for the given section is shown in Fig. Ex. 3.9.

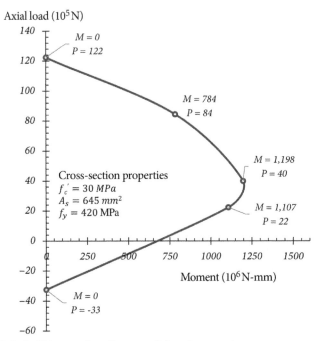

Figure Ex. 3.9 *P–M* interaction diagram of the given section.

The Demand-to-Capacity (*D/C*) Ratio

Question 10: Calculate the capacity ratio based on true *P–M* vector capacity and vector moment capacity at P_u for the following sections. Use CSiCOL, CSI Section Designer or any other computer program to generate *P–M* and *M–M* interaction diagrams.

Given

1. The *PM* interaction diagrams of both cases.
2. A rectangular section with $P_u = 2000$ kN, $M_{ux} = 480$ kN–m, and $M_{uy} = 420$ kN–m.
3. A T-section with $P_u = 1023$ kN, $M_{ux} = 200$ kN–m, and $M_{uy} = 100$ kN–m.

Solution

a. A Rectangular Section

   D/C ratio based on vector moment capacity.
   Fig. Ex. 3.10C shows the $M_{nx}–M_{ny}$ diagram generated using CSiCOL. This can be viewed as a section cut of *PMM* surface at

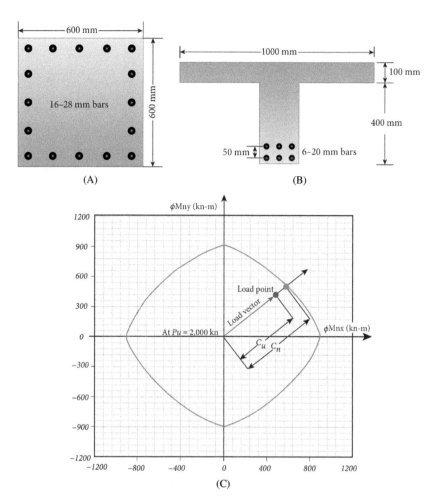

Figure Ex. 3.10 (A) A rectangular RC section. (B) A T-section. (C) Moment capacity diagram for given section at $P_u = 2000$ kN. (D) *PM* interaction diagram for given rectangular section. (E) Moment capacity diagram for given T-section at $P_u = 1023$ kN. (F) *PM* interaction diagram for given T-section.

$P_u = 2000$ kN. Using this figure, we can plot the given load point and can determine corresponding M_{nx} and M_{ny}.

$$C_u = \sqrt{M_{ux}^2 + M_{uy}^2} = \sqrt{480^2 + 420^2} = 637.81 \text{ kN}-\text{m}$$

$$C_n = \sqrt{M_{nx}^2 + M_{ny}^2} = \sqrt{580^2 + 500^2} = 765.76 \text{ kN}-\text{m}$$

Now capacity ratio is calculated as:

$$\frac{C_u}{C_n} = \frac{637.81}{765.76} = 0.833$$

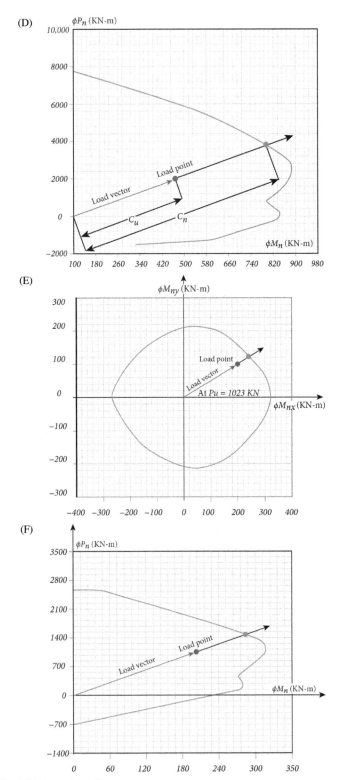

Figure Ex. 3.10 (Continued).

D/C ratio based on true $P-M$ vector capacity.

Again, plotting the load point on true $P-M$ vector capacity curve (obtained using CSiCOL) as shown in Fig. Ex. 3.10D.

$$C_u = \sqrt{P_u^2 + M_{ux}^2} = \sqrt{480^2 + 2000^2} = 2056.8 \text{ kN}-\text{m}$$

$$C_n = \sqrt{P_n^2 + M_{nx}^2} = \sqrt{800^2 + 3800^2} = 3883.3 \text{ kN}-\text{m}$$

Now capacity ratio is calculated as:

$$\frac{C_u}{C_n} = \frac{2056.8}{3883.3} = 0.53$$

b. T-Section

D/C ratio based on vector moment capacity

$$C_u = \sqrt{M_{ux}^2 + M_{uy}^2} = \sqrt{200^2 + 100^2} = 223.61 \text{ kN}-\text{m}$$

$$C_n = \sqrt{M_{nx}^2 + M_{ny}^2} = \sqrt{240^2 + 120^2} = 268.33 \text{ kN}-\text{m}$$

Now capacity ratio is calculated as:

$$\frac{C_u}{C_n} = \frac{223.61}{268.33} = 0.833$$

D/C ratio based on true $P-M$ vector capacity.

$$C_u = \sqrt{P_u^2 + M_{ux}^2} = \sqrt{200^2 + 1023^2} = 1042.37 \text{ kN}-\text{m}$$

$$C_n = \sqrt{P_n^2 + M_{nx}^2} = \sqrt{280^2 + 1470^2} = 1496.43 \text{ kN}-\text{m}$$

Now capacity ratio is calculated as:

$$\frac{C_u}{C_n} = \frac{1042.37}{1496.43} = 0.696$$

Result

1. Capacity ratio for rectangular section based on vector moment capacity $= 0.833$
2. Capacity ratio for rectangular section based on true PM vector capacity $= 0.53$

3. Capacity ratio for T-section based on vector moment capacity = 0.833

4. Capacity ratio for T-section based on true *PM* vector capacity = 0.696

Effect of Reinforcement Ratio on *PM* Interaction Curves

Question 11: Consider a square section having dimensions 400 mm × 400 mm and arrangement of reinforcement is as shown in Fig. Ex. 3.11A. Analyze the effect of total reinforcement ratio on its interaction surface by varying the diameter of reinforcing bars. Use CSiCOL, CSI Section Designer or any other computer program to generate $P-M$ interaction diagrams.

Solution

There are eight bars in the above section and for the given section $f_c' = 25\ \text{N/mm}^2$ and $f_y = 400\ \text{N/mm}^2$. Now we will vary the diameter of bar from 16 to 28 mm as shown in the table below and plot the $P-M$ interaction diagram using CSI Section Builder.

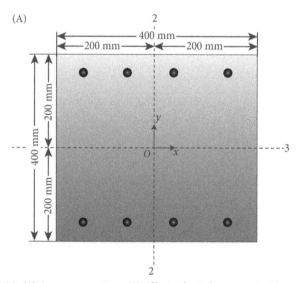

Figure Ex. 3.11 (A) A square section. (B) Effect of reinforcement ratio on *PM* interaction diagram.

(B) Axial load (KN)

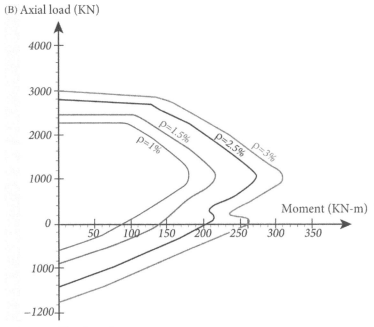

Figure Ex. 3.11 (Continued).

Diameter of Bar d(mm)	Reinforcement Ratio $\rho = A_s/bd$(%)
16	1.0
20	1.5
25	2.5
28	3.0

Result

It can be observed that axial flexural capacity of the given rectangular RC section increased as we increase the reinforcement ratio from 1% to 3%.

Effect of Yield Strength on *PM* Interaction Curves

Question 12: Compare the *PM* interaction curves of given RC section (as shown in figure) with two different yield strengths $(275N/mm^2)$ and $(400N/mm^2)$ of rebars. Use CSiCOL, CSI Section Designer or any other computer program to generate $P-M$ and $M-M$ interaction diagrams.

Solution

The above section has size of 400 mm $\times$ 400 mm and concrete strength is $f_c' = 25$ N/mm^2. The interaction diagram is generated using CSI Section Builder and is shown in Fig. Ex. 3.12B.

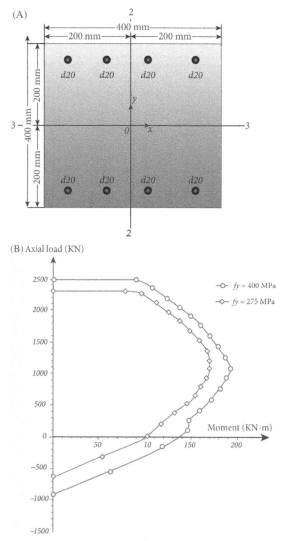

Figure Ex. 3.12 (A) Square section. (B) Effect of strength of reinforcement on *PM* interaction diagram.

Result

It can be observed that axial flexural capacity of given square RC section increased with an increase in yield strength of reinforcing bars from 275 to 400 MPa.

UNSOLVED EXAMPLES

Question 13(a): For the following state of stress shown in Fig. Ex. 3.13, determine the principal stress and maximum shear stress.

Question 13(b): Solve Question 12(a) by Mohr's circle method.

Question 14: A simply supported beam *AB* of 2 m span supports a concentrated load of 150 kN at a distance one-third of span from the left support. The beam is made up of steel and has a rectangular cross-section as shown in Fig. Ex. 3.14. Determine the principal stresses and maximum shear stress at cross-section *PQ* along varying depth and plot them.

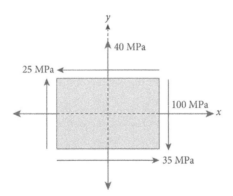

Figure Ex. 3.13 Element in plane stress.

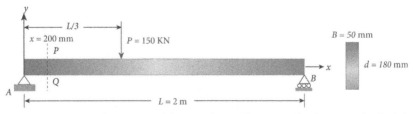

Figure Ex. 3.14 Simply supported beam subjected to point load at one third of the length from the left support.

Question 15: A cantilever beam of rectangular cross-section as shown in figure is subjected to a concentrated load of 100 kN acting at the free end. The beam has width $b = 100$ mm and depth $d = 200$ mm. Determine the principal stresses σ_1 and σ_2 and the maximum shear stress τ_{max} at a cross-section 500 mm from free end at each of the following locations (see Fig. Ex. 3.15).

1. 50 mm from bottom
2. 150 mm from bottom, and
3. The neutral axis.

Question 16: A simply supported beam of rectangular cross-section of width $b = 100$ mm and depth $d = 250$ mm, carries a uniform load of 20 kN/m, its span length is 3 m. Calculate the principal stresses σ_1 and σ_2 and the maximum shear stress τ_{max} at the cross-section 250 mm from left support and midsection at each of the following locations (see Fig. Ex. 3.16).

1. top of the beam
2. 50 mm above and below the neutral axis
3. the neutral axis, and
4. bottom of the beam.

Question 17: A wide flange beam is simply supported with a span length of 3 m as shown in Fig. Ex. 3.17. It supports a concentrated load

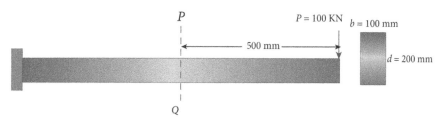

Figure Ex. 3.15 Cantilever beam of rectangular cross-section subjected to point load at free end.

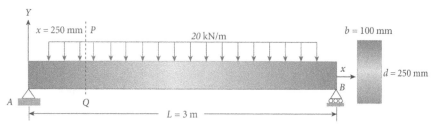

Figure Ex. 3.16 Simply supported beam of rectangular cross-sections subjected to uniform load.

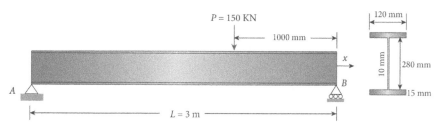

Figure Ex. 3.17 Simply supported beam of wide flange cross-sections subjected to point load at 1 m from right support.

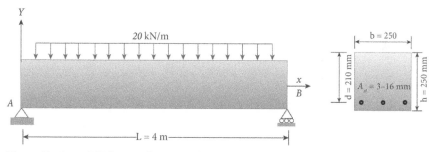

Figure Ex. 3.18 RCC beam of rectangular cross-sections subjected to uniform load.

of 150 kN at 1.0 m from right support. At a cross–section of maximum bending moment and section equals to depth of beam, determine the principal stresses σ_1 and σ_2 and the maximum shear stress τ_{max} at each of the following locations.

1. top of the beam,
2. top of the web, and
3. the neutral axis.

Question 18: A reinforced concrete beam is simply supported in a span of 4 m and has a rectangular cross–section as shown in Fig. Ex. 3.18. It supports a uniformly distributed load of 20 kN/m. Calculate the principal stresses σ_1 and σ_2 and the maximum shear stress τ_{max} at a cross–section located at 0.7 m from the left support at each of the following locations:

1. top of the beam,
2. at a depth equal to half of neutral axis depth from top,
3. the neutral axis, and
4. bottom of the beam.

Question 19: Compute important points on the interaction diagram for the column section shown in Fig. Ex. 3.19. Given: $f_c' = 25\ \mathrm{N/mm^2}$, $f_y = 400\ \mathrm{N/mm^2}$, and $E_s = 2 \times 10^5\ \mathrm{N/mm^2}$.

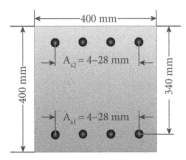

Figure Ex. 3.19 A square section.

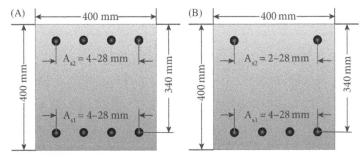

Figure Ex. 3.20 (A) Symmetric square section. (B) Unsymmetric square section.

Question 20(a): Compare the interaction surfaces of the following sections using CSiCOL, CSI Section Builder or any other section analysis computer program (see Fig. Ex. 3.20).

Question 20(b): Compute the capacity ratio based on true PM vector capacity and vector moment capacity at P_u, for the sections in Question 19 (a) for $P_u = 1000$ kN, $M_{ux} = 200$ kN$-$m and $M_{uy} = 100$ kN$-$m. Use CSiCOL, CSI Section Designer or any other computer program to generate $P-M$ and $M-M$ interaction diagrams.

Question 21: Compare the interaction surfaces of the given section below for the following conditions (a) and (b). Use CSiCOL, CSI Section Designer or any other computer program to generate $P-M$ and $M-M$ interaction diagrams (see Fig. Ex. 3.21).

1. Increase in yield strength of rebars: $f_y = 400$ and 250 N/mm^2.
2. Increase in diameter of rebars (with constant $f_y = 250$N/mm^2): 16 and 20 mm reinforcing bars.

Question 22: Determine the interaction surfaces, and $P-M$ and $M-M$ interaction diagrams for the following sections and compare. Given: $f_c' = 24$ N/mm^2 and $f_y = 400$ N/mm^2 (see Fig. Ex. 3.22).

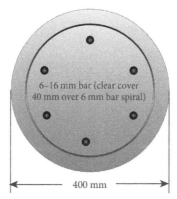

Figure Ex. 3.21 A circular section.

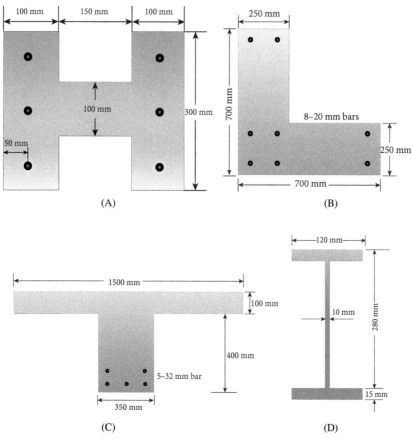

Figure Ex. 3.22 Example sections to practice the generation of *PM* interaction diagrams in CSiCOL. (A) A reinforced concrete H-shaped section. (B) A reinforced concrete L-shaped section. (C) A reinforced concrete T-shaped section. (D) A steel I-section.

SYMBOLS AND NOTATIONS

ε_b strain at the bottom most fiber of the cross-section

ε_{cc} strain at peak stress in concrete

ε_{cu} ultimate strain capacity of concrete

ε_{su} strain in steel at ultimate strength

ε_{sh} strain in steel at the onset of strain hardening

ε_t strain at the top most fiber of the cross-section

ε_y strain in steel at yielding

$\varepsilon(x, y)$ axial strain distribution as a function of x and y

$\varepsilon(y)$ axial strain distribution as a function of y

σ_1 maximum axial stress

σ_2 minimum axial stress

σ_x axial stress in the x-direction at any point on the cross-section

σ_y axial stress in the y-direction at any point on the cross-section

θ_n angle of the resultant moment vector with respect to the x-axis

τ_{max} maximum shearing stress

τ_{xy} shearing stress in the x-plane in the y-direction

A area of the cross-section

A_i area of a point (rebar) on the cross-section

e_x eccentricity in the x-direction

e_y eccentricity in the y-direction

f_a actual axial stress

F_a allowable axial stress

f_{bx} actual stress due to bending in the x-direction

F_{bx} allowable stress due to bending in the x-direction

f_{by} actual stress due to bending in the y-direction

F_{by} allowable stress due to bending in the y-direction

f'_c compressive strength of unconfined concrete

f'_{cc} compressive strength of confined concrete

f_i stress at a point area

f_{su} ultimate strength of steel

f_{sy}, f_y yielding strength of steel

I_{xx} moment of inertia about x-axis

I_{yy} moment of inertia about y-axis

M_x applied moment vector about the x-axis

M_y applied moment vector about the y-axis

M_x bending moment about x-axis

M_y bending moment about y-axis

NA neutral axis

P_z axial force along the longitudinal axis of the member

r combined stress ratio

V_i volume of the stress triangle

x_i distance from the origin to the point on a cross-section along x-direction

y_i distance from the origin to the point on a cross-section along y-direction

REFERENCES

Chen, W.-F. (1997). *Handbook of structural engineering*. Boca Raton, FL: CRC Press LLC. ISBN 0-8493-2674-5.

Chen, W.-F. (2007). *Plasticity in reinforced concrete*. Plantation, FL: J. Ross Publishing. ISBN-13: 978-1-932159-74-5.

Chen, W.-F., & Saleeb, A. F. (1994). *Constitutive equations for engineering materials* (Vol. 1). Amsterdam: Elsevier Science B. V.

Den Hartog, J. P. (2014). *Advanced strength of materials*. New York: Courier Corporation, Dover Publications Inc.

Hassoun, M. N. (1985). *Design of reinforced concrete structures*. PWS Engineering, Original from the University of California. ISBN: 0534037593, 9780534037598.

MacGregor, J. G., Wight, J. K., Teng, S., & Irawan, P. (1997). *Reinforced concrete: Mechanics and design* (Vol. 3). Upper Saddle River, NJ: Prentice-Hall.

Nawy Edward, G. (2005). *Reinforced concrete: A fundamental approach* (5th ed.). New Jersey: Prentice Hall.

Nilson, A. H., Darwin, D., & Dolan, C. W. (2005). *Design of concrete structures*. India: McGraw-Hill Education. ISBN: 0070598541, 9780070598546.

Samali, B., Attard, M. M., & Song, C. (2013). *From materials to structures: Advancement through innovation*. London: CRC Press. Taylor & Francis Group.

Singer, E. L., & Pytel, A. (1980). *Strength of materials*. New York: Harper and Row Publishers, Inc.

Timoshenko, S. (1930). *Strength of materials*. New York: Van Nostrand Company.

Timoshenko, S. (1953). *History of strength of materials: With a brief account of the history of theory of elasticity and theory of structures*. New York: Courier Corporation. Dover Publications Inc.

Varghese, P. C. (2010). *Limit state design of reinforced concrete*. New Delhi: PHI Learning. ISBN-13: 978-8120320390.

Volterra, E., & Gaines, J. H. (1971). *Advanced strength of materials*. Englewood Cliffs, NJ: Prentice Hall.

Wight, J. K., & MacGregor, J. G. (2012). *Reinforced concrete: Mechanics and design* (6th ed.). Upper Saddle River, NJ: Pearson Education, Inc. ISBN-13: 978-0132176521, ISBN-10: 0132176521.

CHAPTER FOUR

Response and Design for Shear and Torsion

INTRODUCTION

The response of a beam or column to the independent effects of shear and torsion as well as combined effects of shear, torsion, moment, and axial load cannot be determined purely based on the response of its cross-section. This is unlike the flexural response to axial load and bending moments, which can almost entirely be defined based on the properties of the cross-section components. This is due to the fact that the nature of stresses generated by flexure and shear—torsion is different. The flexural stresses are generated normal to the cross-section and their summation provides the internal resultant to balance the external actions, independent of adjacent sections. However, in the case of shear, the stresses "between" adjacent sections provide the resistance to shear, and hence requires the behavior of member along the length to be included in analysis. For the case of torsion, the combined effect of stresses, both across and around the cross-section, provides the resistance. This difference if graphically represented in Fig. 4.1.

Although this book is about the behavior and response of the cross-sections, and shear and torsion is more of member responses, the design for shear and torsion however is mostly carried out at cross-section level. The discussion on cross-section design will therefore not be complete without including at least a basic treatment of this subject.

Another aspect that makes the discussion on the shear and torsion behavior a bit more complex than flexural response, especially the reinforced concrete members, is the fact that the response to shear is heavily coupled with torsion as well as flexure. On the other hand, flexural response is not effected directly by the presence of shear and torsion, although the final reinforcement provided for shear, do contribute towards confinement which ultimately affects the flexural response.

Structural Cross Sections
DOI: http://dx.doi.org/10.1016/B978-0-12-804443-8.00004-X
251

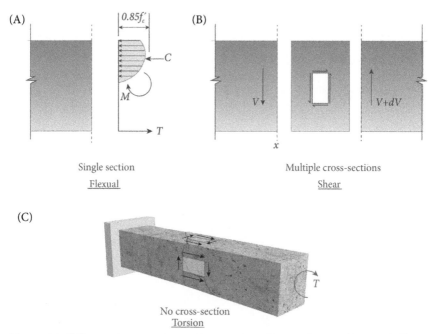

Figure 4.1 Difference between internal response to flexure, shear and torsion effects. (A) Single section, (B) multiple cross-sections, and (C) no cross-section.

Beside shear stresses, torsion also results in warping of cross-section causing axial stresses along longitudinal direction (same as axial stresses caused by bending). Warping stresses point towards out of cross-section plane and may interfere with already existing bending stresses. Therefore, beside shear—torsion, another interaction exists between torsion and axial—flexural response. It is interesting to note that both effects of torsion (shear and axial stresses) are perpendicular in direction to each other. Shear stresses are developed between adjacent cross-sectional planes, however axial stresses due to warping are out-of-plane. Therefore, in reinforced concrete members, sometimes the reinforcement needed for torsion may also need to be combined with flexural rebars.

This chapter presents the fundamentals of elastic response of cross-sections under shear and torsion (see Fig. 4.2), as well as the analysis and practical code design of reinforced concrete (RC) and steel sections (under both shear and torsion). Some special design considerations for deep beams will also be included. Unlike axial—flexural response, there is no unified theory for general cross-section analysis under shear and torsion which can take care of several interactions involved in this problem.

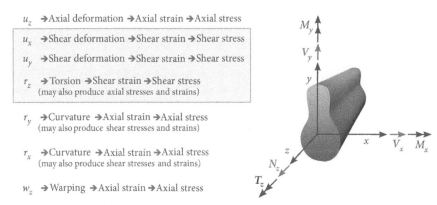

u_z →Axial deformation →Axial strain →Axial stress

u_x →Shear deformation →Shear strain →Shear stress

u_y →Shear deformation →Shear strain →Shear stress

r_z →Torsion →Shear strain →Shear stress
(may also produce axial stresses and strains)

r_y →Curvature →Axial strain →Axial stress
(may also produce shear stresses and strains)

r_x →Curvature →Axial strain →Axial stress
(may also produce shear stresses and strains)

w_z →Warping →Axial strain →Axial stress

Figure 4.2 Strains and stresses corresponding to each degree of freedom.

As discussed above, these interactions may include shear and torsion, shear and axial load, and torsion and moment.

Several theories have been proposed but most could not gain wide acceptance in practice and in terms of inclusion in building codes. The overall understanding about phenomena related to shear and torsion seems to be somewhat less, compared to flexural actions; and this is reflected in lesser capacity reduction factors for design against shear compared to flexure. This chapter will address some of common stress interactions involved however, accounting for all of them simultaneously, is an onerous task. Additional complexities arise from practical situations, e.g., the effect of confinement and several possible reinforcement configurations for each action. We will start now with basic elastic response under shear stresses produced by both shear forces as well as torsion.

BASIC ELASTIC RESPONSE
Shear Stresses

Shear stresses across a beam cross-section (see Fig. 4.3) are produced due to:
- the direct shear deformation due to shear force along the principal axis,
- the indirect shear deformation due to torsion,
- the variation of axial stresses along the depth of the section.

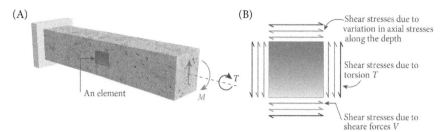

Figure 4.3 The development of shear stresses in a cross-section. (A) A beam under shear, moment, and torsion. (B) Shear stresses in an element due to various actions.

Therefore the shear stress at every location on the cross-section is affected by the magnitude and direction of resultant shear force and the simultaneous presence of torsion. It is also of interest to note, that the shear stress due to shear force will be additive to the shear stress due to torsion at some locations in the cross-section and subtractive at other locations. The properties needed for the determination of shear stress include the shear area along the direction of shear force, S_a and the torsional constant, J. For elastic, uncracked sections, the average shear stress can be computed by a combined stress equation for shear and torsion similar to that for the axial stress. The combined shear stress due to shear and torsion can be obtained as:

$$q = \frac{V_y}{S_{ay}} + \frac{V_x}{S_{ax}} \pm \frac{\rho T}{J} \tag{4.1}$$

where q is the average shear stress, V_x and V_y are the shear forces corresponding to x- and y-directions, S_{ax} and S_{ay} are the shear areas along x- and y-directions, ρ is the distance from center of section to the extreme fiber, T is the applied torque, and J is the torsional constant of the section.

As in the case of axial stresses, for composite sections the stress at each point should be multiplied by the modular ratio of the shape at that point. However, the above equation cannot be used to obtain stress at a particular location on the cross-section. The shear stress distribution function for shear stress due to torsion in the above equation is only defined in an explicit form for a circular section. Therefore it is difficult to derive closed-form equations for determining the combined shear stress due to biaxial shear and torsion in a general section made up of several shapes. The only practical way is to compute the shear stress from shear force by using the general shear area equation described in Chapter 2, Understanding Cross-Sections, and shear stress due to torsion by using the finite element solution. The stress plots for various shapes determined using these numerical solutions are shown in the subsequent sections.

Shear Stress Due to Shear Force

The shear stress distribution due to shear force alone is governed by the Q factor (described in chapter: Understanding Cross-Sections), used for computation of shear area. This primarily depends on the direction of shear force and the shape of the cross-section. We will demonstrate this using the example of a simple elastic beam element. A beam member primarily resists the applied actions by means of internal stress resultants. Consider an elastic beam member as shown in Fig. 4.4A. If we cut two

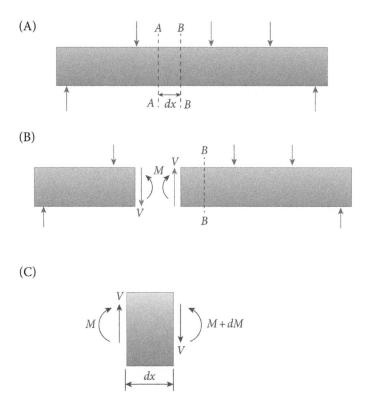

Figure 4.4 Relation between moment and shear for an elastic beam section. (A) Beam, (B) internal forces on section A–A, and (C) internal forces on portion between section A–A and B–B.

sections A and B and draw free-body diagram (see Fig. 4.4C) of a small portion between both sections, the relation between both the internal moment and shear can be easily observed. The equilibrium condition for this small portion implies the following relationship:

$$\frac{dM}{dx} = V \tag{4.2}$$

This simply means that the shear forces and stresses exist in those parts of the beam where the moment is changing. In fact, those shear forces can be defined by the rate at which moments are changing from section to section along the length of member. At a particular section the shear stress distribution along the cross-section depth is given by

$$v = \frac{VQ}{Ib} \tag{4.3}$$

where V is the shear force on that section, I is second moment of area, Q is the first moment (about the neutral axis) of the section area enclosed by extreme fiber and the fiber where shear stresses are being calculated, and b is the width of fiber at which stresses are being calculated. Note the inverse relation between section width and shear stress. Fig. 4.5A—E shows the shear stress distribution in various cross-sections due to applied shear force or shear deformation.

It is worth noting that for T- and I-shaped cross-sections, shear stresses are mostly taken by the webs and the contribution of flanges in shear resistance is minimal. On the other hand, the flexural resistance is mostly contributed by flanges and not the web. This observation makes the box section, one of the most efficient cross-sections in terms of combined resistance against shear—torsion and flexural actions. It is also one of the most economical options for bridge girders due to its ability to optimize for each stress condition as well as has less material requirement compared to solid filled sections.

Shear Stress Due to Torsion

The shear stress distribution for torsion is even more complex than that for shear force. In a homogeneous linear elastic member, the mechanical behavior for torsion can be estimated by using elasticity theory. The angle of twist per unit length is as follows:

$$\theta = \frac{TL}{JG} \tag{4.4}$$

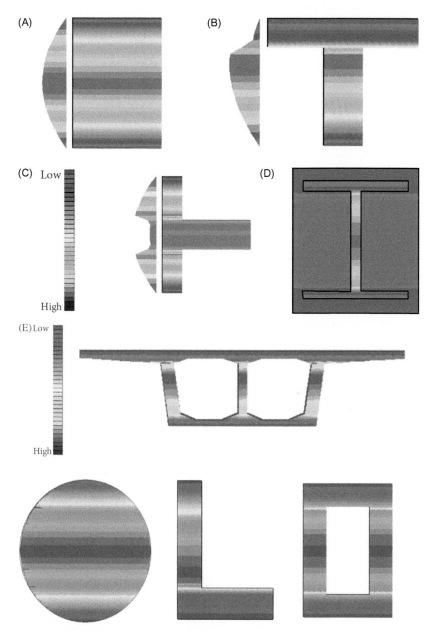

Figure 4.5 Elastic shear stress distribution in various cross-sections due to applied shear force (CSI Section Builder). (A) The shear stress distribution in a rectangular section is of parabolic shape, and is maximum at the mid-depth. (B) The shear stress in a flanged section is the highest in the web. However, the total shear force contributed by the flange could be significant because of large area. (C) The shear stress in a T-section for shear force acting along the flange is not the maximum at the mid-depth. This is due to the large area of the web to resist the shear force at that location. (D) The shear force in a concrete–steel composite section is mostly resisted by the shear stress in the steel web. (E) Shear stresses in a box-girder bridge section, a solid circular, L-shape and hollow box sections.

Table 4.1 The Values of μ With Varying Aspect Ratios for a Rectangular Section

c/b	1.0	1.5	1.75	2.0	2.5	3.0	4.0	6.0	8.0	10.0	∞
μ	0.208	0.231	0.239	0.246	0.258	0.267	0.282	0.299	0.307	0.313	0.333

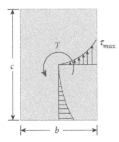

Figure 4.6 Shear stresses in a rectangular section, due to torsion T. Note that the maximum values are at the edges of section.

where T is the twisting moment, G is the shear modulus of material, L is the length of member, and J is the torsional constant. For a rectangular section, J is given as

$$J = cb^3 \left[\frac{1}{3} - 0.21 \frac{b}{c} \left(1 - \frac{b^4}{12c^4} \right) \right]$$ (4.5)

where c and b are the two sides of the rectangle with $b \leq c$ (see Fig. 4.6). The maximum shear stress is at the middle of the longer side c and its value is given in the following equation:

$$\tau_{\max} = \frac{T}{\mu cb^2}$$ (4.6)

where μ is a dimensionless coefficient which varies with the aspect ratio c/b as given in Table 4.1.

For simple, convex shapes, the shear stresses due to torsion form a shear flow or stress loop around the centroid. For flanges sections, several shear flows may be formed, providing the required stress resultant. Fig. 4.7 shows the shear stress distribution in various cross-sections due to applied torsional moment.

Combined Shear Stress Due to Shear Force and Torsion

As discussed earlier, the shear stress at every location on the cross-section is effected by both the magnitude and direction of resultant shear force and the simultaneous presence of torsion. This combination of stress distribution for some common sections is shown in Fig. 4.8 to demonstrate this behavior.

Figure 4.7 Shear stress distribution in various cross-sections due to applied torsional force (CSI Section Builder). (A) In solid, convex shapes, the shear flow due to torsion is formed near the outer edges. The highest stress occurs on the edge nearest to the centroid. (B) In flanged shapes, the shear flow due to torsion tends to form in thicker parts first, and then in thinner parts. For constant thickness, a uniform flow may be formed (greenlight gray indicates high stress, dark gray and medium gray indicate negligible stress). (C) The shear flow in thin-walled open sections is localized in each rectangular element. So the torsional constant can be computed by summation of torsional constant of all rectangular strips (light gray indicates high stress). (D) The shear flow in a steel tube embedded in a concrete section is concentrated in the steel tube. This observation is relevant to concrete sections with hoop reinforcement, where almost the entire torsion is resisted by the reinforcement (light gray indicates high stress). (E) The maximum shear flow in closed shapes forms a loop near the outer surface. (F) The shear flow in square and box sections is developed near the outer edges (light gray indicates high stress).

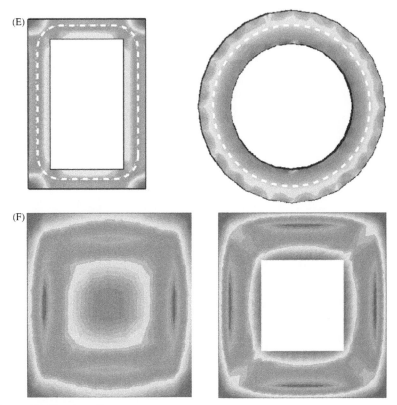

Figure 4.7 (Continued).

Interaction of Shear, Torsion and Bending Stresses

Similar to the axial stress interaction described in Chapter 3, Axial–Flexual Response of Cross-Sections, a shear stress interaction failure surface can be also obtained (between stresses from shear and stresses due to torsion) for any limiting value of shear stress. Such interaction was earlier used by ACI 318-89 to obtain the design capacity of concrete section, leading to a fairly complicated procedure for combined effects of shear and torsion. However, from ACI 318-95 and later versions, this interaction has been neglected and the shear stress for shear force and for torsion are obtained independently considering the other as zero. The interaction between shear stress due to any one component of shear force, say V_y, and shear stress due to torsion will produce an elliptical curve for a rectangular section. The typical shape of torsion–shear interaction for a

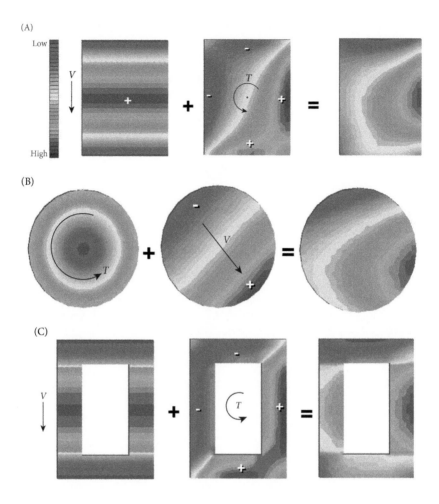

Figure 4.8 The shear stress due to shear force and the shear stress due to torsion is combined together to obtain the total shear stress distribution. The shear stress due to torsion is additive to shear stress from shear force on one edge and subtractive on the other edge (CSI Section Builder). (A) The shear stress due to shear force and the shear stress due to torsion are combined together to obtain the total shear stress distribution. The shear stress due to torsion is additive to shear stress from shear force on one edge and subtractive on the other edge. The stress distribution shown in the middle is the signed shear stress distribution due to torsion indicating the direction as well as the magnitude of the stress. (B) Shear stress distributions for a circular section. From left to right, absolute shear stress due to shear force V, signed shear stress due to torsion T, combined shear stress due to torsion and shear force. (C) Shear stress distributions for a tubular section. From left to right, absolute shear stress due to shear force V, signed shear stress due to torsion T, combined shear stress due to torsion and shear force.

regular section is shown in Fig. 4.9. Similarly, Figs. 4.10 and 4.11 are showing shear–moment and torsion–moment interactions. The exact form of these interaction curves for other section types and shapes may be more complex and generally not documented in references.

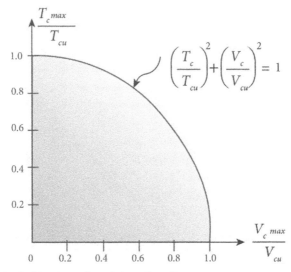

$$\left(\frac{T_c}{T_{cu}}\right)^2 + \left(\frac{V_c}{V_{cu}}\right)^2 = 1$$

Figure 4.9 A typical torsion–shear interaction diagram.

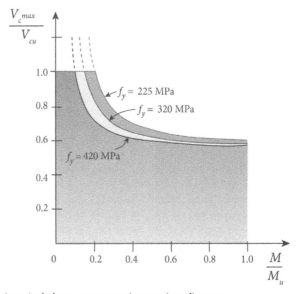

$f_y = 225$ MPa

$f_y = 320$ MPa

$f_y = 420$ MPa

Figure 4.10 A typical shear–moment interaction diagram.

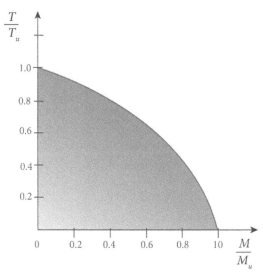

Figure 4.11 A typical torsion–moment interaction diagram.

RESPONSE OF REINFORCED CONCRETE (RC) SECTIONS

As discussed in Chapter 2, Understanding Cross-Sections, reinforced concrete is a heterogeneous and anisotropic material. Individual behaviors of both steel and concrete components are very dissimilar. The response of such a section made of such a material, under various interacting stresses requires an understanding of a wide range of complex phenomena. This makes it very difficult to formulate a unified analytical theory to predict the response of RC section under combined shear and torsion and including the effects of axial load and moment. Some of the current approaches include solid mechanics (principle stress, Mohr's circle), truss mechanism, strut and tie models, diagonal tension and compression field theories, and fracture mechanics. Some key concepts for response evaluation will be discussed in the next section, which will help later in understanding the design rational adopted by various design codes and standards for cross-sections of reinforced concrete and prestressed concrete members. However, the detailed discussion of various theories and behavior of concrete in beyond the scope of this book and is available in many text books and references (Wight and MacGregor, 2011; Nilson, Darwin, & Dolan, 2004; Park & Paulay, 1975).

Principal Stresses and Shear Capacity of RC Members

As discussed in Chapter 3, Axial—Flexual Response of Cross-Sections, the
concept of principal stresses and their representation using Mohr's stress circle provides the relationship between axial and shear stresses as well as the
direction of principal plane. Consider a small rectangular element under
pure shear stress state taken from a loaded RC section. For a simply supported beam, the material fibers experience pure shear stresses with minimal
flexural stresses near the support ends. Pure shear will tend to deform this
element into a parallelogram shape. It is intuitive to consider that the material of this element will be under tension along one diagonal plane and
compression along the other diagonal. The net effect of this pure shear
along the edges of this element would be pure compression and tension
along the diagonals. In this case, these diagonal planes are principal planes
experiencing only direct stresses. Therefore the shear stress problem in one
direction boils down to a direct stress problem in another rotated direction.

If we take similar elements from different locations (see Fig. 4.12)
along the length and depth of the beam, we may also have bending

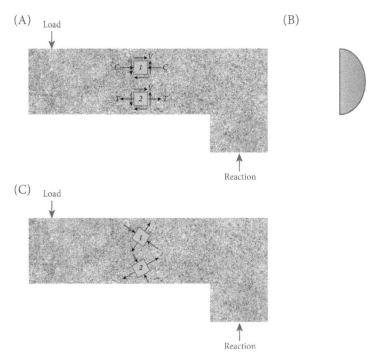

Figure 4.12 Transformation of all actions into principal stresses. (A) Flexural and
shear stresses on elements in a beam. (B) Distribution of shear stresses along the
depth of section. (C) Principal stresses acting on elements.

stresses beside shear stresses (depending upon the location of element). So in a beam member, every location has a unique stress state corresponding to a unique combination of axial and shear stresses. Both the magnitude and direction are changing at all points as the relative values of axial and shear stresses is changing. For pure shear case, the direction of principal stresses will be at an angle of 45 degrees with horizontal. As we start adding axial stresses the principal plane starts to rotate. For pure axial forces, the principal plane will be perfectly vertical (at the center of a simply supported beam). As we move towards support, it starts rotating and reached 45 degrees close to supports, causing diagonal cracking. A more complex case arises if we add torsion also which will produce additional stresses (adding or subtracting depending upon the direction). The additional longitudinal stresses will interact with already existing axial stresses from bending. In case of column members, we may also have pure axial compression. All these combination of stresses (pure axial, bending which is producing variable tension and compression at each section along the length, shear and torsion which is producing both additional shear and axial stresses) can be transformed into principal stresses (compression and tension) acting on the principal planes. The complex problem can be simplified into a combination of tensile and compressive stresses.

Since the concrete tension capacity is far less than (around 10% of the) compression capacity, the governing effect of shear stresses is basically the limit of tensile principal stress along one of the diagonals, often referred to as diagonal tension in standard RC design literature. In short, for RC sections, failure in shear basically means failure in tension (caused by that shear).

The shear and torsion capacities of concrete section are primarily governed by the tension capacity of concrete. Shear reinforcement is provided to reinforce the section against diagonal tension cracking. In old practice, diagonal shear reinforcements were provided in beams to traverse the shear cracks directly, however soon it was realized that vertical shear reinforcement is more practical and improves the flexural performance also by providing confinement (see Fig. 4.13). In order to determine the amount of shear reinforcement, it is necessary to know the shear capacity of plain concrete. Shear reinforcement is designed to resist the component of applied shear stresses remaining after subtracting shear capacity of plain concrete. It is worth noting that while designing reinforcement against bending moment, the tensile capacity of concrete is generally ignored. However, for shear design, ACI 318-14 recommends to consider the shear capacity of plain concrete.

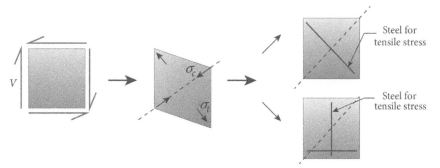

Figure 4.13 Principal stresses, diagonal tension, and shear reinforcement for an RC section.

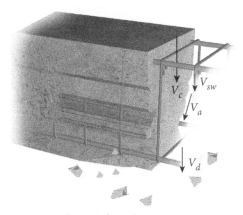

Figure 4.14 Components of a reinforced concrete beam resistance against shear forces.

The total shear resistance offered by a reinforced concrete section can be divided into several components. Fig. 4.14 is showing these components on a cracked RC beam section. The component V_c represents the shear resistance of intact/uncracked concrete. Various codes and standards provide guidelines to estimate it and are primarily based on tensile capacity of concrete. V_a represents aggregate interlocking along the crack. This interlocking provides shear resistance depending upon crack width and other factors, e.g., aggregate texture, size, etc. V_{sw} is the shear resistance provided by shear reinforcement (provided in the form of stirrups). It is believed that flexural reinforcement while traversing the shear crack, also provides some shear resistance and is referred to as dowel action. This component is denoted by V_d. There is another component which arise from tied-arch type of behavior which exists in rather deep beams, and will be discussed later. The total shear resistance for a reinforced concrete beam is a summation of all these components, though most practicing engineers prefer to neglect dowel action and aggregate interlocking, being difficult to estimate. The most important component is V_s and many times in design for earthquake effects, the reinforcement demand by confinement requirements for joints comes out to be more than shear requirement. In such cases confinement detailing governs the amount and configuration of shear reinforcement. Detailed procedure to design sections for shear is discussed later in this chapter.

Average Shear Stresses in Cracked RC Beams

Soon after initial cracking in reinforced concrete members, the applied actions are transferred to reinforcing steel. In fact, it is, a sort of, compulsory for an RC section to crack in order to fully utilize its capacity. Consider a cracked RC beam member as shown in Fig. 4.15A. If we take a small uncracked portion between any two consecutive cracks and draw its free-body diagram, as shown in Fig. 4.15B, we can develop a relationship for an average shear stress based on equilibrium conditions. The tension in longitudinal reinforcing bars is represented by T and the internal

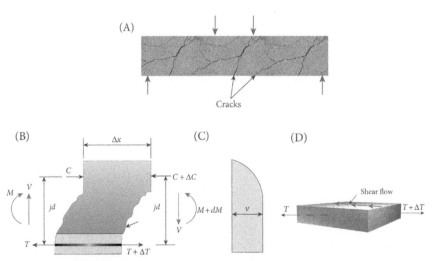

Figure 4.15 The average shear stresses on a cracked RC section. (A) A cracked beam. (B) Portion of beam between two cracks. (C) Average shear stress. (D) Bottom part of beam. *Based on Wight and MacGregor (2011).*

compressive resistance offered by the uncracked part of section is denoted with C (with jd as the lever arm between them which is assumed to be constant here).

$$T = \frac{M}{jd} \quad \text{and} \quad T + \Delta T = \frac{M + \Delta M}{jd} \tag{4.7}$$

Subtracting the two, we get

$$\Delta T = \frac{\Delta M}{jd} \tag{4.8}$$

The equilibrium implies

$$\Delta M = V \Delta x \quad \text{and thus,} \quad \Delta T = \frac{V \Delta x}{jd} \tag{4.9}$$

Fig. 4.15D shows a small segment of beam cut from the bottom at the location of reinforcement. The average value of horizontal shear can be computed by dividing ΔT with corresponding area under shear. Therefore

$$v = \frac{\Delta T}{b_w \Delta x} \quad \text{and hence,} \quad v = \frac{V}{b_w jd} \tag{4.10}$$

If we include the possibility of varying lever arm "jd" instead of considering it as constant, the actual relationship between shear and moment will become

$$V = \frac{d}{dx}(Tjd) \quad \text{or} \quad V = \frac{d(T)}{dx}jd + \frac{d(jd)}{dx}T \tag{4.11}$$

Eq. (4.11) implies that if the change in rebar tension shown in Fig. 4.15D is completely neglected (an extreme case when the inclined cracks prevent the shear flow), the shear is transferred by arch action instead of beam action.

Response Under Torsion

Torsion is a moment acting about the longitudinal axis of a member. In structures, torsion may result from the eccentric loading of beams or from deformations resulting from the continuity of beams or similar members that join at an angle to each other. From the design point of view, torsional moments are classified into equilibrium moments and compatibility torsional moments. Equilibrium torsion is the torsional moment against which a structure must provide internal resistance in order to keep the force equilibrium in the overall structural system. If this torsional moment is neglected in the calculation of the force equilibrium of the structure the stability of overall structure is destroyed. Few examples are shown in Fig. 4.16.

Compatibility torsion is a moment caused by the compatibility between the members meeting at a joint (composing statically indeterminate structure), and provides an influence mainly on the elastic deformation of the structure. Few examples are shown in Fig. 4.17.

In general, the torsional rigidity of a concrete member is greatly reduced after cracking and plastic deformation by torsion. Hence, the torsional moment action on the member becomes relatively small when a concrete member of a statically indeterminate structure reaches such a state. In a reinforced concrete member, the mechanical behavior before torsional crack can be estimated by using elasticity theory, assuming that the gross concrete section is effective.

Torsional Cracks in Unreinforced Concrete

In an unreinforced concrete beam, pure torsional moment act on a member resulting in pure shear stresses. The principal stress angle is 45 degrees. Using Mohr's stress circle, we can get principal stresses (σ_1, σ_2) as follows:

$$\sigma_1 = \sigma_2 = |\tau_{\max}| = \frac{M_t}{K_t}(tension: +) \tag{4.12}$$

It is not difficult to visualize the direction of torsional crack propagation in such a case. Since the direction of principal tension remains same on all faces, all cracks, AB, BC, CD, and DE propagate diagonally. Fig. 4.18 shows the cracks caused by σ_1 when it exceeds the tensile strength of concrete material.

Let us consider the same unreinforced concrete beam subjected to combined torsion and shear. In this case, the two shearing stress components add on one side face and counteract each other on other side. As a result, inclined cracking starts on the face where stresses add (crack AB) and extends across the flexural tensile face of the beam. If the bending moments are sufficiently large, the cracks will extend almost vertically across the back face (crack CD) as shown in Fig. 4.19.

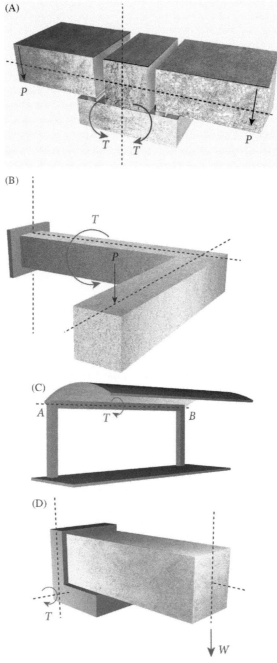

Figure 4.16 Some examples of equilibrium torsional moments. (A) Section through a beam supporting precast floor slab. (B) Cantilever with eccentrically applied load. (C) A canopy supported by a beam *AB*. (D) A rectangular beam resting on L-shaped section.

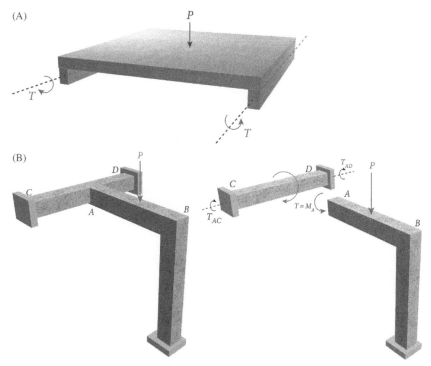

Figure 4.17 An example of compatibility torsional moment. (A) A slab supported by beams. (B) A beam supporting an another beam.

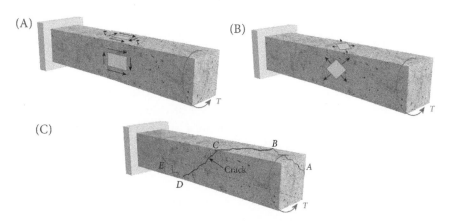

Figure 4.18 Torsional cracks for pure torsion case in an unreinforced concrete beam. (A) Shear stresses. (B) Principle stresses. (C) Torsional crack. *Based on Wight and MacGregor (2011).*

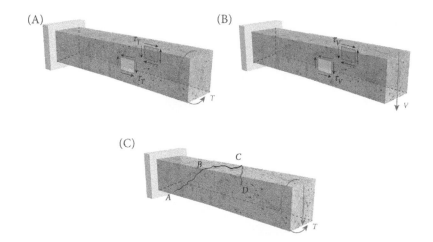

Figure 4.19 Torsional cracks for combined torsion and shear in an unreinforced concrete beam. (A) Shear stresses due to torsion. (B) Shear stresses due to shear. (C) Crack pattern due to torsion and shear. *Based on Wight and MacGregor (2011).*

In case of combined torsion and axial force, the net tensile principal stress (σ_1) can be determined using Mohr's stress circle as shown in Eqs. (3.13) and (3.14)

$$\sigma_1 = \frac{\sigma_n}{2} + \sqrt{\left(\frac{\sigma_n}{2}\right)^2 + \tau^2} \tag{4.13a}$$

where σ_n is the normal stress and τ is the shear stress. At the onset of cracking, σ_1 will be a tensile stress denoted by f_t. Using $\sigma_1 = f_t$ in above equation and solving for τ, we get,

$$\tau = f_t \sqrt{1 - \sigma_n/f_t} \tag{4.13b}$$

$$M_{tc} = K_t \tau = K_t f_t \sqrt{1 - \sigma_n/f_t} \tag{4.13c}$$

It can be observed from the equation above that due to presence of tensile stresses, the torsional capacity (M_{tc}) of overall member is reduced. This reduction is governed by σ_n/f_t ratio. On the face subjected to negative axial force (compression), we replace σ_n with $-\sigma'_n$ in Mohr's stress equations and can get similar expressions as follows:

$$\sigma_1 = \frac{-\sigma'_n}{2} + \sqrt{\left(\frac{-\sigma'_n}{2}\right)^2 + \tau^2}; \quad Crack:\ \sigma_1 = f_t \tag{4.14a}$$

$$\tau = f_t \sqrt{1 + \sigma'_n/f_t} \tag{4.14b}$$

$$M_{tc} = K_t f_t \sqrt{1 + \sigma'_n/f_t}$$

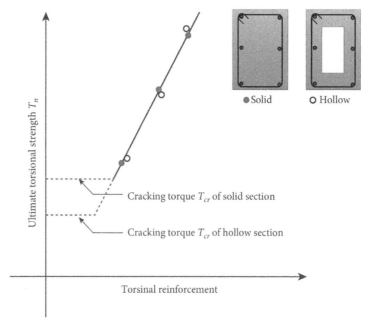

Figure 4.20 Ultimate torsional strength of solid and hollow sections of the same size. *Based on Shihada (2011).*

Above expression shows that the torsional capacity of member is increased on the face subjected to compressive stresses.

Fig. 4.20 presents the ultimate torsional capacity of two reinforced concrete members (solid and hollow) for various levels of torsional reinforcements. It can be seen that capacities of both members are nearly the same and increasing with reinforcement provided that outer dimensions and rebar configuration are same once the concrete has cracked. As the maximum shear stresses caused by tension are around the edges of member, there is no significant reduction in torsional capacity if we take out the material from center. This shows that both solid and hollow members can be considered as tubes for various practical design purposes. The concrete at the center of member has minimal effect on torsional capacity and can be ignored. The stress contours of hollow and solid sections shown earlier in the chapter also help to understand and confirm this phenomenon.

Space Truss Analogy for Reinforced Concrete Members for Torsion

The concept of space truss analogy assumes that concrete resists only compressive stresses and no tensile stresses, and the principal tension is

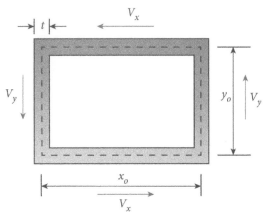

Figure 4.21 Shear flow in a rectangular tube.

resisted by reinforcement. In this model, the torsional capacity of a rectangular RC section can be derived from the reinforcement and the concrete surrounding the steel only. For this purpose, a thin-walled section is assumed which acts as a space truss in which the inclined spiral concrete strips between adjacent cracks resist the compressive stresses and torsional moment is resisted by longitudinal bars and stirrups. Fig. 4.21 shows the shear flow in a rectangular concrete cross-section which is represented by a hollow tube. The value of shear can be determined as follows:

$$M_t = K_t \tau_{max} \tag{4.15}$$

where K_t is the section modulus for torsion and is determined based on the shape and size of cross-section. For rectangular tube, it can be determined as follows:

$$K_t = 2A_m t \tag{4.16}$$

where A_m is the area enclosed by the centerline of wall thickness ($=x_o y_o$) and t is the thickness of tube. Using the above relation between shear stresses and torsional moment capacity, we can have

$$\tau = \frac{M_t}{2A_m t} \tag{4.17}$$

therefore

$$V_x = \tau \times t x_o = \frac{M_t}{2A_m} x_o \quad \text{and} \quad V_y = \tau \times t y_o = \frac{M_t}{2A_m} y_o \tag{4.18}$$

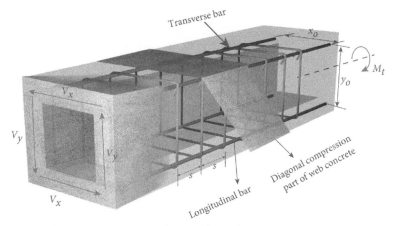

Figure 4.22 Torsion space truss analogy, with reinforcement.

Fig. 4.22 shows various elements of a space truss assumed for torsion analysis of a reinforced concrete beam. We will now develop the relationships between shear stresses and corresponding torsional moment capacity using equilibrium conditions, as follows.

Let us assume that the total area of longitudinal bars and stirrups are A_l and A_w, respectively, and the corresponding stresses are represented by σ_1 and σ_w. The yield strengths of both types of reinforcements is denoted by f_y and the spacing between stirrups is "s" (see Fig. 4.23). The concrete stresses and diagonal compressive capacity is represented by σ_c' and f_{wc}', respectively.

Longitudinal force equilibrium implies the following expression:

$$\sum A_l \sigma_1 = 2V_x \cot\theta + 2V_y \cot\theta = \frac{M_t}{2A_m} 2(x_o + y_o)\cot\theta \qquad (4.19)$$

Rearranging, we can get

$$\sigma_1 = \frac{M_t}{2A_m} \frac{2(x_o + y_o)\cot\theta}{\sum A_l} \qquad (4.20)$$

where $\sum A_l$ is the total cross-sectional area of longitudinal bars. Similarly the force equilibrium in transverse direction yields the following:

$$A_w \sigma_w \frac{y_o \cot\theta}{s} = V_y, \quad A_w \sigma_w \frac{x_o \cot\theta}{s} = V_x, \quad \text{and} \quad \sigma_w = \frac{M_t}{2A_m} \frac{s \tan\theta}{A_w}$$

$$(4.21)$$

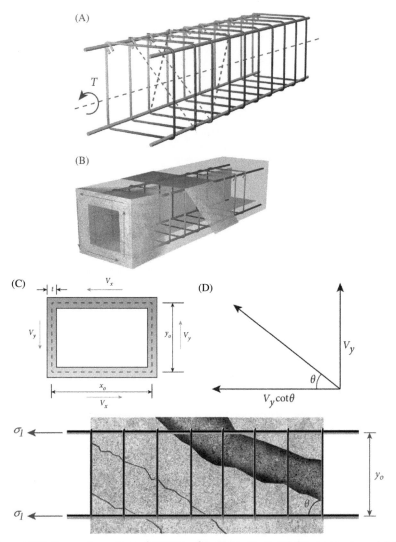

Figure 4.23 Forces in various elements of a space truss. (A) Space truss model for torsion. (B) Space truss in an RC beam member. (C) Shear flow around the tube. (D) Two components of shear force. (E) Diagonal compression struts (uncracked concrete portion) and tension ties (steel rebars).

Applying the force equilibrium in diagonal direction, we get

$$\sigma'_c \cdot ty_o \cos\theta = \frac{V_y}{\sin\theta}, \quad \sigma'_c \cdot tx_o \cos\theta = \frac{V_x}{\sin\theta}, \quad \text{and} \quad \sigma'_c = \frac{M_t}{2A_m} \frac{1}{t\cos\theta\sin\theta}$$

$$(4.22)$$

By replacing σ_1 and σ_w with f_y, we can get torsional capacity (M_{ty}) corresponding to yielding of longitudinal bar and stirrup. The resulting relations are shown in the following equations:

$$\sigma_1 = \frac{M_t}{2A_m}\frac{2(x_o + y_o)\cot\theta}{\sum A_l}, \quad \tan\theta = \frac{M_t}{2A_m}\frac{2(x_o + y_o)}{\sum A_l f_y}, \quad \text{and} \quad \sigma_w = \frac{M_{ty}}{2A_m}\frac{s\tan\theta}{A_w}$$

$$(4.23)$$

$$f_y = \left(\frac{M_{ty}}{2A_m}\right)^2 \frac{2s(x_o + y_o)}{A_w \sum A_l f_y}, \quad M_{ty} = 2A_m f_y \sqrt{\frac{A_w \sum A_l}{2s(x_o + y_o)}},$$

$$(4.24)$$

$$\text{and} \quad \tan\theta = \sqrt{2(x_o + y_o)\frac{A_w}{s\sum A_l}}$$

The diagonal compressive capacity (M_{tcu}) can be determined by replacing σ'_c with f'_{wc} as follows:

$$f'_{wc} = \frac{M_{tuc}}{2A_m}\frac{1}{t\cos\theta\sin\theta} \quad \text{and} \quad M_{tuc} = 2A_m t\cos\theta\sin\theta \cdot f'_{wc} \quad (4.25)$$

For a balance condition of failure, the reinforcement yields at the same instant as concrete is crushed in diagonal compression. We can equate the corresponding capacities $(M_{ty} = M_{tuc})$ to get the following expressions:

$$2A_m f_y \sqrt{\frac{A_w A_s}{2s(x_o + y_o)}} = 2A_m t\cos\theta\sin\theta \cdot f'_{wc}, \quad \tan\theta = \sqrt{2(x_o + y_o)\frac{A_w}{sA_s}},$$

$$\text{and} \quad \frac{A_w}{ts} = \rho_w = \frac{f'_{wc}}{f_y}\sin^2\theta$$

$$(4.26)$$

These expressions can be used to estimate the amount and spacing required to ensure adequate torsional capacity for reinforced concrete beams using space truss analogy.

CODE-BASED SHEAR AND TORSION DESIGN OF RC SECTIONS

Design codes provide simplified procedures and expressions to determine required cross-sectional size, amount, configuration of reinforcement and section capacities for various actions and demands. Using the most widely used construction material, i.e., reinforced concrete, we will provide an overview of code-based cross-sectional design procedure for beams and column sections subjected to shear and torsion. This procedure applies to members that have no explicit design requirement for high seismic demands and are considered to be parts of ordinary moment resisting frames. The design of members from intermediate and special moment resisting frames requires special seismic considerations in terms of ductility and is beyond the scope of this book, and readers are referred to Wight and MacGregor (2011), Nilson et al. (2004), and Park and Paulay (1975).

Fig. 4.24 shows the basic design process for a typical RC beam member. Note that shear design is the successor of design against bending moment and for a member to pass the design criteria, it should not only satisfy strength requirements (for both shear and moment) but also has to satisfy serviceability requirements.

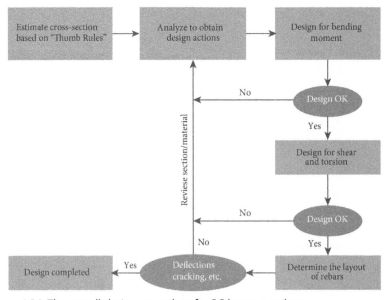

Figure 4.24 The overall design procedure for RC beam members.

We will now separately consider code-based design, based on simplified behavior of RC members for both shear and torsion.

Design of RC Beams for Shear

Design for RC beams for shear primarily consists of the following steps:
- Compute cross-sectional properties
- Compute shear capacity of concrete alone
- Compute minimum shear reinforcement
- Determine shear to be resisted by steel rebars
- Compute shear reinforcement area
- Select rebars and determine stirrup spacing
- Check spacing limits on stirrups

ACI 318, 2014 provides simplified procedure to determine shear capacity of RC beams as follows.

Total shear capacity:

$$V_n = V_c + V_s \tag{4.27}$$

Considering effects of longitudinal reinforcements, moments, and shear magnitudes, the following equation for determining the shear strength provided by concrete (V_c) can be used.

$$V_c = \left(1.9\lambda\sqrt{f_c'} + 2500\rho_w \frac{V_u d}{M_u} \right) b_w d \quad \text{(lb − in units)} \tag{4.28}$$

where f_c' is 28-day compressive strength of concrete and b_w and d are the width and depth of rectangular beam cross-section. ρ_w is the reinforcement ratio ($\rho_w = A_s/b_w d$) and, V_u and M_u are the ultimate shear force and bending moment. λ is a modification factor accounting for the reduction in mechanical properties of lightweight concrete, relative to normalweight concrete of the same compressive strength.

A more widely used, conservative and simplest expression is as follows:

$$V_c = 2\lambda\sqrt{f_c'}b_w d \tag{4.29}$$

For axial compressive force, N_u:

$$V_c = 2\left(1 + \frac{N_u}{2000A_g} \right) \sqrt{f_c'}b_w d \tag{4.30}$$

For axial tensile force, N_u:

$$V_c = 2\left(1 + \frac{N_u}{500A_g}\right)\sqrt{f_c'}\,b_w d \qquad (4.31)$$

Shear taken by stirrups:

$$V_s = A_v f_{ys}(\sin\alpha + \cos\alpha)\frac{d}{s} \qquad (4.32)$$

where α is the angle of stirrups with horizontal and s is the spacing of stirrups. A_v and f_{ys} are the cross-sectional area and tensile yield strength of stirrups, respectively. For $\alpha = 90°$, the expression for V_s can be simplified as follows:

$$V_s = A_v f_y \frac{d}{s} \qquad (4.33)$$

If $V_u \geq \phi V_n$, the stirrups should be provided as follows:

$$\phi V_n \geq V_u - \phi V_c \quad \text{or} \quad V_s \geq \frac{V_u}{\phi} - V_c \qquad (4.34)$$

Fig. 4.25 shows a graphical interpretation of ACI 318 requirements for shear design of beams. The region A corresponds to $V_u \leq \phi V_c/2$,

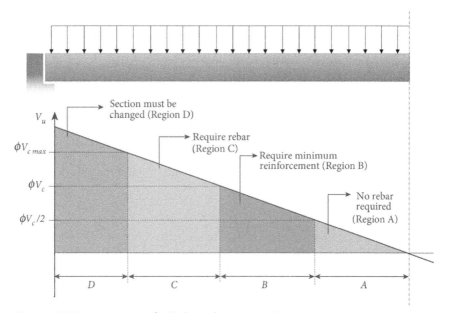

Figure 4.25 Interpretation of ACI shear design equations.

where no shear reinforcement is required $(A_v/s = 0)$. Region B is provided with minimum shear reinforcement and corresponds to the condition $\phi V_c/2 < V_u \leq \phi V_c$. The region C corresponds to $\phi V_c < V_u \leq \phi V_{cmax}$. Here the reinforcement is needed corresponding to the different of applied shear force and shear capacity provided by concrete only $[A_v/s = (V_u - \phi V_c)/(\phi f_{ys} d)]$. For region D $(V_u \geq \phi V_{cmax})$, a change in section is recommended. The maximum shear reinforcement is limited to $8\sqrt{f_c'} b_w d$ since a beam shear strength cannot be increased indefinitely by adding more and more shear reinforcing. The greater the shear transferred by shear reinforcing to the concrete, the greater will be the chance of a combination shear and compression failure. The minimum shear reinforcement is given by,

$$A_{v,min} = 0.75\sqrt{f_c'}\frac{b_w s}{f_{yt}} < \frac{50 b_w s}{f_{yt}} \tag{4.35}$$

The ACI 318-14 shear design procedure for rectangular, T- and L-shaped beams is presented here in the form of a flowchart (see Fig. 4.26). The required inputs include the section dimensions, rebar sizes, material capacities and elastic moduli, and ultimate shear forces. ACI 318-14 can be referred further for detailed explanation of all notations and parameters.

On similar lines, Figs. 4.27 and 4.28A,B show the procedure for the design of beams for shear, as prescribed in BS 8110 (1997) and EC 2 (2004), respectively, which can be referred further for detailed explanation of all notations and parameters.

Design of RC Beams for Torsion

The current design procedure in ACI 318-14 for torsion is based on the following assumptions:

1. Torsion has no effect on shear strength of concrete. The design of RC beams for torsion is carried out almost independently of the design for shear.

2. Torsion stress determination is based on thin-walled tube, space truss analogy. Both solid and hollow members are considered as tubes before and after cracking, and resistance is assumed to be provided by the outer part of the cross-section centered around the stirrups.

3. No interaction exists between moment, shear, and torsion. Reinforcement for each of the three forces is calculated separately and then combined.

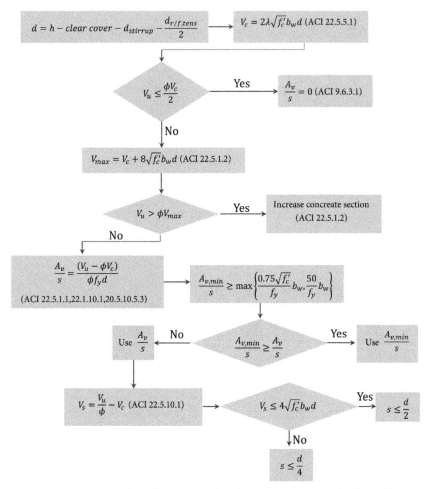

Figure 4.26 ACI 318-14 shear design procedure for rectangular, T- and L-shaped beams.

The overall procedure for torsion design can be summarized under the following steps:

- Determine the factored torsion, T_u.
- Determine special section properties.
- Determine critical torsion capacity, based on capacity of concrete to resist torsion.
- Determine the reinforcement steel, if required.
- Check the minimum reinforcement

Due to thin-walled tube and space truss analogy, special section properties are required in the determination of torsional stresses and member

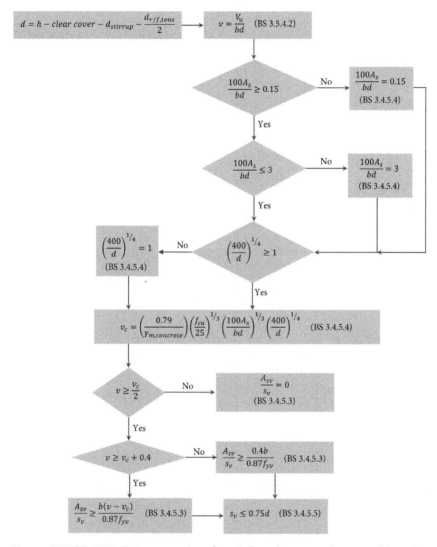

Figure 4.27 BS 8110 design procedure for reinforced concrete beams subjected to shear.

capacities (instead of actual dimensions of section). These special properties are defined as follows:

A_{cp} = Area enclosed by outside perimeter of concrete cross-section

A_{oh} = Area enclosed by centerline of the outermost closed transverse torsional reinforcement

A_o = Gross area enclosed by shear flow path

(A) Shear R/F checks

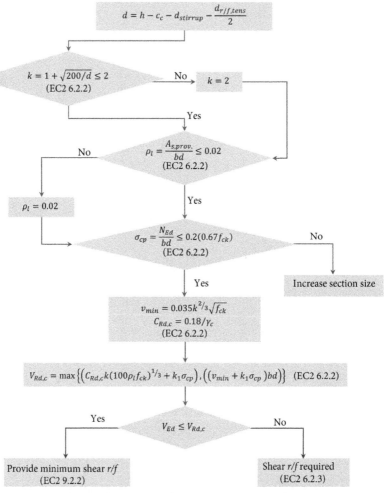

Figure 4.28 (A) Shear reinforcement checks according to EC 2 design procedure for reinforced concrete beams subjected to shear. (B) Determination of shear reinforcement and spacing according to EC 2 design procedure for reinforced concrete beams subjected to shear.

p_{cp} = Outside perimeter of concrete cross-section

p_n = Perimeter of centerline of outermost closed transverse torsional reinforcement

Fig. 4.29 shows the determination these properties for rectangular and T-shape RC sections.

(B) Shear reinforcement

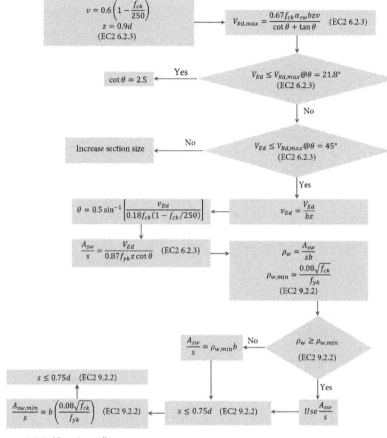

Figure 4.28 (Continued).

Based on ACI 318 procedure, which is analogous to shear design, if the factored applied torsion is less than a critical torsion value, its effects can be ignored. Critical torsion can be determined (in lb–in units) as follows.

For nonprestressed members:

$$T_{cr} = \phi\lambda\sqrt{f_c'}\,\frac{A_{cp}^2}{p_{cp}} \qquad (4.36)$$

For prestressed members:

$$T_{cr} = \phi\lambda\sqrt{f_c'}\,\frac{A_{cp}^2}{p_{cp}}\sqrt{1 + \frac{f_{pc}}{4\lambda\sqrt{f_c'}}} \qquad (4.37)$$

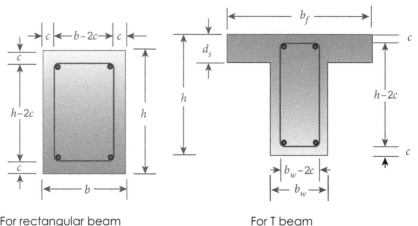

For rectangular beam

$A_{cp} = bh$

$A_{oh} = (b - 2c)(h - 2c)$

$A_o = 0.85A_{oh}$

$p_{cp} = 2b + 2h$

$p_n = 2(b - 2c) + 2(h - 2c)$

For T beam

$A_{cp} = b_w h + (b_f - b_w)d_s$

$A_{oh} = (b_w - 2c)(h - 2c)$

$A_o = 0.85A_{oh}$

$p_{cp} = 2b_f + 2h$

$p_n = 2(b_w - 2c) + 2(h - 2c)$

Figure 4.29 Special sectional properties for rectangular and T-beams.

For nonprestressed members subjected to an axial tensile or compressive force:

$$T_{cr} = \phi\lambda\sqrt{f_c'}\frac{A_{cp}^2}{p_{cp}}\sqrt{1 + \frac{N_u}{4A_g\lambda\sqrt{f_c'}}} \tag{4.38}$$

If the factored torsion is greater than critical $(T_u > T_{cr})$, the reinforcement is required to be designed. The required longitudinal rebar area is

$$A_l = \frac{T_u p_h}{\phi 2A_o f_y \tan\theta} \tag{4.39}$$

The required closed stirrup area is

$$\frac{A_t}{s} = \frac{T_u \tan\theta}{\phi 2A_o f_y s} \tag{4.40}$$

Minimum reinforcement is determined as follows.

Closed stirrup area:

$$A_v + 2A_t = 0.75\sqrt{f_c'}\frac{b_w s}{f_{yt}} \tag{4.41}$$

Minimum longitudinal rebar area:

$$A_{l,min} = 5\sqrt{f_c'}\frac{A_{cp}}{f_y} - \left(\frac{A_t}{s}\right)p_h\frac{f_{yt}}{f_y} \tag{4.42}$$

ACI 318 limits the size of cross-section to reduce unsightly cracking and compressive failure due to shear and torsion. The maximum torsional moment strength can be checked using Eqs. (4.43) and (4.44).

For solid sections:

$$\sqrt{\left(\frac{V_u}{b_w d}\right)^2 + \left(\frac{T_u p_h}{1.7A_{oh}^2}\right)^2} \leq \phi\left(\frac{V_c}{b_w d} + 8\sqrt{f_c'}\right) \tag{4.43}$$

For hollow sections:

$$\frac{V_u}{b_w d} + \frac{T_u p_h}{1.7A_{oh}^2} \leq \phi\left(\frac{V_c}{b_w d} + 8\sqrt{f_c'}\right) \tag{4.44}$$

The AC1 318-14 torsion design procedure for rectangular, T- and L-shaped beams are presented in Fig. 4.30 in the form of a flowchart. The required inputs include the section dimensions, rebar sizes, material capacities and elastic moduli, and ultimate torsion.

For torsion design of T and L sections, it is generally assumed that placing torsion reinforcement in the flange area is inefficient. With this assumption, the flange is ignored for torsion reinforcement calculation.

Figs. 4.31 and 4.32 show the procedure for the design of beams for torsion, as prescribed in BS 8110 (1997) and EC 2 (2004), respectively, which can be referred further for detailed explanation of all notations and parameters.

Design of RC Sections for Combined Shear–Torsion

After carrying out both shear and torsion designs, the overall output is combined to come up with the most efficient detailing scheme. Fig. 4.33 provides an overview and details of the process for the combined shear–torsion design of RC sections.

It is important to note that combined stirrup area for both shear and torsion (A_v) is the sum of area required by shear (A_{vs}) and two times the area required by torsion (A_{vt}). It is due to the fact that A_{vs} is the cross-sectional area of two legs of stirrups as both legs together provide the

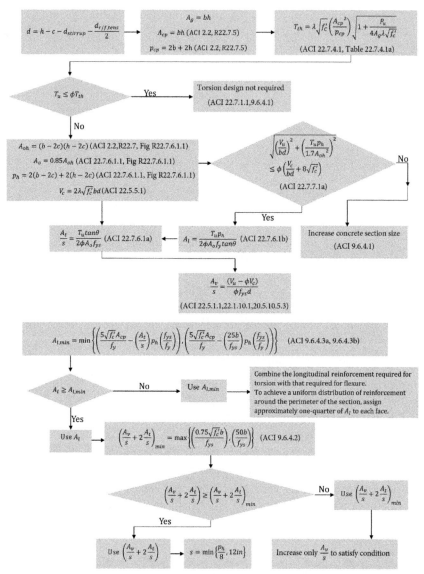

Figure 4.30 ACI 318-14 torsion design procedure for rectangular, T- and L-shaped beams.

shear resistance against inclined shear cracks. Generally, for the purpose of determining stirrup spacing (s) for combined shear and torsion, the amount of stirrups is expressed in terms of A_{vs}/s or A_{vt}/s. Since A_{vt} refers to the cross-sectional area of single leg of stirrup, it is multiplied with two before adding to A_{vs}, for consistent spacing of both individual and combined stirrup areas. A_v, therefore, refers to cross-sectional area of both legs of stirrups required for combined shear and torsion.

Torsion reinforcement checks

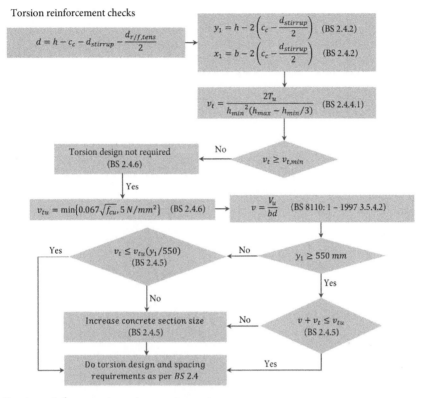

Torsion reinforcement, spacing requirements

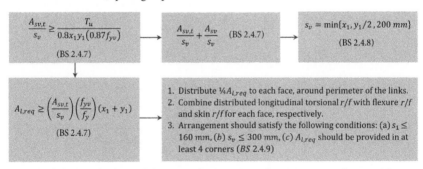

Figure 4.31 Determination of shear reinforcement and spacing according to BS 8110 design procedure for reinforced concrete beams subjected to shear.

Special Considerations for Deep Beams

Deep beams are a type of beams in which a significant amount of applied load is carried to the supports by a compression force combining the load and the reaction. As a result, the strain distribution is no longer

Torision reinforcement checks

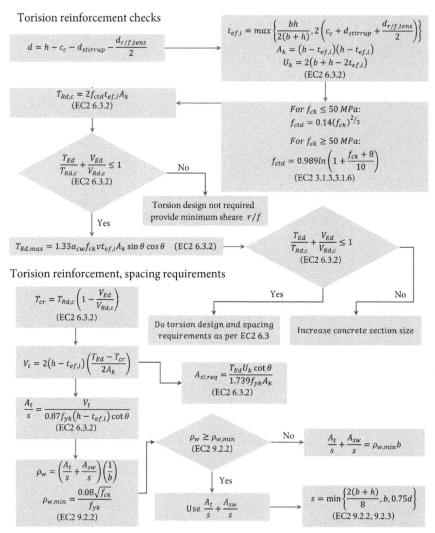

Torision reinforcement, spacing requirements

Figure 4.32 Determination of shear reinforcement and spacing according to EC 2 design procedure for reinforced concrete beams subjected to shear.

considered linear, and the shear deformations become significant when compared to pure flexure. Floor slabs subjected to lateral loading, short span beams subjected to high vertical loads, and transfer girders are examples of deep beams. The depth/thickness ratio of deep beams is large and shear span depth ratio is less than 2.5 for concentrated load and 5 for uniform load.

The most widely used framework to design deep beams is to employ strut-and-tie model. This is based on the same truss analogy which was

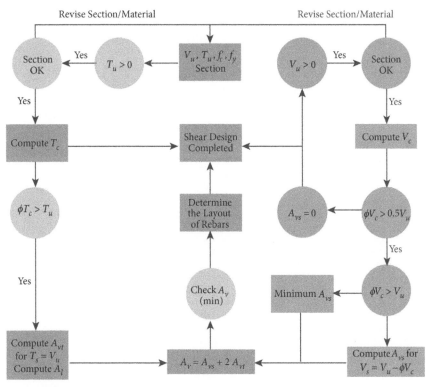

Figure 4.33 Combined shear–torsion design of RC sections.

discussed earlier in torsion design. It is assumed that a hidden truss is developed (see Fig. 4.34) inside the real RC beam which carries the applied stresses using its axially loaded compression and tension members.

Truss models are also included in some building codes (e.g., the strut-and-tie method of ACI 318, Appendix A). The designer is required to first visualize the stress flow within a structural member in order to come up with an adequate truss model. Past experience from similar design examples or a simple exercise of linear structural analysis can help in identifying the stress patterns. The directions of principal compressive stress govern the directions of compression struts. In flexural members, the required reinforcement is determined from flexural tensile stresses in ties however, additional reinforcement is placed to establish the equilibrium with compression struts and form a stable truss. For a given design example, generally there is no unique truss model and more than one solution

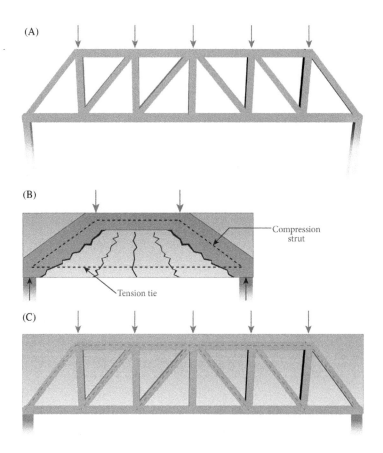

Figure 4.34 Truss model for design of deep beams. (A) A real truss. (B) Development of compression struts and tension ties. (C) An RC beam with a hidden truss mechanism.

can produce acceptable results. It requires expertise and iterations to develop the most adequate truss models for a particular member and design situation.

An important issue related to design and detailing of deep beams is mid-depth cracks (see Fig. 4.35). Various experimental works showed that at or close to mid-depth of deep beams, cracks with significant width arise which not only are undesirable for esthetic reasons, but also can cause potential corrosion problem to reinforcement. To safeguard against these cracks, skin reinforcement is provided on the sides of deep beams at a distance measured two-third depth from the tension face. Although the

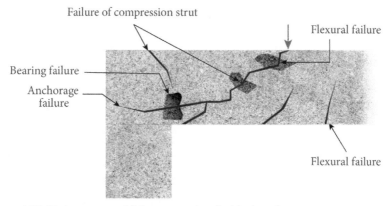

Figure 4.35 Various types of failures associated with deep beams.

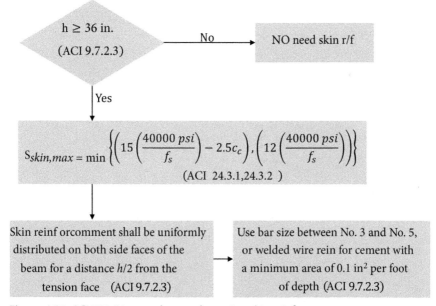

Figure 4.36 ACI 318-14 procedure to determine skin reinforcement.

primary purpose of skin reinforcement is to control crack width, it may be used for providing additional bending resistance. Fig. 4.36 shows the procedure to determine skin reinforcement for deep beams according to ACI 318-14.

Reinforcement Requirement for Combined Axial–Flexural and Shear–Torsion

As mentioned earlier, the reinforcement requirement of an RC section to resist various applied actions is different depending upon both the magnitude and direction of actions as well as cross-section geometry. Tensile reinforcement should be placed near the bottom of a simply supported beam where the tensile bending stresses are maximum. Similarly the shear caused by pure torsion will require a hoop reinforcement near the edges of section. In order to understand the role of rebars and their relative locations, let us assume a reinforced concrete T-shaped section subjected to various actions. Figs. 4.37–4.42 show the location of required reinforcement against various individual and combined actions. This provides a clear understanding of "where the section resistance is coming from?" which is a key concept for RC cross-section design.

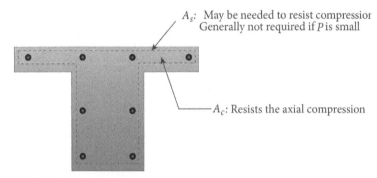

A_s: May be needed to resist compression
Generally not required if P is small

A_c: Resists the axial compression

Figure 4.37 Rebars for axial load—P.

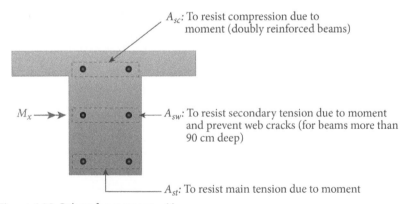

A_{sc}: To resist compression due to moment (doubly reinforced beams)

M_x →

A_{sw}: To resist secondary tension due to moment and prevent web cracks (for beams more than 90 cm deep)

A_{st}: To resist main tension due to moment

Figure 4.38 Rebars for moment—M_x.

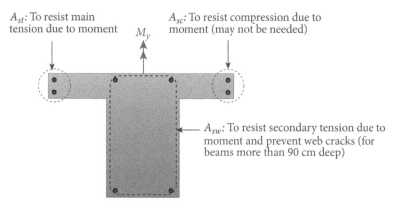

A_{st}: To resist main tension due to moment

M_y

A_{sc}: To resist compression due to moment (may not be needed)

A_{sw}: To resist secondary tension due to moment and prevent web cracks (for beams more than 90 cm deep)

Figure 4.39 Rebars for moment—M_y.

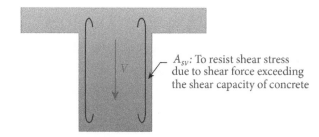

A_{sv}: To resist shear stress due to shear force exceeding the shear capacity of concrete

V

Figure 4.40 Rebars for shear—V.

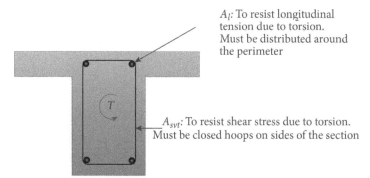

A_l: To resist longitudinal tension due to torsion. Must be distributed around the perimeter

T

A_{svt}: To resist shear stress due to torsion. Must be closed hoops on sides of the section

Figure 4.41 Rebars for torsion—T.

A_{st}: To resist tension due to M_y

$A_{sc} + A_l/4$: To resist compression due to moment M_x (doubly reinforced beams) and tension due to torsion

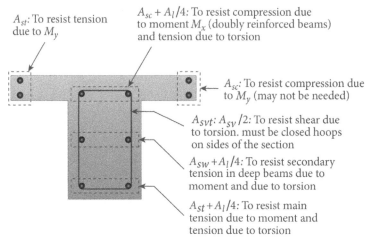

A_{sc}: To resist compression due to M_y (may not be needed)

A_{svt}: $A_{sv}/2$: To resist shear due to torsion. must be closed hoops on sides of the section

$A_{sw} + A_l/4$: To resist secondary tension in deep beams due to moment and due to torsion

$A_{st} + A_l/4$: To resist main tension due to moment and tension due to torsion

Figure 4.42 Rebars for $P + M_x + M_y + V + T$.

SOLVED EXAMPLES

Horizontal Shear Stress in a Beam Section

Problem 1: A simply supported rectangular beam having span length of beam is 2 m, width 50 mm, and depth 1000 mm as shown in Fig. Ex. 4.1. It carries a concentrated load of 60 kN at midpoint. Determine the maximum horizontal shear stress develops in the beam due to the loading. Also investigate the shear stress along the depth of beam.

Given
Length of beam = 2 m
Width of beam = 50 mm
Depth of beam = 1000 mm
Point load at midpoint = 60 kN

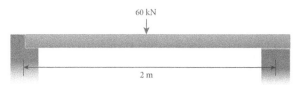

60 kN

2 m

Figure Ex. 4.1 A simply supported beam.

Solution

Reaction at the support (see Fig. Ex. 4.2).

$$R_A = \frac{60}{2} = 30 \text{ kN}$$

$$R_B = \frac{60}{2} = 30 \text{ kN}$$

Maximum shear force from shear force diagram = 30 kN.
Maximum Horizontal Shear Stress Will Occur at Neutral Axis
Neutral axis will occur at mid-depth

$$y = \frac{100}{2} = 50 \text{ mm}$$

Horizontal shear stress at N.A.

$$Q = \frac{VAy'}{Ib}$$

where

$$V = 30 \text{ kN}$$

$$b = 50 \text{ mm}$$

$$y' = \frac{50}{2} = 25 \text{ mm}$$

$$A = 50 \times 50 = 2500 \text{ mm}^2$$

$$I = \frac{bd^3}{12} = \frac{50 \times 100^3}{12} = 4.167 \times 10^6 \text{ mm}^4$$

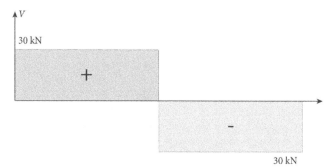

Figure Ex. 4.2 Shear force diagram.

Hence:

$$Q = \frac{VAy'}{Ib} = \frac{30 \times 1000 \times 2500 \times 25}{4.167 \times 10^6 \times 50} = 9 \text{ N/mm}^2$$

The Horizontal Shear Stress at Different Depth of Beam:

$$V = 30 \text{ kN}$$

$$b = 50 \text{ mm}$$

The depth of neutral axis = 50 mm

$$I = \frac{bd^3}{12} = \frac{50 \times 100^3}{12} = 4.167 \times 10^6 \text{ mm}^4$$

At $y = 50$ mm

$$Q = 9 \text{ N/mm}^2$$

At $y = 60$ mm

$$A = 50 \times 40 = 2000 \text{ mm}^2$$

$$y' = \frac{60}{2} = 30 \text{ mm}$$

$$Q = \frac{VAy'}{Ib} = \frac{30 \times 1000 \times 2000 \times 30}{4.167 \times 10^6 \times 50} = 8.64 \text{ N/mm}^2$$

At $y = 70$ mm

$$A = 50 \times 30 = 1500 \text{ mm}^2$$

$$y' = \frac{70}{2} = 35 \text{ mm}$$

$$Q = \frac{VAy'}{Ib} = \frac{30 \times 1000 \times 1500 \times 35}{4.167 \times 10^6 \times 50} = 7.56 \text{ N/mm}^2$$

At $y = 80$ mm

$$A = 50 \times 20 = 1000 \text{ mm}^2$$

$$y' = \frac{80}{2} = 40 \text{ mm}$$

$$Q = \frac{VAy'}{Ib} = \frac{30 \times 1000 \times 1000 \times 40}{4.167 \times 10^6 \times 50} = 5.76 \text{ N/mm}^2$$

At $y = 90$ mm

$$A = 50 \times 10 = 500 \text{ mm}^2$$

$$y' = \frac{90}{2} = 45 \text{ mm}$$

$$Q = \frac{VAy'}{Ib} = \frac{30 \times 1000 \times 500 \times 45}{4.167 \times 10^6 \times 50} = 3.24 \text{ N/mm}^2$$

At $y = 100$ mm

$$A = 50 \times 0 = 0 \text{ mm}^2$$

$$y' = \frac{100}{2} = 50 \text{ mm}$$

$$Q = \frac{VAy'}{Ib} = \frac{30 \times 1000 \times 0 \times 50}{4.167 \times 10^6 \times 50} = 0 \text{ N/mm}^2$$

Results

1. Horizontal shear stress at neutral axis $= 9 \text{ N/mm}^2$
2. Given below are the values of shear stress at different depths.

y (mm)	d' (mm)	y' (mm)	A (mm)	Q (MPa)
50	50	25	2500	9
60	40	30	2000	8.64
70	30	35	1500	7.56
80	20	40	1000	5.76
90	10	45	500	3.24
100	0	50	0	0

Principle Stresses and Maximum Shear Stresses

Problem 2: A rectangular steel beam is simply supported over a span length 1.8 m and has a cross-section width of $b = 100$ mm and height $h = 200$ mm. It supports a concentrated load $P = 150$ kN at midpoint. Investigate the principal stresses and maximum shear stresses at cross-section PQ, located at a distance $x = 200$ mm from left support of the beam and at mid-span. Consider only the in-plane stresses (see Fig. Ex. 4.3).

Given

Span length $= 2.0$ m
Cross-section width $b = 100$ mm
Cross-section height $h = 200$ mm
Concentrated load $P = 150$ kN

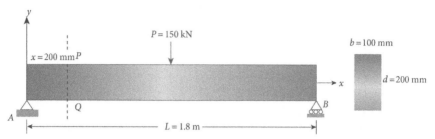

Figure Ex. 4.3 Beam with a rectangular cross-section subjected to point load at center.

Solution

First we calculate the flexure and shear stresses acting on cross-section. Knowing the applied stresses, the principal stresses can be determined from the equations of plane stress. We can also plot the distribution of principal stresses over the depth of beam.

$$R_A = \frac{P}{2} = 75 \ \text{kN}$$

Maximum bending moment at PQ

$$M_{PQ} = R_A \times 0.2 = 75 \times 0.2 = 15 \ \text{kN-m}$$

Maximum bending moment at mid

$$M_{mid} = R_A \times \frac{L}{2} = 75 \times 1 = 75 \ \text{kN-m}$$

Shear force $V = R_A = 75$ kN at both sections
Moment of inertia of the section

$$I = \frac{bh^3}{12} = \frac{100 \times 200^3}{12} = 6.667 \times 10^7 \ \text{mm}^4$$

Normal Stress on Cross-Section

At section PQ

$$\sigma_x = -\frac{My}{I} = \frac{15 \times y \times 10^6}{6.667 \times 10^7} = 0.225y \ \text{N/mm}^2$$

At mid of section

$$\sigma_x = -\frac{My}{I} = \frac{75 \times y \times 10^6}{6.667 \times 10^7} = 1.125y \ \text{N/mm}^2$$

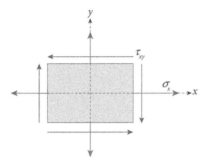

Figure Ex. 4.4 Beam element in plane stress.

The stresses calculated from the above equations are positive when in tension (i.e., when y is negative) and negative when in compression. A stress element cut from the beam at cross-section PQ is shown in Fig. Ex. 4.4. For reference purposes, a set of xy-axis is associated with the element. The normal stress σ_x and the shear stress τ_{xy} are shown acting on the element in their positive directions.

Shear Stress Calculations

First moment for rectangular cross-section

$$Q = b\left(\frac{h}{2} - y\right)\left(y + \frac{\frac{h}{2} - y}{2}\right) = \frac{b}{2}\left(\frac{h^2}{4} - y^2\right)$$

Shear stress

$$\tau = \frac{VQ}{Ib} = \frac{12V}{bh^3(b)}\frac{b}{2}\left(\frac{h^2}{4} - y^2\right) = \frac{6V}{bh^3}\left(\frac{h^2}{4} - y^2\right)$$

The shear stresses τ_{xy} acting on the x face of the stress element are positive upward, whereas the actual stresses τ act downward. Therefore the shear stresses τ_{xy} are given by the following formula:

$$\tau = -\frac{6 \times 75 \times 1000}{100 \times 200^3} \times \left(\frac{200^2}{4} - y^2\right) = -5.625\left(10,000 - y^2\right) \times 10^{-4}$$

Calculation of Principal Stresses

Let us divide the depth of the beam into eight equal intervals and label the corresponding point from A to I. The calculation of stresses, principal stresses and maximum shear stresses are shown in table below. Principal stresses and maximum shear stresses at each of the points from A through I may be determined from equations below:

$$\sigma_{1,2} = \frac{\sigma_x + \sigma_y}{2} \pm \sqrt{\left(\frac{\sigma_x - \sigma_y}{2}\right)^2 + \left(\tau_{xy}\right)^2}$$

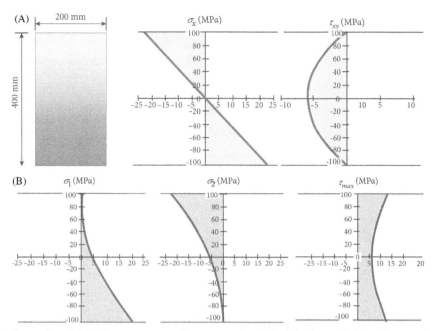

Figure Ex. 4.5 Various stress profiles at section *PQ*. (A) The variation of normal and shear stresses along the depth of cross-section *PQ*. (B) Variation of principal and maximum shear stresses along the depth of cross-section *PQ*.

Since there is no normal stress in y-direction, this equation simplifies to

$$\sigma_{1,2} = \frac{\sigma_x}{2} \pm \sqrt{\left(\frac{\sigma_x}{2}\right)^2 + \left(\tau_{xy}\right)^2}$$

Also the maximum shear stresses are (see Fig. Ex. 4.5)

$$\tau_{max} = \sqrt{\left(\frac{\sigma_x - \sigma_y}{2}\right)^2 + \left(\tau_{xy}\right)^2} = \sqrt{\left(\frac{\sigma_x}{2}\right)^2 + \left(\tau_{xy}\right)^2}$$

Stress Variation at Section *PQ*

Point	y (mm)	σ_x (MPa)	τ_{xy} (MPa)	σ_1 (MPa)	σ_2 (MPa)	τ_{max} (MPa)
A	100	−22.5	0	0	−22.5	11.25
B	75	−16.875	−2.46094	0.351563	−17.2266	8.7890625
C	50	−11.25	−4.21875	1.40625	−12.6563	7.03125
D	25	−5.625	−5.27344	3.164063	−8.78906	5.9765625
E	0	0	−5.625	5.625	−5.625	5.625
F	−25	5.625	−5.27344	8.789063	−3.16406	5.9765625
G	−50	11.25	−4.21875	12.65625	−1.40625	7.03125
H	−75	16.875	−2.46094	17.22656	−0.35156	8.7890625
I	−100	22.5	0	22.5	0	11.25

Stress variation at the middle of beam (see Fig. Ex. 4.6):

Point	y (mm)	σ_x (MPa)	τ_{xy} (MPa)	σ_1 (MPa)	σ_2 (MPa)	τ_{max} (MPa)
A	100	−112.5	0	0	−112.5	56.25
B	75	−84.375	−2.46094	0.071716	−84.4467	42.2592164
C	50	−56.25	−4.21875	0.314646	−56.5646	28.4396462
D	25	−28.125	−5.27344	0.956257	−29.0813	15.0187566
E	0	0	−5.625	5.625	−5.625	5.625
F	−25	28.125	−5.27344	29.08126	−0.95626	15.0187566
G	−50	56.25	−4.21875	56.56465	−0.31465	28.4396462
H	−75	84.375	−2.46094	84.44672	−0.07172	42.2592164
I	−100	112.5	0	112.5	0	56.25

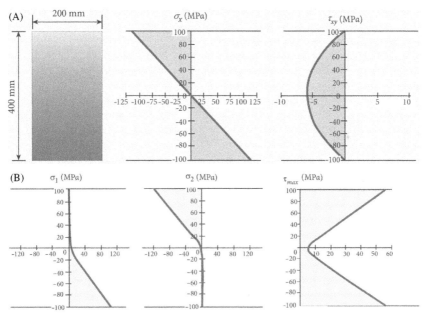

Figure Ex. 4.6 Various stress profiles at a section at mid-span of beam. (A) Variation of normal and shear stresses along depth at mid of cross-section. (B) Variation of principal and maximum shear stresses along the depth at mid of beam.

Shear Stresses Due to Torsion

Problem 3: A solid bar of circular cross-section having diameter of 50 mm and length 1.2 m. Shear modulus of material is 20 GPa. The bar is subjected to twisting moment of 300 N−m. Determine (a) the maximum shear stress, τ_{max} and angle of twist, ϕ. (b) If the allowable shear stress $\tau_{all} = 40$ MPa and $\phi = 2.0°$, determine the twisting moment, T (see Fig. Ex. 4.7).

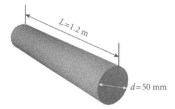

Figure Ex. 4.7 Solid bar of circular cross-section.

Given
Diameter of solid bar = 50 mm
Length of solid bar = 1.2 m
Shear modulus = 20 GPa
Twisting moment of solid bar = 300 N−m

Solution

$$\tau_{max} = \frac{16T}{\pi d^3}$$

$$\tau_{max} = \frac{16 \times 300 \times 10^3}{\pi \times 50^3} = 12.22 \ \text{MPa}$$

$$I_p = \frac{\pi \times d^4}{32} = \frac{\pi \times 50^4}{32} = 6.13 \times 10^5 \ \text{mm}^4$$

$$\phi = \frac{TL}{GI_P} = \frac{300 \times 10^3 \times 1.2 \times 1000}{80 \times 10^3 \times 6.13 \times 10^5} = 0.00734 \ \text{rad}$$

$$T_1 = \frac{\pi \times d^3 \times \zeta_{all}}{16} = \frac{\pi \times 50^3 \times 40}{16} = 981,747 \ \text{N−mm}$$

$$T_1 = 981,747 \ \text{N−mm} = 981.747 \ \text{N−m}$$

$$\phi_{all} = 2.0° = \frac{2 \times \pi}{180} = 0.035 \ \text{rad}$$

$$T_2 = \frac{GI_P\phi_{all}}{L} = \frac{80 \times 10^3 \times 6.13 \times 10^5 \times 0.035}{1.2 \times 1000} = 1,430,333.33 \ \text{N−mm}$$

$$T_2 = 1,430,333.33 \ \text{N−mm} = 1430.333 \ \text{N−m}$$

Thus T_{all} is the maximum of T_1 and T_2.
Hence, $T_{all} = 1430.333 \ \text{N−m}$.

Results

1. $\tau_{max} = 12.22$ MPa and $\phi = 0.00734$ rad
2. $T_{all} = 1430.333$ N—m

Problem 4: A rectangular steel bar 30 mm wide and 40 mm deep is subjected to torque of 500 N—m. Evaluate maximum shear stress in the material and angle of twist/unit length given shear modulus 80 GN/m².

Given

Width of steel bar, $b = 30$ mm
Depth of steel bar, $c = 40$ mm
Torque $= 500$ N—m.
Shear modulus $= 80$ GN/m².

Solution

$$J_1 = cb^3 \left[\frac{1}{3} - 0.21 \frac{b}{c} \left(1 - \frac{b^4}{12c^4} \right) \right]$$

$$J_1 = 40 \times (30)^3 \left[\frac{1}{3} - 0.21 \times \frac{30}{40} \left(1 - \frac{30^4}{12(40)^4} \right) \right]$$

$$J_1 = 1,080,000 \times 0.179$$

$$J_1 = 193,320 \text{ mm}^4$$

$$\theta = \frac{TL}{J_1 G_c} = \frac{500 \times 10^3 \times 1000}{80 \times 10^3 \times 1.93 \times 10^5} = 0.032 \text{ rad}$$

Now maximum shear stress is calculated as

$$\tau_{max} = \frac{T}{\mu c b^2}$$

$$\frac{c}{b} = \frac{40}{30} = 1.33$$

Since the ratio is 1.33, from Table 4.1, we can get,

$$\mu = 0.2218$$

$$\tau_{max} = \frac{500 \times 10^3}{0.2218 \times 40 \times 30^2} = 62.6 \text{ N—mm}^2$$

Results

1. Angle of twist is 0.032 rad
2. Maximum shear stress is 62.6 $N-mm^2$

Determination of Allowable Torque

Problem 5: A solid shaft having its outer diameter of 100 mm. Maximum shear stress of 80 MPa is applied on the shaft. Calculate the torque that can be applied on the shaft for the given shear stress. Also evaluate the torque considering the hollow shaft of 100 mm inner diameter.

Given

Shear stress, $\tau_{max} = 80$ MPa
Diameter of solid shaft, $d = 100$ mm
Inner diameter of hollow shaft, $d' = 100$ mm

Solution
For Solid Shaft

$$d = 100 \text{ mm} = 0.1 \text{ m}$$

$$C = \frac{1}{2}d = \frac{1}{2} \times 0.1 = 0.05 \text{ m}$$

$$\frac{J}{C} = \frac{\pi}{2}C^3 = \frac{\pi}{2} \times 0.05^3 = 1.963 \times 10^{-4} \text{ m}^3$$

$$T = \frac{\tau_{max}J}{C} = \left(80 \times 10^6\right) \times \left(1.963 \times 10^{-4}\right) = 15,704 \text{ N}-\text{m}$$

$$T = 15.70 \text{ kN}-\text{m}$$

For Hollow Shaft

$$C_1 = \frac{1}{2}d' = 0.05 \text{ m}$$

$$C_2 = \sqrt{C_1^2 + C^2} = \sqrt{0.05^2 + 0.05^2} = 0.0707$$

$$J = \frac{\pi}{2}\left(C_2^4 - C_1^4\right)$$

$$J = \frac{\pi}{2}\left(0.07^4 - 0.05^4\right) = 27.9 \times 10^{-6} \ \text{m}^4$$

$$J = 27.9 \times 10^{-6} \ \text{m}^4$$

$$T = \frac{\tau_{max}J}{C_2}$$

$$T = \frac{80 \times 10^6 \times 27.9 \times 10^{-6}}{0.0707} = 31,570 \ \text{N}-\text{m}$$

$$T = 31.57 \ \text{kN}-\text{m}$$

Result
Torque for solid shaft under given shear stress, $T = 15.70$ kN$-$m
Torque for hollow shaft under given shear stress, $T = 31.57$ kN$-$m

Shear Stress of Circular Shaft

Problem 6: Determine the maximum shearing stress in the hollow shaft subjected to torsion, shown in Fig. Ex. 4.8. Also determine the diameter of the solid shaft considering the same maximum shearing stress evaluated in the first part of the problem.

Given
Outer diameter, $d_2 = 70$ mm
Inner diameter, $d_1 = 50$ mm
Torque, $T = 2600$ N$-$m

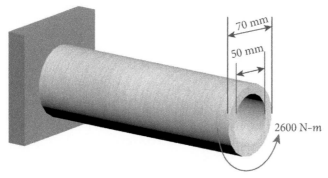

Figure Ex. 4.8 Circular hollow shaft.

Solution

$$C_1 = \frac{1}{2} d_1 = 0.025 \ \text{m}$$

$$C_2 = \frac{1}{2} d_2 = 0.035 \ \text{m}$$

$$c = 0.035 \ \text{m}$$

$$J = \frac{\pi}{2} \left(C_2^4 - C_1^4 \right)$$

$$J = \frac{\pi}{2} \left(0.035^4 - 0.025^4 \right) = 1.74 \times 10^{-6} \ \text{m}^4$$

Maximum Shear Stress

$$\tau_{max} = \frac{Tc}{J} = 2600 \times \frac{0.035}{1.74 \times 10^{-6}} = 52.3 \ \text{MPa}$$

Diameter of Solid Shaft

$$\tau_{max} = \frac{TC}{J}$$

$$J = \frac{\pi}{2} C_3{}^4$$

$$\tau_{max} = \frac{2T}{\pi C_3^3}$$

$$C_3^3 = \frac{2 \times 2600}{\pi \times (52.3 \times 10^6)} = 3.165 \times 10^{-5} \ \text{m}^3$$

$$C_3 = 0.032 \ \text{m}$$

$$d_3 = 2 \times C_3 = 63.3 \ \text{mm}$$

Results

1. Maximum stress of hollow shaft is 52.3 MPa
2. Diameter of solid shaft for maximum shear stress of 52.3 MPa is 63.3 mm

Problem 7: Evaluate torsional stiffness and maximum shear stress of the following cross-sections. Also compare the computed values.

1. Hollow tube with 3 mm wall thickness and 45 mm diameter
2. Hollow tube with 2.5 mm wide cut along the length.
3. A rectangular solid bar side ratio 3 to 1 and have same area
4. Leg angle section with same thickness and perimeter
5. Box section (square) with same thickness and perimeter

Solution

1. **Hollow tube with 3 mm wall thickness and 45 mm diameter**

   Equation for torsion is:

   $$\frac{T}{J} = \frac{G\theta}{L} = \frac{\tau}{r}$$

   Polar moment of area J for the tube is:

   $$J = 2\pi r^3 t$$

   $$\frac{T}{\theta} = \frac{GJ}{L} = \frac{2\pi \times (22.5 \times 10^{-3})^3 \times (3 \times 10^{-3})G}{1}$$

   $$\frac{T}{\theta} = 214.71 \times 10^{-9}G$$

   Now shear stress,

   $$\tau_{max} = \frac{TR}{J} = \frac{T \times (22.5 \times 10^{-3})}{2\pi(22.5 \times 10^{-3})^3 \times (3 \times 10^{-3})}$$

   $$\tau_{max} = 0.105 \times 10^6 \ T$$

2. **Hollow tube with 2.5 mm wide cut along the length**

   $$\frac{T}{\theta} = \frac{K_2(2\pi r - x)t^3 G}{L}$$

   $$\frac{T}{\theta} = \frac{0.333 \times \left[(2\pi \times 22.5 \times 10^{-3}) - (2.5 \times 10^{-3})\right] \times (3 \times 10^{-3})^3 G}{1}$$

   $$\frac{T}{\theta} = 124.07 \times 10^{-12}G$$

$$\tau_{max} = \frac{T}{K_1 \, db^2}$$

$$\tau_{max} = \frac{T}{0.33 \times 138.87 \times 10^{-3} \times (2.5 \times 10^{-3})} = 3.46 \times 10^6 \ T$$

3. A rectangular solid bar side ratio 3 to 1 and have same area

$$A_{Hollow \ bar} = \pi(22.5 \times 10^{-3})^2 = 1.59 \times 10^{-3}$$

$$A_{rectangular \ bar} = 3b^2 = 1.59 \times 10^{-3}$$

$$b^2 = 10.6 \times 10^{-4} \ \text{m}$$

$$b = 32.6 \ \text{mm} \approx 33 \ \text{mm}$$

$$d/b = 3/1 = 3$$

$$d = 3b = 99 \ \text{mm}$$

By using this value of d/b, we will get k_1 and k_2

$$k_1 = 0.267$$

$$k_2 = 0.263$$

So torsional stiffness will be

$$\frac{\theta}{L} = \frac{db^3 G}{L} = \frac{0.263(99 \times 10^{-3})(3.26 \times 10^{-2})^3 G}{1}$$

$$\frac{T}{\theta} = 902.07 \times 10^{-9} G$$

Maximum shear stress

$$\tau_{max} = \frac{T}{K_1 \, db^2} = \frac{T}{0.267 \times 99 \times 10^{-3} \times (3.26 \times 10^{-2})^2}$$

$$\tau_{max} = 0.0355 \times 10^6 \, T$$

4. **Same perimeter and thickness as with tube but with equal leg angle**

Perimeter of angle = perimeter of tube

Perimeter of angle = $2\pi \times 22.5 \times 10^{-3}$

Length for one side = 0.0706 m

For torsional stiffness

$$\frac{\theta}{L} = \frac{3T}{G \sum db^3} = \frac{3T}{G(0.0706 \times 2) \times (3 \times 10^{-3})^3}$$

$$\frac{\theta}{L} = \frac{3T}{G(3.8124 \times 10^{-9})}$$

$$\frac{T}{\theta} = 1.2708 \times 10^{-9} G$$

Maximum shear stress

$$\tau_{max} = \frac{3T}{\sum db^2} = \frac{3T}{2 \times 0.0706 \times (3 \times 10^{-3})^2}$$

$$\tau_{max} = 2.36 \times 10^6 T$$

5. **Square box closed**

Perimeter can be calculated as:

$$P = 2\pi \times (22.5 \times 10^{-3})$$

Length of side of square box

$$S = \frac{2\pi}{4}(22.5 \times 10^{-3}) = 0.0353 \text{ m}$$

So bounded area

$$= A = 0.0353^2 = 1.25 \times 10^{-3}$$

Torsional stiffness

$$\theta = \frac{TLS}{4A^2 Gt}$$

$$\frac{T}{\theta} = \frac{4A^2 Gt}{LS}$$

$$\frac{T}{\theta} = \frac{4 \times 0.035^2 \times G \times 3 \times 10^{-3}}{2\pi \times 22.5 \times 10^{-3}} = 131.8 \times 10^{-9} G$$

Maximum shear stress

$$\tau_{max} = \frac{T}{2At} = \frac{T}{2 \times 0.0353 \times 3 \times 10^{-3}}$$

$$\tau_{max} = 0.136 \times 10^6 \, T$$

Results

S. No	Section	Torsional Stiffness	Maximum Shear Stress
1	Hollow tube with 3 mm wall thickness and 45 mm diameter	214.71×10^{-9} G	0.105×10^6 T
2	Hollow tube with 2.5 mm wide cut along the length	124.07×10^{-12} G	3.46×10^6 T
3	A rectangular solid bar side ratio 3 to 1 and have same area	902.07×10^{-9} G	0.0355×10^6 T
4	Leg angle section with same thickness and perimeter	1.2708×10^{-9} G	2.36×10^6 T
5	Box section (square) with same thickness and perimeter	131.8×10^{-9} G	0.136×10^6 T

Design of a Rectangular Beam for Shear

Problem 8: Rectangular beam with an effective span of 8 m and is subjected to 120 kN/m of working live load and no other dead load except itself. Design necessary shear reinforcement given the compressive strength of concrete is 27.6 MPa, yield strength of steel is 414 MPa, depth of the cross–section is 700 mm, width and height of the cross–section are 360 and 750 mm, respectively. Reinforcement provided is 69 bars ($A_s = 3850$ mm^2)

Given
$f_c' = 27.6$ MPa
$f_y = 414$ MPa
$b = 360$ mm
$d = 700$ mm
$h = 750$ mm
$A_s = 69$ bars(diameter = 28.6 mm) = 3850 mm^2

Solution
Beam self-weight

$$\text{D.L} = 360 \times 750 \times 23.6 \times 10^{-3} = 6372 \text{ kN/mm} = 6.37 \text{ kN/m}$$

Total factor load

$$T_L = 1.2 \text{ D.L} + 1.6 \text{ L.L}$$
$$T_L = 1.2 \times 6.37 + 1.6 \times 120 = 199.644 \text{ kN/m}$$

Factored Shear Force at Face of Support

$$V_u = \frac{199.644 \times 8}{2} = 798.576 \text{ kN}$$

$$\frac{L}{2} = \frac{800}{2} = 4000 \text{ mm}$$

V_u at the distance d from the support is calculated as:

$$V_u = \frac{4000 - 700}{4000} \times 798.576 = 658.83 \text{ kN}$$

$$V_n = \frac{V_n}{\phi} = \frac{658.83}{0.75} = 878.44 \text{ kN}$$

$$V_c = \frac{\lambda \sqrt{f_c'} bd}{6} = \frac{1.0 \times \sqrt{27} \times 360 \times 700}{6} = 218,238.40 \text{ N} = 218.24 \text{ kN}$$

Check for Adequacy of Section

$$V_c + \left(\frac{2}{3}\sqrt{f_c'}\right) bd = 218.24 \times 1000 + \left(\frac{2}{3}\sqrt{27} \times 360 \times 700\right)$$

$$V_c + \left(\frac{2}{3}\sqrt{f_c'}\right) bd = 10,911,991.60 \text{ N} = 1091.19 \text{ kN} > 798.57$$

Hence Section is OK.

Web Steel Reinforcement
Try No. 04 stirrups, $A_V = 2 \times 100 \text{ mm}^2 = 200 \text{ mm}^2$

$$s = \frac{A_v f_y d}{V_u - V_c}$$

$$s = \frac{200 \times 414 \times 700}{(798,576 - 218,238.40)} = \frac{57,960,000}{580,337.6} = 99.87 \text{ mm}$$

Plane X_1 at $s = d/4$, maximum spacing

$$V_n - V_c = 798.576 - 218.240 = 580.336 \text{ kN}$$

$$\frac{1}{3}\sqrt{f_c'}bd = \frac{\sqrt{27.6}}{3} \times 360 \times 700 = 441,299.89 \text{ N} = 441.299 \text{ kN} < 580.336 \text{ kN}$$

Finding plane for $s = d/4$ at a distance x_1 from mid-span

$$V_{n1} = V_c + 441 = 218 + 441 = 659 \text{ kN}$$

$$x_1 = \frac{(4000 - 700) \times 659}{878} = 2476.87 \text{ mm}$$

Finding plane x_2 for $s = d/2$ maximum spacing

$$s = \frac{d}{2} = \frac{A_v f_y d}{V_{n2} - V_c}$$

$$V_{n2} = V_c + \frac{A_v f_y d}{s/2}$$

$$V_{n2} = 218 + \frac{200 \times 411 \times 700}{350} \times \frac{1}{1000} = 382.4 \text{ kN}$$

$$x_2 = \frac{(4000 - 700) \times 382.4}{878} = 1437.26 \text{ mm}$$

Finding plane x_3 at shear force V_c

$$V_C = 218 \text{ kN}$$

$$x_3 = \frac{(4000 - 700) \times 218}{798} = 901 \text{ mm}$$

Discontinue Stirrups at Plane

$$A_{v(min)} = \frac{360 \times \dfrac{700}{2}}{3 \times 414} = 101.45 \text{ mm}^2$$

$$< A_v = 200 \text{ mm}^2 (\text{OK})$$

$$x_4 = \frac{901}{2} = 450.4 \text{ mm}$$

Results
See Fig. Ex. 4.9.

Design of a T-Beam for Shear

Problem 9: Consider simply supported T-beam section as in Fig. Ex. 4.10. The beam carries factored dead load of 25 kN/m including self-weight and service live load of 30 kN/m. Normal-weight concrete is used with

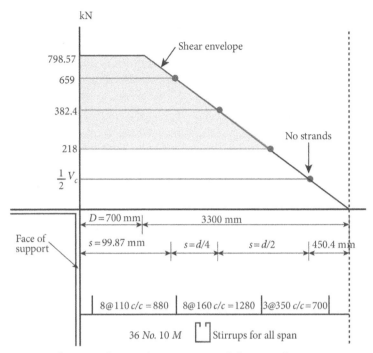

Figure Ex. 4.9 Shear envelope and arrangement of shear reinforcement.

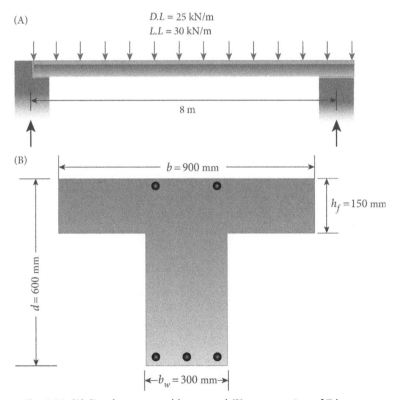

Figure Ex. 4.10 (A) Simply supported beam and (B) cross-section of T-beam.

compressive strength of 20 MPa. Yield strength of flexural reinforcement is 420 MPa whereas yield strength of stirrups is 300 MPa. Design vertical stirrups following ACI 318.

Given

D.L $= 25$ kN/m
L.L $= 30$ kN/m
$f_c' = 20$ MPa
$f_y = 420$ MPa
$f_{ys} = 300$ MPa

Solution
Factored Shear Force:

$$W_u = 1.2 \text{D.L} + 1.6 \text{L.L}$$

$$W_u = (1.2 \times 25 + 1.6 \times 30) = 78 \text{ kN/m}$$

Now shear force diagram is shown in Fig. Ex. 4.11.

$$\frac{V_u}{\phi} \text{ at } d = 416 - \frac{0.60}{4}(416 - 64) = 353.6 \text{ kN}$$

$$V_c = \frac{\lambda \sqrt{f_c'} bd}{6}$$

For normal-weight concrete, $\lambda = 1.0$

$$V_c = \frac{1 \times \sqrt{20} \times 300 \times 600}{6} = 134 \text{ kN}$$

$$\frac{V_u}{\phi} > 0.5 \ V_c$$

Hence stirrups are required.

Check for Cross-Section:

Maximum shear in stirrups

$$V_{s(max)} = \frac{2}{3}\sqrt{f_c'} bd$$

$$V_{s(max)} = \frac{2}{3}\sqrt{20} \times 300 \times 600 = 536.65 \text{ kN}$$

$$\left(\frac{V_u}{\phi}\right)_{max} = V_c + V_{s(max)} = 134 + 536.65 = 670.65 \text{ kN} > 353.6 \text{ kN}$$

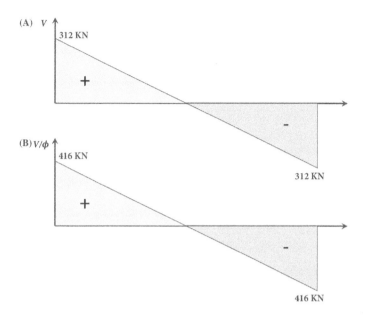

Figure Ex. 4.11 (A) Shear force diagram and V/ϕ diagram for factored loads. (B) Shear force diagram for factored loads. (C) V/ϕ diagram for factored loads.

Hence section is OK.

$$s_{max} = \frac{3A_v f_y}{b} = \frac{3 \times 142 \times 300}{300} = 426 \text{ mm}$$

Hence maximum spacing based on depth of beam $= 300$ mm.

Stirrups Spacing to Resist Shear Forces:

$$s = \frac{A_v f_y d}{\left(\frac{V_c}{\phi}\right) - V_c} = \frac{2 \times 71 \times 300 \times 600}{(353.6 - 134) \times 1000}$$

$$s = \frac{25,560,000}{229,200} = 116.39 \text{ mm}$$

Since this spacing is quite small considering depth hence, we revise procedure using No.13 double leg bars.

Trail No. 02 (No.13 Double Leg Stirrups)

$$A_V = 2 \times 129 = 258 \text{ mm}^2$$

$$s = \frac{258 \times 300 \times 600}{(353.6 - 134) \times 1000} = 211.48 \text{ mm}$$

The spacing of 211.48 mm is quite reasonable. Since the shear force decreases linearly along the length of beam. Next, we will find a point where we can reduce shear reinforcement. Using 10 double leg stirrups at 300 mm, we have

$$V_s = \frac{A_v F_y d}{s} = \frac{142 \times 300 \times 600}{300} = 85.2 \text{ kN}$$

$$V_n = V_c + V_s = 134 + 85.2 = 219.2 \text{ kN}$$

Using shear force diagram, and similar triangle concept, the above shear stress occurs at

$$x_1 = \frac{416 - 219.2}{416} \times 4000 = 1892.31 \text{ mm}$$

Hence, the shear stress occurs at a distance of 1900 mm from the center.

Stirrups can be disconnected where

$$\frac{V_c}{2} \geq \frac{V_u}{\varnothing} = 67 \text{ kN}$$

Again using similar triangles concept, we can locate this point

$$x_2 = \frac{416 - 67}{416} \times 4000 = 3355.77 \approx 3350 \text{ mm}$$

Results

Hence, the final design is listed below
- Provide the first 13 stirrup at 100 mm from the edge.
- Provide 913 stirrups at 200 mm extend to 1900 mm.
- Provide 610 stirrups at 300 mm extend to 3800 mm.

Design of a T-Beam for Equilibrium Factored Torsion Moment

Problem 10: A T-beam cross-section is shown in Fig. Ex. 4.12. The beam is subjected to equilibrium factored torsional moment of 55 kN−m and 200 kN factored shear force. Design web reinforcement, given bending reinforcement $A_s = 2190 \text{ mm}^2$, concrete compressive strength, $f_c' = 27.5 \text{ MPa}$, and yield strength of steel $f_y = 345 \text{ MPa}$.

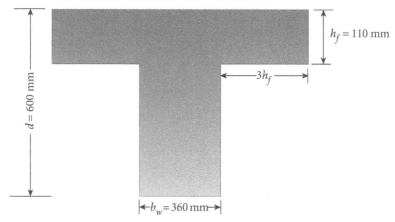

Figure Ex. 4.12 T-beam cross-section.

Given

$A_s = 2190 \ \text{mm}^2$

$f_c' = 27.5 \ \text{N/mm}^2$

$f_y = 345 \ \text{N/mm}^2$

Factored torsional moment $= 55 \ \text{kN-m}$

Factored shear force $= 200 \ \text{kN}$

Solution

$$A_{cp} = b_w \times h = 360 \times 650 = 234{,}000 \ \text{mm}^2$$

$$P_{cp} = 2(b_w + h) = 2(360 + 650) = 2020 \ \text{mm}$$

Torsional Moment

$$T_u = \frac{\phi \sqrt{f_c'}}{12} \frac{A_{cp}^2}{P_{cp}}$$

$$T_u = \frac{0.75\sqrt{27.5}}{12} \times \frac{234{,}000^2}{2020} = 8.80 \times 10^6 \ \text{N-mm}$$

$$= 8.8 \ \text{kN-m} < 55 \ \text{kN-m}$$

Hence, the above calculated torsional moment can be neglected. For design

$$T_n = \frac{T_u}{\phi} = \frac{55}{0.75} = 73.33 \ \text{kN-m}$$

Section Properties

$$A_o = 0.85\, A_s h$$

Assume 40 mm clear cover and No. 10 bars

$$d = 11.3 \text{ mm}$$

$$A_s = 100 \text{ mm}^2$$

$$x_1 = 360 - 2\left(40 + \frac{11.3}{2}\right) = 268.7 \text{ mm}$$

$$y_1 = 650 - 2\left(40 + \frac{11.3}{2}\right) = 558.7 \text{ mm}$$

$$A_o = x_1 \times y_1 = 268.7 \times 558.7 = 150{,}122.69 \text{ mm}^2$$

$$d = 650 - \left(40 + 11.3 + \frac{11.3}{2}\right) = 593 \text{ mm} \approx 590 \text{ mm}$$

$$P_n = 2(x_1 + y_1) = 2 \times (268.7 + 558.7) = 1654.8 \text{ mm} \approx 1655 \text{ mm}$$

Using $\theta = 45°$; $\cot\theta = 1.0$.

Check for Adequacy of Section:

$$\sqrt{\left(\frac{V_u}{bd}\right)^2 + \left(\frac{TP_n}{1.7A_o^2}\right)} < \phi\left(\frac{V_c}{bd} + \frac{8\sqrt{f_c'}}{12}\right)$$

$$V_c = \lambda\sqrt{f_c'}\,\frac{bd}{6} = 1.0\frac{\sqrt{27.5}}{6} \times 360 \times 590 = 183.94 \text{ kN}$$

Now LHS of above equation can be calculated as:

$$\sqrt{\left(\frac{V_u}{bd}\right)^2 + \left(\frac{TP_n}{1.7A_o^2}\right)} = \sqrt{\left(\frac{183{,}943}{360 \times 590}\right)^2 + \left(\frac{55 \times 10^6 \times 1655}{1.7 \times 150{,}122.69^2}\right)^2} = 0.88 \text{ N/mm}^2$$

$$\phi\left(\frac{V_c}{bd} + \frac{8f_c'}{12}\right) = 0.75\left(\frac{183{,}943}{360 \times 600} + \frac{8\sqrt{27.5}}{12}\right)$$

$$\phi\left(\frac{V_c}{bd} + \frac{8f_c'}{12}\right) = 0.75 \times (0.851 + 3.46) = 3.23 \text{ N/mm}^2$$

$3.23 > 0.88$, hence section is OK.

Torsional Reinforcement:

$$T_n = 73.33 \ \text{kN-m} = 73.3 \times 10^6 \ \text{N-mm}$$

$$\frac{A_s}{\rho} = \frac{T_n}{2A_o F_{yt} \cot\theta} = \frac{73.3 \times 10^6}{2(150, 122.69 \times 345 \times 1)}$$

$$\frac{A_s}{\rho} = 0.707 \ \text{mm}^2/\text{mm/one leg}$$

Shear Reinforcement:

$$V_c = \lambda\sqrt{f_c'}\frac{bd}{6} = 183,943 \ \text{N}$$

$$V_n = \frac{V_c}{\varnothing} = \frac{183,943}{0.75} = 245,257.33 \ \text{N}$$

$$V_s = V_n - V_c = 61.31 \ \text{kN}$$

$$\frac{A_v}{\rho} = \frac{V_s}{f_y d} = \frac{61.31 \times 10^3}{345 \times 600} = 0.30 \ \text{mm}^2/\text{mm/two legs}$$

$$\frac{A_{vt}}{\rho} = \frac{2A_t}{\rho} + \frac{A_v}{\rho} = 2 \times 0.707 + 0.25 = 1.664 \ \text{mm}^2/\text{mm/two legs}$$

Try No. 10M closes stirrups ($d = 11.3$ mm, $A_s = 100$ mm^2). Area of two legs $= 2 \times 100 = 200$ mm^2

$$S = \frac{200}{1.664} = 120.19 \approx 120 \ \text{mm}$$

Maximum Allowable Spacing

Maximum allowable spacing s_{max} will be smaller of 125 mm or

$$\frac{1}{8}f_h = 2(x_1 + y_1) = \frac{2020}{8} = 252.5 \ \text{mm}$$

$$252.5 \ \text{mm} > 125 \ \text{mm}$$

$$\frac{1}{16}\sqrt{f_c'} = \frac{1}{16}\sqrt{27.5} = 0.32 < 0.35$$

Hence

$$A_{vt} = \frac{0.35bs}{f_y} = \frac{0.35 \times 360 \times 125}{345}$$

$$A_{vt} = 45.65 \text{ mm}^2$$

So, use No. 10M closed stirrups at 125 mm center to center.

$$A = \frac{A_t}{\rho} \rho_h \frac{f_{yt}}{f_y} \cot^2\theta = 0.707 \times 2020 = 1428.14 \text{ mm}^2$$

$$A_{min} = \frac{5\sqrt{f_c'} A_{cp}}{12f_y} - \frac{A_t}{\rho} \rho_h \frac{f_{yt}}{f_y} < 0.707$$

$$A_{min} = \frac{5\sqrt{27.5} \times 234,000}{12 \times 345} - 1480.458$$

$$A_{min} = 1468.477 - 1428.14 = 40.337 \text{ mm}^2$$

$$\frac{A_t}{\rho} \geq \frac{0.175b_w}{345} = \frac{0.175 \times 360}{345} = 0.182$$

$$0.182 < 0.707 \text{ OK}$$

Hence

$$A_f = 1480.458 \text{ mm}^2$$

It is assumed that $\frac{1}{4}A_l$ goes to top and that $\frac{1}{4}A_l$ goes to bottom of stirrups to balance flexural bars.

$$\Sigma A_s = \frac{1}{4}A_l + A_s = \frac{1480.458}{4} + 2190 = 2560 \text{ mm}^2$$

Results

Final design is illustrated below:

- Provide 525 longitudinal bars $A_s = 2500$ mm^2 at the bottom.
- Provide 215 bar at the top corner of stirrups (400 mm^2).
- Provide 215 bars at each vertical face of the web.

Design of T-Beam for Compatibility Factored Torsion Moment

Problem 11: Design web reinforcement needed for the same section as mentioned in Problem 10 for compatibility factored torsion $T_u = 32$ kN−m.

Given

$T_u = 32$ kN−m
$f'_c = 27$ N/mm²
$f_y = 345$ N/mm²

Solution

Since $T_u = 32 > 8.8$ kN−m

Hence, closed stirrups are needed to be provided for compatibility torsion

$$T_u = \frac{\phi \sqrt{f'_c}}{3} \frac{A_{cp^2}^2}{P_{hp}} = 4 \times 8.8 = 35.2 \text{ kN−m} > 32 \text{ kN−m}$$

Hence, use $T_u = 32$ kN−m for torsional design

$$T_n = \frac{T_u}{\phi} = \frac{32}{0.75} = 42.67 \text{ kN−m}$$

Torsional Reinforcement

$$A_o h = 150, 122.69 \text{ mm}^2$$

$$P_n = 1655 \text{ mm}$$

$$\frac{A_t}{\rho} = \frac{T_n}{A_o h f_y \cot\theta} = \frac{42.67 \times 10^6}{2 \times 150, 122.69 \times 345}$$

$$\frac{A_t}{S} = 0.411 \text{ mm}^2/\text{mm/one leg}$$

From previous example

$$\frac{A_v}{S} = 0.30 \text{ mm}^2/\text{mm/two leg}$$

$$\frac{A_{vt}}{S} = 2\frac{A_t}{S} + \frac{A_v}{S} = 2(0.411) + 0.3 = 1.122 \text{ mm}^2/\text{mm/two leg}$$

Using No. 10M closed stirrups

$$S = \frac{2 \times 100}{1.12} = 178.25 \text{ mm}$$

This is less than $\frac{1}{8} P_h$ or 300 mm. Hence, use No. 10M with 180 mm at critical sections,

$$A_t = \frac{A_t}{S} P_h \frac{f_{yt}}{f_y} \cot^2 \theta = 0.411 \times 1655 = 680.205 \ \text{mm}^2$$

$$A_{t(min)} = \frac{5\sqrt{f_c'} A_{cp}}{f_y} - \frac{A_t}{\rho} P_h \frac{f_{yt}}{f_y}$$

$$A_{t(min)} = 1468.477 - 680.205 = 788.272 \ \text{mm}^2$$

Use $A_t = 948 \ \text{mm}^2$

Distribution of Torsional Longitudinal Bars

$$A_l = \frac{948}{4} = 237 \ \text{mm}^2$$

Results

Final design is illustrated below:
- Provide bottom area $A_s = 2500 \ \text{mm}^2$ with 5 No. 25M bars.
- Provide 2 No. 15 $M(A_s = 400 \ \text{mm}^2)$ bar at top corners and each of two vertical faces.

UNSOLVED PROBLEMS

Problem 12: A circular rod with an outer diameter of 100 mm and internal diameter 80 mm is shown in Fig. Ex. 4.13. Calculate maximum tensile, compressive and shear stresses provided $T = 6$ kN$-$m and $G = 27$ GPa.

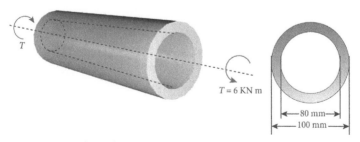

Figure Ex. 4.13 A circular rod.

Problem 13: A rectangular shaft of 50 mm × 30 mm is subjected to torque of 2 kN−m. Calculate maximum shear stress in the shaft. Also determine twisting angle per unit length. Use $G = 80$ GN/m².

Problem 14: Consider a long tube having double cell and effective cross-section as shown in Fig. Ex. 4.14. Determine maximum shear stress in each wall of the tube when the tube is subjected to a torque of 150 N−m. Assume no buckling occurs and tube has constant twist per unit length.

Problem 15: Consider an I-section as shown in Fig. Ex. 4.15. Calculate maximum torque if the section is subjected to shear stress of 60 MN/m². The twist per meter length is 9 degrees. Assume $G = 80$ GN/m².

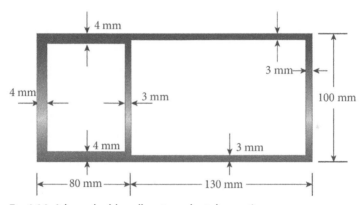

Figure Ex. 4.14 A long double-cell rectangular tube section.

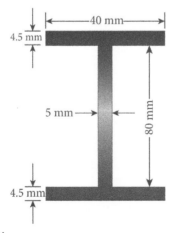

Figure Ex. 4.15 An I-section.

Problem 16: A T-beam cross-section is shown in Fig. Ex. 4.16. The beam is subjected to a point load of 3 kN and the length of the beam is 2.5 m. Calculate maximum shear stress. Also determine shear flow and shear stress at the joint between flange and wed. Finally, also evaluate bending stresses in the cross-section.

Problem 17: Simply supported rectangular beam have clear span of 7 m. The beam carries uniform dead load of 15 kN/m and live load of 20 kN/m. Assuming no torsion exists, design tension reinforcements given: $b_w = 300$ mm, $d = 430$ mm, $h = 500$ mm, $A_s = 3800$ mm^2, $f_c' = 27.5$ MPa, and $f_y = 414$ MPa.

Problem 18: The shear force envelope of the beam is shown in Fig. Ex. 4.17. Design necessary vertical stirrups provided, $b_w = 360$ mm, $d = 500$ mm, $V_{u1} = 334$ kN, $V_{u2} = 267$ kN, $V_{u3} = 200$ kN, $f_c' = 27.5$ MPa, and $f_y = 414$ MPa.

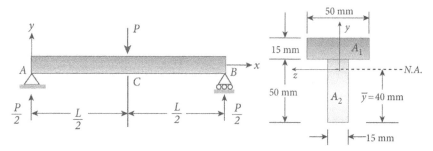

Figure Ex. 4.16 A simply supported beam with having T-shaped cross-section.

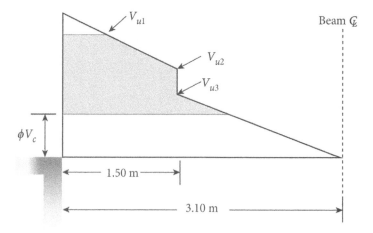

Figure Ex. 4.17 Shear envelope of given beam.

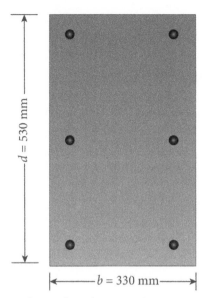

Figure Ex. 4.18 A rectangular reinforced concrete beam cross-section.

Problem 19: A cantilever beam carries a live load of 120 kN/m acting at a distance of 1 m from the support. The dimensions of the cross-sections are 0.25 m × 0.50 m. The effective depth of the beam is 450 mm. Design necessary stirrups reinforcement if compressive strength of lightweight concrete is 20 MPa and yield strength of for all reinforcements is 270 MPa.

Problem 20: Rectangular cross-section is shown in Fig. Ex. 4.18. Design torsional reinforcement provided $f_c' = 27.5$ MPa, $f_y = 420$ MPa, torsional moment $T_u = 12$ kN/m, shear force is 170 kN, and $A_s = 1150$ mm^2.

REFERENCES

ACI 318. (2014). Building code requirements for structural concrete (ACI 318-14) and commentary. American Concrete Institute (ACI), and International Organization for Standardization.

BS 8110. (1997). *Structural use of concrete. Part I: Code of practice for design and construction.* UK: British Standards Institution.

Eurocode 2 (2004). Design of concrete structures. European Committee for Standardization.

Nilson, A., Darwin, D., & Dolan, C. (2004). *Design of concrete structures* (14th ed.). New York: McGraw-Hill Higher Education.

Park, R., & Paulay, T. (1975). *Reinforced concrete structures.* New York: John Wiley & Sons.

Shihada, S.M. (2011). Design of reinforced concrete structures. Lectures at The Islamic University of Gaza.

Wight, J. K., & MacGregor, J. G. (2011). *Reinforced concrete: Mechanics and design* (5th ed.). Englewood Cliffs, NJ: Prentice Hall International, Inc.

FURTHER READING

ACECOM (1997). *General engineering assistant and reference, GEAR, released by Asian Center for Engineering Computations and Software, ACECOMS*. Thailand: Asian Institute of Technology (AIT).

CSI (2001). *Section Builder, released by Computers and Structures Inc*. Berkeley, CA: CSI.

CSI (2008). *CSICOL 8.4.0 released by Computers and Structures Inc*. Berkeley, CA: CSI.

CSI (2012). *Section Designer ETABS 2013, released by Computers and Structures Inc*. Berkeley, CA: CSI.

MacGregor, J. G. (1992). *Reinforced concrete: Mechanics and design* (2nd ed.). Englewood Cliffs, NJ: Prentice Hall International, Inc.

Wang, C.-K., & Salmon, C. G. (1985). *Reinforced concrete design* (4th ed.). New York: Harper & Row Publishers.

CHAPTER FIVE

Response and Design of Column Cross-Sections

Design of column cross-sections signifies the importance of interaction of axial load and biaxial load bending and efficiency of cross-section shape and reinforcement layout.

INTRODUCTION

Columns are one of the most important structural components. Historically the symbolic value of columns or pillars has been used in the literature and art as a symbol of strength, stability, and support in many contexts. Columns have been and are one of the most important and visible features in ancient construction masterpieces of various historical civilizations including Roman, Greek, and Arabic architecture. The structural significance of columns is evident from their use in almost every type of structural system. The columns can range from a single masonry pillar in a single-story house to a large bridge pylon supporting a long cable-stayed or suspension bridge. Irrespective of the size, type, or usage, they all share one characteristic; they all carry axial load, mostly in compression after in conjunction with bending moment and almost always support other structural elements or members.

Columns cover a very wide spectrum of usage and applications. They can be constructed in a variety of shapes and materials, and can exhibit several of structural response. Their applications range from a single signpost to columns in buildings, to piers and pylons in bridges to piles embedded in soil. This wide spectrum makes the column design problem a fairly complex one. Fig. 5.1 shows the overall view of the column design problem. It can be seen that the column design deals with the overall framing, cross-section

Structural Cross Sections
DOI: http://dx.doi.org/10.1016/B978-0-12-804443-8.00005-1

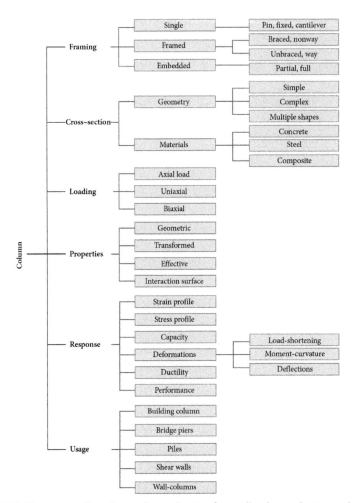

Figure 5.1 The scope, diversity, and complexity of overall column design problem.

shape and materials, loading types, geometric properties, various kinds of responses and finally, a varied range of usage. Each of these main categories has several specializations making the overall column design problem fairly diverse and complex. This chapter intends to cover some key aspects of the behavior and response of column cross-sections and provides a brief overview of some practical considerations necessary for their design.

Before discussing basic issues related to column design problem, let us first answer this question, "what is a column, really?" It may seem like a simple and obvious question, but the answer is not so straightforward. Generally, it is understood that horizontal load-carrying members are

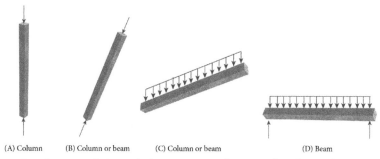

(A) Column   (B) Column or beam   (C) Column or beam   (D) Beam

Figure 5.2 Can we distinguish between a column and a beam based on its orientation?

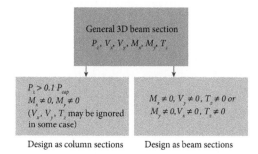

Figure 5.3 Definition of a column and beam for design purpose.

beams and vertical members are columns; but what about, say, an inclined member at an angle of 45 degrees, which is subjected to a variety of load conditions? Should this be designed as a column or a beam? The geometrical transition from a generally understood so-called column to a beam is shown Fig. 5.2. It is obvious that the behavior and thus, definition of columns is not merely a matter of orientation.

If orientation cannot be used to define a column, then what other definition can be used? One common definition of a column is that if the applied axial load is greater than 10% of the nominal axial load capacity of the cross-section, then the member should be designed as a column, otherwise as a beam (see Fig. 5.3). This distinction has also used by ACI 318, 2014 while determining the design moment strength of cross-sections. However, in a real structure, the members are subjected to several load combinations. Some of these combinations may have higher axial load than 10% of their axial capacity (P_{no}) and some may have less. Therefore strict adherence to this definition may require the same member to be designed as a column for some load combinations and as a beam for some. From a

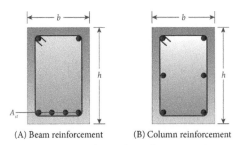

(A) Beam reinforcement   (B) Column reinforcement

Figure 5.4 Difference between detailing of reinforcement between beams and columns: (A) section designed as a beam and (B) same section designed as a column for same moment.

strength and capacity point of view, this may produce different values. The requirements for minimum and maximum reinforcement, development lengths, bar curtailment, transverse reinforcement, and other detailing requirements are so different for beams and columns and the final member design can be significantly affected by the initial selection of the beam or column type. The rebar arrangement suddenly changes from beam type (bars on one face) to column type (bars on all faces) when the load crosses $0.1 P_{no}$ threshold, as shown in Fig. 5.4.

This code-based definition can be used as a general guideline and may be reasonable for members subjected to no axial loads (pure bending) and those subjected to fairly high axial loads. A great deal of engineering judgment is needed in borderline cases where the distinction between beam and column behavior is not clear.

As a more general guideline, members with significant bending moments, axial force, shear force, and torsion, should be designed to satisfy the more restrictive requirements for both, as a beam and as a column. For example, in such a case, the longitudinal reinforcement may be determined by using the general capacity interaction curves, whereas transverse reinforcement may be determined from beam–shear and torsion design procedures. The rebar layout may be determined from the predominant direction of moments in the governing load combinations. Another difference that is often used to distinguish between beam design and column design is the presence of $P-\Delta$ effects. The $P-\Delta$ effects increase the bending moments due to the presence of axial loads. So, if significant moment magnification due to an axial load is expected, the member should obviously be designed as a column.

This chapter, therefore, focuses on the issues that are relevant reinforced concrete and composite to members that are subjected to significant axial loads, either tension or compression, in addition to the presence of significant bending moments about one or both axes and accompanied shear forces and torsional moments. The members are assumed to be long enough to be considered for slenderness effects.

COMPLEXITIES INVOLVED IN ANALYSIS AND DESIGN OF COLUMNS

As mentioned earlier, the strength and stability of buildings and structures depend heavily on the strength and performance of columns. Failure of a column is generally more catastrophic for a structure than failure of a beam.

Generally, five types of columns are used in different structures, as shown in Fig. 5.5:

1. Free standing single piers that are used in bridges and sometimes in industrial buildings, etc.
2. Columns that are part of a moment resisting framing system and are connected to several members at each end.
3. Columns that are placed as stiffeners in concrete walls or connected to reinforced masonry walls.
4. Columns that are embedded in soil and used as piles or caissons.
5. Sometimes, large structural members, such as shear walls, are also designed as columns.

The behavior and overall design of each type of these columns varies considerably, especially in terms of load application, load transfer, stability, cross–section type and detailing of reinforcement, etc.

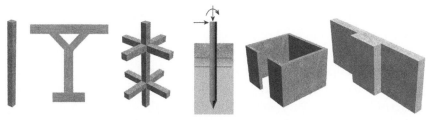

Figure 5.5 Free standing single columns/piers, column in building frame, embedded columns (piles), shear walls as columns, and wall-columns.

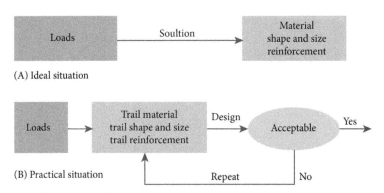

Figure 5.6 The iterative design process for most of practical cases.

Irrespective of the usage or type of column, the main design problem is to determine the appropriate dimensions, cross-section shape, material characteristics, reinforcement amount and distribution based on a set of applied actions, column geometry and framing conditions. There appears to be no direct solution available that can independently provide the desired ideal solution, i.e., provide a final design purely based on the available input parameters (Fig. 5.6 A). It is rather an iterative process, requiring considerable "initial design" information. In fact, it will not be unreasonable to say that "the column must be designed before it can be designed." Fig. 5.6 (B) shows this iterative procedure associated with most of practical design situations. An ideal situation corresponds to the direct solution which can provide cross-sectional shape, dimensions, and reinforcement configurations. While in practical cases, the trial sizes need to be tested iteratively until an optimum option is found.

We will first discuss the complexities involved in the typical column design problem, and later present general procedure and considerations for design of column sections, and discuss some of the simplifications for practical design process. Generally, compared to design of beams and slabs, column design is more complex due to various parameters and affecting phenomena, the most critical of which include the following:

- Loading type and level of load
- Cross-section materials, shape, and layout
- Column length, bracing characteristics, and stability issues

Other complexities may arise from the consideration of ductility, performance, axial shortening, shear demand, etc. Fig. 5.7 shows a general 3D column subjected to combined axial load (P_x) and biaxial bending moments (M_y and M_x) at both ends. These loading and moments

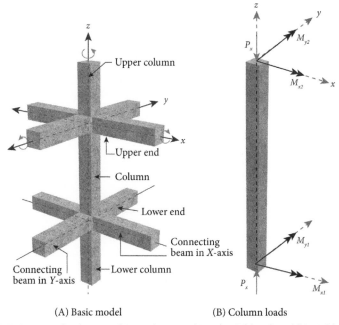

(A) Basic model   (B) Column loads

Figure 5.7 A general column subjected to combined axial load and biaxial bending.

are transferred to column depending upon different framing conditions which may vary from simple cases of point load transfer to complex cases of connecting members with variable stiffnesses, etc.

Based on all possible variations in this general case shown in Fig. 5.7, a "complexity space" can be constructed to show the effect of various parameters on the level of complexity involved in the design process. The first level of complexity space, as shown in Fig. 5.8, considers the primary factors, i.e., cross-section shape, loading type, and slenderness effects for column design. The complexity increases as we move along any of these axes. As can be seen in Fig. 5.8, the simplest case is that of a circular or rectangular short column subjected to only an axial load. The design for this case can be carried out by using a single formula involving the cross-section area and the material strengths. However, the design of a simple reinforced concrete rectangular column subjected to axial load and uniaxial moment cannot be performed directly by using closed–form formulae or equations. Similarly the design of a long, rectangular column subjected to axial load may involve consideration for slenderness effects which makes the design more complex. It can be seen further that the design of even a simple rectangular column subjected to several biaxial bending

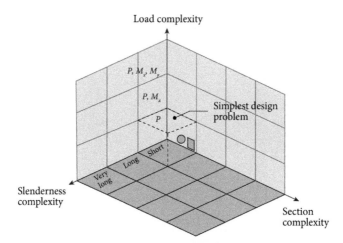

Figure 5.8 Overall complexity of column design problem. The complexity depends on the type of loads, slenderness, and cross-section. The design becomes more difficult as we move along each axis.

load combinations can be quite complicated. Most practicing engineers, however, deal with the design of columns that are assumed to be in the simple to medium complexity range.

New structural systems and innovative design of buildings are resulting in various new cross-sectional shapes for columns which may further add the complexity in terms of determining the axial—flexural capacity. The shape is often governed by architectural, esthetic, or structural requirements. Columns of rectangular, circular, oval, polygonal, flanged, and even triangular shapes can be seen in several buildings and bridges. These sections may be solid or hollow, or may have an arbitrary arrangement of shapes. The cross-section material may be simple concrete, reinforced concrete or composite made up from concrete, steel, and rebars. Even in the case of simple reinforced concrete columns, the concrete confined within the rebar cage behaves differently than the concrete outside the confining cage, thus creating further complexities in proper evaluation of cross-sectional capacity, especially at high strain levels. This complexity, in relation to the loads, is shown in Fig. 5.9. It can be seen that a further complication may arise due to the interaction of applied actions.

Columns may be subjected to axial load, combined axial load and bending about one axis, or combined axial load and bending about two axes. This interaction makes it difficult to determine the strain profile on

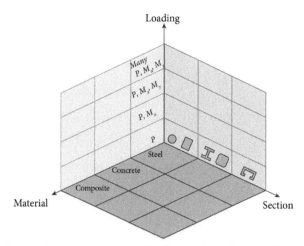

Figure 5.9 Complexity space between loading and section. Section is subdivided into geometry and material.

a cross-section. This means, that it is not easy to find the direction and location of neutral axis for a set of axial loads, and bending moments. Due to this, the capacity calculations cannot be performed using manual or closed-form solutions and hence, the overall adequacy and safety of the cross-section cannot be directly determined. This problem is further complicated by the fact that a column cross-section must be designed for several independent load combinations, each with its own set of actions. The maximum and minimum values from the action envelopes cannot be used for column design, rather all the load combinations need to be checked.

Another level of complexity is due to the slenderness effects related to the length, framing and loading conditions of the columns and the structures as a whole. The slenderness effects modify the actions, and hence the capacity in a highly nonlinear manner. This nonlinearity in reinforced concrete columns arises, both from geometric changes ($P-\Delta$ effect) and material changes (cracking, creep, nonlinear stress—strain curves, etc.) thus, making the exact evaluation of slenderness effects fairly complicated.

The procedures available in most text books (Leet, 1982; MacGregor, Wight, Teng, & Irawan, 1997; McCormac & Brown, 2015; Nilson, 1997) and design charts generally deal with rectangular and circular cross-sections with regular rebar distribution, subjected to uniaxial bending and without considering the slenderness effects. The

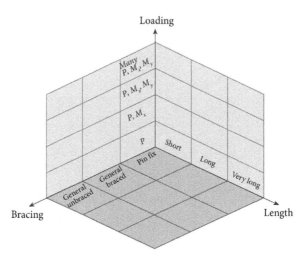

Figure 5.10 Complexity space between loading and slenderness. Slenderness effects are subdivided into bracing and length.

biaxial bending is generally considered using approximate methods, such as Load Contour or Breslar Approximation. Computer programs can be effectively used for determining the capacity of general shapes, and can provide greater insight into their behavior, when combined with graphical visualization techniques. The complexity space between slenderness and loading space can be further expanded to the type of loading, the dimensions and length of the column itself, and the bracing and connectivity conditions as shown in Fig. 5.10. It is again obvious that the determination of slenderness effects for pin- or fix-ended single column is much easier than for a column that is part of an unbraced frame subjected to sway loading.

SLENDERNESS AND STABILITY ISSUES IN COLUMNS

Before discussing the actual design procedure, it is important to understand some of the key behavioral aspects of columns. Stability is one of the most important consideration in analysis and design of column members. This section deals with various issues related to slenderness and stability, and presents basic concepts and applications to various cases of practical nature.

What Is Slenderness?

As discussed in Chapter 3, Axial—Flexual Response of Cross-Sections, the interaction between the axial load and moments significantly affects various aspects related to design of cross-sections of axial—flexural members. This interaction is dependent upon the cross-section properties as well as the level of loads. It can be seen from the capacity interaction surface, as well as from load—moment interaction diagrams (presented in chapter 3) that the moment capacity increases by increasing the axial load in some load range and decreases with increasing load in another range. This interaction relates to the internal stress resultants or the capacity of cross-section and is not affected by the external actions.

There exists another type of interaction between axial load and moment which is a result of column deflection. Let us consider two columns; one, a very short one, with a length less than three times the least dimension of the column, and the other column which is long, having length more than, say 15 times the least dimension of the column cross-section, as shown in Fig. 5.11.

Both columns have the same cross-section and reinforcement, and both are subjected to an axial load (P) with an initial eccentricity of "e." The load—moment interaction curve for this cross-section is shown on

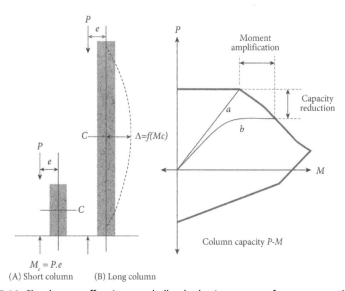

Figure 5.11 Slenderness effect in a nutshell—the basic concept of moment amplification and capacity reduction in long columns.

the right side. Now the moment diagram due to the eccentric load for both columns will show a constant moment at, $M = P.e$. Due to this moment, there will be some deflection in both the columns. It is obvious that the deflection in the short column will be much smaller than in the longer column due to a lower flexural stiffness of long column. The deflection Δ and corresponding moment M are related as shown in the following equation:

$$\Delta = \int_a^b \frac{M}{EI} dx \qquad (5.1)$$

Let us further assume that the deflection in short column is so small that it can be ignored. If the axial load is increased in the short column, the moment will increase proportionally and the load—moment relationship will be a straight-line which will meet the capacity interaction curve at point "A" and the column cross-section would fail or reach its theoretical capacity. On the other hand, it is obvious that the deflection due to moment will be significant in long column. This deflection will increase the eccentricity of the load with respect to the deformed shape of the column. This increased deflection will produce additional moment on the column, and the total moment diagram will be a sum of the initial moment ($M = Pe$) and the additional moment ($M = P\Delta$). This phenomenon, however, does not stop here. This additional moment due to deflection (also referred to as second-order moment) is expected to produce additional deflection, which will in turn produce additional moment. Now if the initial axial load was small, the initial moment would be small, the corresponding deflection would also be small and the additional moment due to this deflection would not be large enough to cause significant additional deflection and hence, this phenomenon would die out quickly.

This means that the load—moment relationship for a long column will almost be a straight-line for small axial loads, similar to short column. However, when the axial load is high, the initial moment will be large, deflection will be large, and therefore additional moment can be significant. The deflection due to additional moment will produce further additional moment and this process will continue. In fact, a stage will soon reach, when the additional moment will keep increasing by progressive deflections without any increase in the axial load and the applied load—moment curve will meet the capacity interaction curve at point B, at much lower load than the short column. It is quite clear that the axial load capacity has decreased and the moment has been magnified. So the net effect of the slenderness is that the moment is increased due to the presence of axial load and the axial load-carrying capacity is reduced due to the load—moment interaction. It is, however, interesting to note that if the initial load is very small and accompanied by a very large eccentricity, the moment magnification due to slenderness may in fact increase the axial load-carrying capacity due to the nature of the load—moment interaction curve. This interaction of axial load and the increase in elastic moment can be termed as the slenderness effects. If the magnification of elastic moments is significant, then the slenderness effects need to be considered in the design of columns as well as their impact on overall structure.

It may also be emphasized here that as long as the applied moment is well below the yield moment, the relationship between the moment and curvature is nearly linear, so the shape of the M/EI diagram will be similar to the moment diagram (see Fig. 5.12A). However, when the moment exceeds the yield moment then the curvature increases rapidly and therefore, the shape of the M/EI diagram can be significantly different compared to the bending moment diagram. Fig. 5.12A shows the bending moment and curvature diagrams for a uniform column cross-section. Since both the end moments M_1 and M_2 are below the yield moment as indicated by moment—curvature response of this cross-section (see Fig. 5.12B), the curvature diagram can simply be obtained by dividing bending moment diagram with EI.

There can be several ways to consider slenderness effects in column design. One possible option is elastic second-order analysis which takes

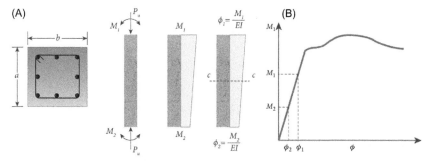

Figure 5.12 The curvature distribution along the length of column remains proportional to bending moment as long as moment remains below yield point. (A) Bending moment and curvature diagrams of a uniform reinforced concrete column cross-section. (B) Moment—curvature relationship of square RC section.

into account the additional moments caused by axial load eccentricity developed due to lateral deflection. A more convenient option is to use code-based procedure, e.g., moment magnification method as prescribed in ACI 318, which will be discussed later.

Buckling and Slenderness Ratio

Buckling is a sudden lateral failure of an axially loaded member in compression, under a load value less than the compressive load-carrying capacity of that member. The axial compressive load corresponding to this mode of failure is referred to as critical buckling load. A load greater than critical load results in unpredictable and sudden deformation of member in lateral direction. This may further result in complete loss of member's capacity to withstand applied load. If the member is part of a complex assembly of members, the load will be redistributed to other connected and intact members. Buckling is a major concern in design of axially loaded members in compression, especially the ones with significant slenderness effects. The extent of these effects is characterized by a parameter referred to as slenderness ratio. In its simplest form, it is the ratio of effective length of column to the least radius of gyration of its cross-section as shown in the following equation:

$$\lambda = \frac{kL}{r} \tag{5.2}$$

where k is the effective length factor and its values depend on support conditions of columns. Fig. 5.13 shows the values of effective length factor (k) for various support conditions.

Based on slenderness ratio, columns are usually classified as short or slender. The critical or limiting values of this ratio are used to decide whether slenderness effects should be considered in a column or not. It is therefore an important design parameter for columns. Generally the steel columns with slenderness ratio less than 50 are considered as short columns; those having between 50 and 200 are intermediate columns, and those with greater than 200 are considered as slender columns. Generally the behavior of intermediate steel columns is controlled by the strength of material it is made of, and long columns are governed by its modulus of elasticity. For reinforced concrete, slenderness is sometimes characterized as l/d ratio, where l is the unsupported length of column and d is the least

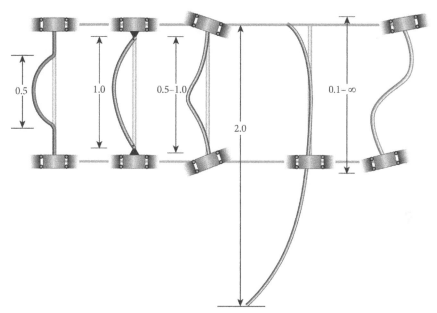

Figure 5.13 Effective length factor, k for various end support conditions.

dimension of the cross-section. The columns with l/d ratios less than 10 are considered as short columns and those with $l/d > 10$ are considered as long or slender columns.

In 1757 a Swiss mathematician Leonhard Euler derived a simple formula to determine critical buckling load for a homogeneous and ideally straight column with no lateral forces. This famous Euler's buckling formula is shown in the following equation:

$$P_{cr} = \frac{\pi^2 EI}{(kL)^2} \tag{5.3}$$

where P_{cr} is the critical buckling load, L is the length of column, k is the effective length factor, and EI is the cross-sectional stiffness. An interesting observation that can be made from this formula is that there is no role of compressive strength of material in determining the critical buckling load. It is only dependent on elastic modulus of material, moment of inertia of cross-section and effective length of column. Effective length is again dependent on boundary conditions of column, the effect of which may become significant in some cases. In fact, as shown in Fig. 5.13, the boundary conditions control the deformed shape and the position of inflection points (i.e., the points of zero internal moment or at which curvature changes its sign) on deformed shape of column; and the distance between two inflection points is the effective length (*kL*) of column. It is therefore very important to understand the behavior of practical boundary conditions and their role in slenderness considerations, which will be discussed next.

The Role of Boundary Conditions

As reflected by the effective length factor, boundary conditions play an important role in consideration of slenderness effects in design. Based on degree of fixity or release at the ends, a frame can be classified either as "sway" or "nonsway" (see Fig. 5.14).

1. When the two ends of the column are not significantly braced against lateral movements relative to each other, it is referred to as sway condition.

2. When the two ends of the column are sufficiently braced against the movement in lateral direction, it is termed as "braced" or "nonsway."

The design of columns for sway and nonsway frames will differ due to different assumptions for boundary conditions. The effective length factor for braced columns varies from 0.5 to 1.0, whereas for unbraced columns,

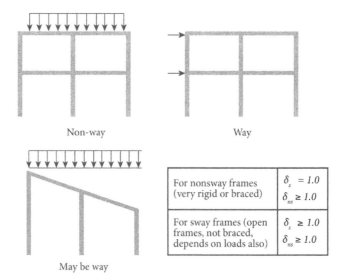

For nonsway frames (very rigid or braced)	$\delta_s = 1.0$ $\delta_{ns} \geq 1.0$
For sway frames (open frames, not braced, depends on loads also)	$\delta_s \geq 1.0$ $\delta_{ns} \geq 1.0$

Figure 5.14 Columns as part of sway and nonsway frames.

it can vary from 1 to infinity. ACI 318 (Section 10) allows the columns to be considered as component of nonsway (braced) frame if the additional moments induced by second-order effects are not greater than 5% of the initial end moments. Practically the distinguishing factors between braced and unbraced columns may not remain constant throughout the intended life of column. In real structures, almost all columns are subjected to normal lateral loads and they undergo some lateral deflections. It is difficult to encounter columns that do not undergo any relative displacement at the two ends in some stage during the lifetime of the structure. However, "very small" lateral displacements may not necessarily make the column unbraced (see Fig. 5.15).

Although the discussion in this section does not promise to answer all the questions, it may help to resolve some of the issues that can affect the determination of the effective length factors. In general, the end boundary conditions for the column in a frame can be represented by a spring system. Each spring represents the stiffness of the members attached to the ends of the column. The true restraining conditions in a 3D building are practically very difficult to evaluate. However, a simplified model can be considered along two principal planes of the structure (or any arbitrary plane for that matter). On this restraining plane, the degrees of freedom at ends can be replaced by springs (see Fig. 5.16). The stiffness of these springs can range from "0" indicating a pure freedom to infinity indicating complete fixity. Both of the extreme values are difficult to achieve in reality, as even a "pure" cantilever column is not really free. It has to be

Figure 5.15 Various practical boundary conditions with varying degrees of end restraining effects for columns.

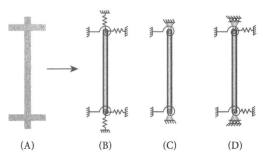

Figure 5.16 Various idealizations for modeling of actual restraining effects of practical boundary conditions: (A) actual column, (B) full spring model, (C) restrained pin model, and (D) roller spring model.

connected to something for any practical load to be transferred to the column. As soon as any structural or even a nonstructural element is connected, some degree of restraint is provided practically. The effective length of the column or the so-called "bent length" will depend on the stiffness of the end restraints in this idealized system. The real question is how to evaluate these end restraints in terms of stiffness? For simplicity, the axial stiffness is assumed to be very high (although axial shortening of columns in tall buildings is often significant) and hence, this degree of freedom is generally locked in the evaluation of effective length factors. The lateral resistance or stiffness is either considered infinite, hence locked and produces the so-called braced or nonsway column condition, or it is considered completely free resulting in the so-called unbraced or sway condition. It is obvious that both of these conditions are extreme bounds and most of the actual columns will have lateral restraints in between. The lateral restraint is provided by the flexural stiffness of other columns or shear stiffness of shear walls and bracing systems. This stiffness can neither be exactly zero nor infinity.

A conventional procedure for computing the end restraining stiffness in columns is to add up the flexural stiffness of the connecting beams and columns. The flexural stiffness computed by the factor EI/L is only a representative and relative value which does not consider the end framing conditions of the connecting members themselves. However, for all practical purposes this value can be considered an adequate representation of the end restraining effect.

A difficult problem of practical nature arises when the lower end of the column is connected to a foundation. There is very little information available in the literature to help the engineers in evaluating the true restraining effect of various types of foundations. To determine the stiffness of the foundation, we need to realize that the foundation and the supporting soil together act as a complex spring system which may be elastic, inelastic, or even plastic. Many engineers use arbitrary values of foundation stiffnesses, or more correctly, the ratio of foundation to column stiffnesses.

Another problem that is often encountered by practicing engineers is related to the determination of "free" length for the columns embedded in soil or in floor slabs. This question may also be asked for columns that change in cross-section along a single unbraced free length. Fig. 5.17 shows some of the practical cases and configurations of columns embedded in soil. Case (A) corresponds to a single column with some embedded length in soil. For this case, depending on the size of columns, type of soil, and presence of lateral loads, the column may become fixed against rotation (or effectively fixed) at a depth of about 3—5 times the average dimension of column cross-section, below the compact or stable soil level. In case of columns or piers in waterways, the soil erosion, silting, and scour may keep changing this level. For case (B) again, the embedded depth less than the fixity depth can be considered as clear

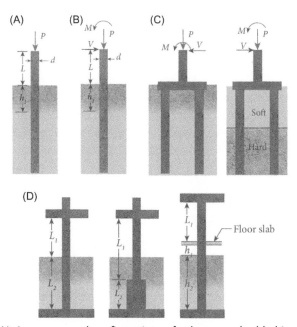

Figure 5.17 Various cases and configurations of columns embedded in soil.

length. Another factor which may make the problem more complex is the amount of restraint or fixity provided by back fill and compacted soils. Case (C) is even more complex where the column size below the soil level is larger than the main column. So the effect of larger cross-section as well as larger restraint due to soil needs to be taken into account. At the same time because of the larger column near the footing, it is not immediately clear as to which column dimensions should be considered in determining the effective length. In this case, one can use the concept of inverse stiffness (or flexibility) to determine the stiffness of the variable column.

Case (D) is further complicated by the presence of a concrete floor which is not rigidly connected to the column but nevertheless provides significant lateral restraint. The well-constrained compact filling under the floor provide further restraint which is numerically difficult to evaluate. In this case, one can consider the column to be hinged for the determination of slenderness effects. However, this hinged condition should also be considered in modeling of the frame to obtain consistent and compatible moment in the column. An even more complex case is that of a column supported on a pile cap and pile foundation. If the piles are driven into soft top layer or are exposed above the firm soil layers, then the column—pile cap—pile—soil system will interact together to determine the effective bending length or effective length of the whole system. The solution to such problems is to carry out a $P-\Delta$ analysis of the complete system or to estimate the effective length from a first-order analysis.

The problem of effective lengths is also complicated by the connection types between steel columns and foundation. Several details are used in practice, each affecting the effective length in a different manner. First, it is always difficult to assess the degree of fixity provided by the connection between the column and the base plate and the connection between the base plate and the footing.

When Are Slenderness Effects Important?

It may seem apparently contradictory, but the slenderness effects in columns of very low rise buildings may be more critical than in medium and high-rise buildings. This may be explained by three reasons. Let us consider a particular building layout, say a column spacing of 6 m × 6 m (20 ft. × 20 ft.) with a story height of say 3 m (10 ft.). For the same layout, for a medium high-rise building say more than 30 floors, the column size would be about 0.8 m (30 in.). For a 3 m height, assume an effective length factor of 1.0, we get a $kl = 20$, which is fairly small. So the column moment magnification is likely to be very small and thus, slenderness effects are negligible. Also the column is most likely to be braced due to the presence of shear wall(s), elevation shafts, and other bracing systems to resist the lateral loads. The lateral drift ratio between two adjacent floors is also likely to be small because of a check and control on overall lateral deflection of the building. The columns are likely to have relatively small moments due to the contribution of shear walls and bracing system and due to small eccentricity the design will be "axial load" controlled.

On the other hand, if we consider same floor plan for a four-story building, the column size is likely to be about 0.4 m (16 in.). The column will most likely to be unbraced in the absence of any shear walls or elevator shafts and the moments due to lateral loads are going to be significant in proportion to the axial load. The effective length ratio will be more than 1, say 1.5 or so, and therefore the moment magnification in this case will be significant and is likely to affect the column design. Due to high eccentricity, the design will be moment controlled and hence directly affected by the moment magnification.

In general, the corner and edge columns in buildings are more affected by slenderness effects than the interior columns and this can be due to several reasons. The corner columns are often subjected to high biaxial moments, due to gravity as well as lateral loads. The axial load is relatively small, so the moment governs the design. The effective length factor is generally larger due to smaller total stiffness of beams connected at the ends of the columns. Therefore the moment magnification may be higher and may affect the design more significantly than for the interior columns. This is especially true for laterally unbraced frames or sway loading conditions. However, for braced frames or nonsway load conditions, the interior columns may experience greater moment magnification due to the presence of high axial loads. In general, the moment magnification of braced or nonsway columns is greatly affected by the ratio between the axial load and the critical buckling load, whereas for sway or unbraced columns, it is also affected by the amount of lateral load and the relative lateral drift ratio.

Effect of Column Slenderness on Overall Frame Behavior

As we have seen, the slenderness effects magnify the moments, with a corresponding magnification of deformations and deflections as well. This additional deformation when translated to the ends of a framed structure will change the elastic or first-order deformation characteristic of the overall frame, including other columns and beams. This will, in turn, change the moments, shear, and axial forces in other members of the frame. If we consider the moment magnification in just one column in a typical frame, the effect propagates with a diminishing intensity. The beams and columns adjacent to the column may also undergo the magnification. However, in a real structure several columns may be undergoing the moment magnification at the same time and hence, the effects are cumulative. The problem is further complicated by the fact that the moments modified by moment magnification are further magnified due to slenderness and affect the moments in other member. So the overall $P-\Delta$ effect in the frame becomes highly nonlinear.

When Should Slenderness Effects Be Considered?

The simple answer to this question would be "almost always." Considering slenderness effects in column design by default is always a good (and safe) thing to do. The worst that can happen is that there is no significant moment magnification, and the additional effort is probably considered as wasted. However, just to check whether slenderness effects need to be considered or not, requires significant amount of computations. Various codes provide limiting values of slenderness ratio to check whether moment magnification should be considered in a column or not. For example, ACI 318 (Section 10) specifies that the slenderness effects may be ignored for compression members braced against side-sway when the slenderness ratio $\lambda = kL/r$ is less than $34-12(M_{1b}/M_{2b})$ and for compression members not braced against side-sway, λ is less than 22. For λ values higher than these prescribed limits, the slenderness effects must be considered. M_{1b} and M_{2b} are the smaller and larger factored end moments, respectively.

To apply these checks, we first need to determine whether the column is braced or unbraced for a particular load combination that itself needs some computations and rather arbitrary checks. Second, to evaluate the kL/r ratio, the effective length factor needs to be determined, requiring considerable stiffness calculations of all members framing into the column in question. After all these computations, if we conclude that slenderness effects do not need to be considered, a considerable effort would already have been spent. On the other hand, if a computer-aided tool is being used for column design, it is probably best to simply evaluate the slenderness effects as part of the routine. If the magnification turns out to be small, so be it. But at least no critical magnification would be missed by following the rather simple and arbitrary division between short and long columns.

Some general guidelines to facilitate the design of columns in buildings are presented here. However, these should be treated as just guidelines, and not general rules.

1. The slenderness effects may be more pronounced in low-rise buildings without shear walls and bracings, and in some cases upper floors of tall buildings.
2. Unbraced frame or sway loading conditions in columns are likely to produce larger slenderness effects than in braced or nonsway columns unless the nominal axial stress is higher than about 50% of buckling stress.
3. Higher slenderness ratio is likely to produce higher slenderness effects, but not necessarily. The actual amount of axial load and moment value and distribution are also important.
4. Most practical column proportion in low to medium rise buildings will fail in flexural buckling mode and will undergo some moment magnification when axial load approaches the cross-section capacity.
5. It is always safer to include the slenderness effects than to exclude them.
6. Columns of narrow cross-sections, or unsymmetrical cross-sections may have moment magnification in the lateral direction in addition to or instead of the primary bending direction being considered.
7. The presence of transverse bracing beams can affect the primary moment magnification if the bracing is connected to lateral bracing systems. However, if the bracing simply connects to similar columns, the effect may by negligible.

COLUMN DESIGN PROCESS AND PROCEDURES

The design process is a general sequence of steps required to complete the design of columns, whereas design procedures are rules specified by the regulatory codes. As discussed earlier, an iterative procedure is almost always involved between the analysis and design phases. It is also important to note that the modeling and analysis phase requires that the structural configuration, the member sizes, their cross-sections, and even the material properties be defined before starting the analysis process. This effectively means that the preliminary or the conceptual design phase must generate a significant amount of design information before the formal design process can begin. It is also important to note that the selection of the appropriate structural system and member sizing at the conceptual design stage is affected by, not only structural considerations, but also by several other, often conflicting, factors. These include architectural requirements, esthetic considerations, environmental consideration, constructability concerns, maintenance considerations and of course, economic considerations. In the case of column design, the selection of shape, its location, length, framing conditions, etc., will often be dictated by nonstructural considerations. The modeling and analysis phase, on the other hand, is purely in the structural engineer's domain. Irrespective of the type of structure, type of load or method and extent of analysis, the outcome of this phase is deformations and the corresponding actions acting on or induced at various locations or in various structural components and members. In the conventional linear-elastic analysis, the output from the analysis phase can be used as an input to the design and detailing phase. The design of members can then be carried out in two ways:

1. For some simple members and sections, the cross-section dimensions and properties are directly determined by the actions. The design of beams and slabs may be carried out in this manner.
2. The capacities of the members or cross-sections are checked against the action; with an appropriate margin of safety. The design of RC columns, however, needs to be carried out in this manner, as the direct design of columns based on applied actions is not practical or even possible in most cases. The design of steel sections may be carried out directly using standard shapes.

This design process can be viewed as a two-step process: Analysis and Design (see Fig. 5.18). This process is typically used when linear-elastic-static behavior is assumed or considered.

Figure 5.18 The two-step process: Analysis and Design.

As discussed in Chapter 1, Structures and Structural Design, ideally, a solution algorithm should be available that can directly provide the optimum design of a particular column in terms of the cross-section shape, cross-section dimensions, material properties, reinforcement amount, and their efficient location based on the sets of axial loads and moments, including the effect of slenderness. Unfortunately, no such algorithm or solution technique is available at this time. In fact, to the best of the author's knowledge, not many computer programs are capable of automatically producing optimized design of reinforced concrete and composite columns of general shape and configuration. CSI column from computer and structures does offer some capability to do this.

The response and capacity of a column to resist external actions depends so much on the column design that it is safe to say that a column should be designed before it can be designed. The determination of cross-section capacity requires that the column size, material properties and reinforcement size, and distribution should be known in advance. It is important to note that the determination of the slenderness effects on the columns requires that the cross-section design must be known, but the design of the cross-section depends on the determination of magnified moments obtained from the slenderness effect calculations. This is especially true when slenderness effects need to be included in the design. The $P-\Delta$ effect that modifies the original elastic actions, and hence affects the column design in a very significant manner, depends heavily on the column cross-section proportions, material characteristics, reinforcement amount, its location, and so on. Practically, all design parameters should be known before slenderness effects can be evaluated. Generally the linear-elastic analysis provides the basic actions, which are modified for slenderness effects and the modified actions are then used in the column design.

One of the options is to carry out a fully integrated nonlinear analysis and design. In this case, the loads are applied gradually, and at each stage of incremental load, the analysis is carried out. The modified properties of members are computed, then the modified response is computed including the $P-\Delta$ effects or geometric effects, the capacity of sections is checked, the sections are modified if required and the next loading stage is applied. At the completion of this process, the final design is obtained including the effects of cracking, $P-\Delta$ and other types of nonlinearities. A modified form of this analysis can be used in performance-based design methodologies, such as the "Pushover Analysis." Fig. 5.19 summarizes the three basic processes for design of columns.

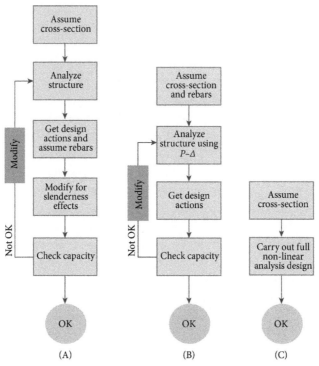

Figure 5.19 The analysis design process for column design. (A) The conventional linear-elastic analysis, (B) $P-\Delta$ analysis, and (C) full nonlinear integrated analysis and design.

The Code-Based Design Procedures

The procedure for the design of a column has undergone several revisions in the recent editions of the building codes, especially ACI 318 code. The general ACI design procedure is outlined in Fig. 5.20. As can be seen, the procedure is based on the usual linear elastic, iterative process. The cross-section size and reinforcement is estimated based on the results of the elastic analysis. The preliminary design is then used to check if slenderness effects need to be considered. If so, the magnified design moments are computed using a fairly comprehensive (but approximate) procedure. The magnified moments are then used to check the capacity and decide the adequacy of the cross-section. If the section is inadequate, the modifications in material, size, or reinforcement are made and the entire process is repeated. It is important to note that the code does not provide any specific procedure for the determination of the biaxial

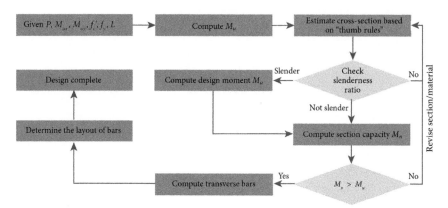

Figure 5.20 An overview of ACI 318 column design procedure.

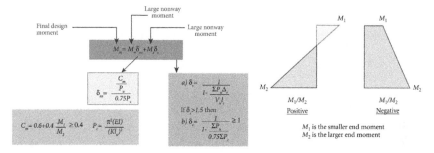

Figure 5.21 Summary of moment magnification according to ACI 318.

capacity of a general cross-section other than providing general guidelines and recommendations. There is, however, a fairly detailed description and justification provided for the determination of magnified moments. There is no specific provision in the procedure that prohibits the application of this procedure to composite cross-sections.

Fig. 5.21 shows the process to determine the total magnified moment due to slenderness effects in long columns. Various notations are explained as follows:

M_{ns} = Larger nonsway moment

M_s = Larger sway moment

δ_{ns} = The moment magnifier for nonsway condition

δ_s = The moment magnifier for sway condition

C_m = Moment correction factor relating the actual moment diagram to that of a uniform equivalent moment diagram having same peak moment as actual diagram. It attempts to estimate the moment

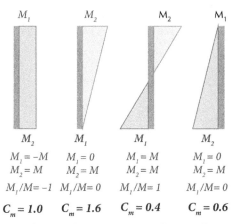

Figure 5.22 The value of C_m and moment magnification for various cases of end moments.

distribution using some form of interpolation function using two, three, or four moment values.

P_u = Total factored vertical load

P_c = Critical buckling load

V_u = Horizontal story shear

The moment variation (or moment correction) factor C_m is smaller for cases where end moments are not equal or when the deflected shape of column is having double curvature. Fig. 5.22 shows the values of C_m for various cases of end moment values and directions. It can be seen that for the same magnitude and opposite directions of maximum moments $M_1 = -M_2$ corresponding to case (A), the moment magnification is highest.

The overall procedure for the design of reinforced concrete column based on BS 8110 is similar to the ACI procedure; however, the concepts behind determination of additional or magnified moment, due to slenderness are significantly different. Also the overall guidelines and recommendations for the determination of biaxial bending capacity, the procedure for the determination of additional moments is relatively simpler compared to the ACI procedure. Unlike ACI, a clear distinction is made between uniaxial bending and biaxial bending in determination of additional moments. The overall procedure is shown in Fig. 5.23.

Most of the basic concepts discussed in the previous sections related to slenderness effects are applicable to both reinforced concrete as well as steel columns. However, the evaluation of slenderness effects in steel columns based on the AISC provisions is simpler than the procedure

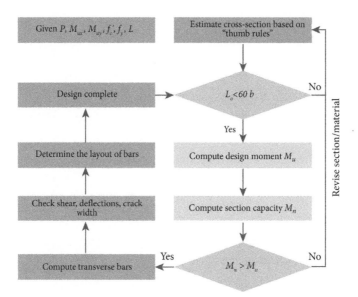

Figure 5.23 An overview of BS 8110 design procedure.

given by ACI for concrete columns. The design of steel columns is governed by the concept of combined steel ratio (R) as shown in Fig. 5.24. The combined stress equation of AISC code is applicable to short columns or when axial stress ratio is less the 0.15. For other cases, the equation is modified (see Fig. 5.24) using the factors δ_x and δ_x which are referred to as flexural stress magnification factors.

This approach is in fact similar to moment magnification method used in ACI code for reinforced concrete columns. However, instead of magnification of moments, the resulting stresses are modified directly. The stress magnification factors depend upon the critical buckling stress (F'_{EX} and F'_{EY}), which in turn depends on the section stiffness, effective length, and moment distribution along the length. Here the cross–section stiffness is more or less constant and hence easier to evaluate and use.

Who Is Responsible for Column Design?

The question as to who is really responsible for the design of columns, is a pertinent and interesting one. The immediate answer to this question appears to be quite obvious; i.e., the structural designers. However, if we look a little bit deeper then we will realize that this may not always be the case, or at least not entirely so. Take the example of columns in a

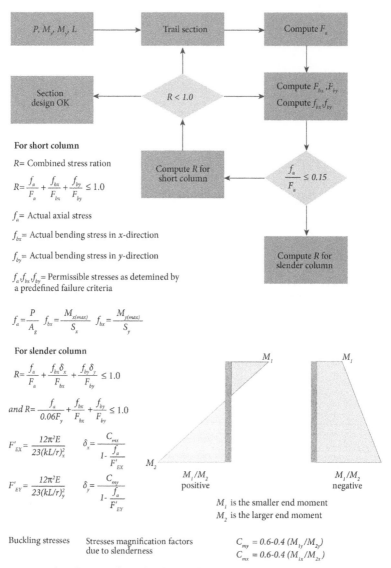

Figure 5.24 The design of steel columns based on combined stress ratio (*R*), AISC code.

multistory building. If the column is located in a parking floor, then the shape of the column cross–section will most likely be specified by the architect to be circular and rounded or elongated depending on the clearance and vehicular traffic considerations. In the case of residential

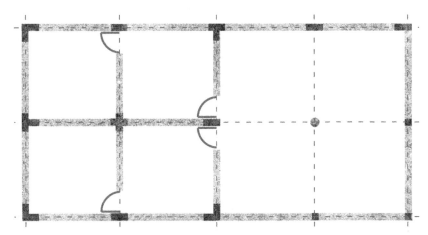

Figure 5.25 The architectural requirements and constraint may completely control the design of columns in buildings.

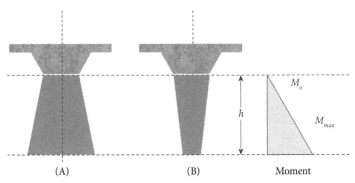

Figure 5.26 The esthetic considerations may override the structural considerations. (A) Design based on moment demand and (B) design based on esthetic demands.

apartment buildings with several partitions and planning constraints, the columns may be required to be aligned with the walls, and in some cases, concealed within the walls. The shape of cross-sections may need to be rectangular, Ell, Tee, or Cross depending on the location of the column. Fig. 5.25 shows various nonstructural constraints in a typical building that govern the design of a column, often to the detriment of the structural strength and performance.

In some cases, the esthetic requirements and considerations may completely overshadow the structural considerations. Fig. 5.26 shows an example of a bridge pier design governed by esthetic demand. It is obvious that based on the moment diagram, the cross-section should be the largest at the base and smaller at the top. But in this case, a completely contrary design is often adopted.

These two examples indicate that the design of columns, and in fact the whole structure, is not really in the hands of structural engineers and designers. It also shows that the design of members such as columns starts long before the analysis has been carried out. The consideration of moments, axial loads, and slenderness effects comes in at a much later stage, when not much can be done to optimize the design or to make it effective and efficient.

PRACTICAL DESIGN CONSIDERATIONS FOR RC COLUMNS

Selecting Column Cross-Section Shape

It is obvious that the cross-section shape, dimensions, proportions, and overall size together with the layout of longitudinal rebars are the most important parameters affecting the behavior of the column section. Other affecting factors are the strength of concrete, the grade and type of rebars and the arrangement of transverse rebars. So the question is, how can we select a cross-section, which will provide the most efficient solution to the given set of actions? The first issue is choosing the predominant, or the critical actions that govern the overall design. If there were only one set of actions, for a single-load combination case, then it would probably be easier to formulate the rules or guidelines. However, in most practical cases the critical action sets exist in pairs, corresponding to positive and negative moments obtained for the predominant direction of the lateral loads. Such cases suggest a symmetrical cross-section and rebar layout. However, there are several cases where a single predominant bending direction exists, such as in the columns of portal frames, columns supporting large cantilevers, or columns in other structures where lateral loads do not govern the design. In such cases, and even for cases of symmetrical moments, it is worthwhile to see how the cross-section and the rebar layout affect the column capacity so that the most efficient cross-section may be selected.

In buildings, the size and shape may be restricted due to architectural and space constraints. Generally a very high axial load is present, especially in lower floors of high-rise buildings requiring the use of high-strength concrete. Consideration of differential axial shortening and slenderness effects, especially in sway (unbraced) frames, is generally important. Presence of biaxial moments in the corner columns due to gravity loads and all columns due to diagonal wind or seismic load direction may make them critical from design point of view.

Effect of Cross-Section Shape on Column Strength

The following guidelines (not rules) may be used to choose the initial shape and dimensions of column cross-section:

- Whenever practical, use square, rectangular, or circular columns. It is obvious that these shapes are the easiest to design and construct. They are often efficient when there is small moment or the relative value of biaxial moment is nearly the same. Circular columns are especially suitable for seismic prone areas where high strength and ductility are needed in all directions. It is much easier to confine the concrete using special reinforcement in circular columns (or square) than in other shapes.

- Avoid shapes requiring complicated formwork unless it can be reused several times. Complicated shapes such as chamfered or rounded rectangles, grooved sections, and complex polygonal sections are mostly used for architectural and esthetic reasons as well as bridge piers. Their use is justified if the columns form the integral part of the architectural design. Sometimes, precast cladding can be attached to the structural columns to obtain the descried effect and sometimes complicated shapes can be used for structural advantage as well as for material saving when several repetitions are needed.

- Use oblong shapes when the predominant moment in one direction is clearly much larger than in the other direction; however, the aspect ratio of the cross-section should be reasonable ($0.25 < h/b < 4$). In this case, the cross-section depth can be larger than the width. It should be noted that the moment magnification due to slenderness effect along the smaller dimensional may be higher and may need to be considered in selection of dimensions.

- Use hollow shapes or I shape or H shape when moment is much larger as compared to axial load, such as in high bridge piers or columns of portal frames. It is obvious that the moment capacity is predominantly provided by the lever arm between the tension and compression stress resultants in the cross-section. Any concrete and rebars in between the compressive and tensile force are not effective in resisting the moment and hence, can be removed. Sometimes even two columns connected by diagonal or horizontal braces may be used.

- Avoid highly unsymmetrical and open shapes (C, Z, L, etc.). Highly unsymmetrical shapes are often structurally inefficient. Their moment capacity depends more heavily on the axial load than for symmetrical sections. In such sections, even loads, applied at the geometric centroid may produce additional biaxial moments. The unsymmetrical and open sections are also prone to twisting, lateral torsional buckling, nonlinear strain distribution, and shear lag between tension-compression couple.

- For parking and no-wall spaces in buildings, use circular or polygonal columns whereas use square or rectangular columns for closed and partitioned spaces. The circular, polygonal, or chamfered edges provide greater clearance and look visually uniform, pleasing, and slender than rectangular or square sections, when used in open spaces. However, the same shapes can complicate the detailing and construction process when used with masonry walls. In such cases either rectangular or slotted columns may be used.

- For high bridge piers, use hollow rectangular, circular, or polygonal box sections and give special consideration to esthetic impact of shape. Provide nosing or spoilers for better aerodynamics and fluid flow considerations.

Selecting Concrete Strength for Columns

The concrete strength affects the column size, which in turn affect several other factors. However, the strength of concrete, mostly affects the axial load capacity and moment capacity in the compression-controlled region. There is almost a direct relationship between the axial load capacity and concrete strength in this region for highly reinforced columns, in tension-controlled region, the increase in load and moment capacity is controlled more by the amount, location, and strength of reinforcing steel. If the axial load is negative (tension) then of course, the concrete strength has little effect on the load—moment capacity of the column. If the eccentricity is high, the concrete strength may not be the governing parameter. It, however, affects the durability, stiffness hardness, elastic shortening, creep, and related long-term effects. Generally, highest strength concrete for columns, in tall buildings is used where the primary action is the axial load. For high bridge piers and other columns subject to small loads and large moments, medium strength concrete is generally adequate. Generally, high-strength concrete is always preferable for columns submerged in water or in aggressive environments.

Selecting Steel Strength for Columns

The yield strength of rebars affects the moment capacity more than the axial load capacity, especially for tension-controlled sections. It is important to note that the higher strength rebars generally results in a section that has lower stiffness than a column using lower strength rebars. This is because the stiffness of a reinforced section depends on the area of steel and the modular ratio between the steel and the concrete, and is independent of the yield strength of rebars. Lower strength steel would generally require more steel and hence would increase the stiffness. So if stiffness is a primary concern, especially in long columns and structures subjected to lateral loads, a lower strength steel may be advantageous, unless a higher than required amount of steel is used for high-strength bars. In general, higher strength steel should be used with higher strength concrete for greatest overall economy in the design of columns. Selecting high-strength steel, however, increases the development length and lap splice length requirements, especially for large diameter bars. Therefore if splices or dowels are needed at short intervals (small story height), high-strength steel may lead to greater wastage.

Selecting Sizes and Layout of Longitudinal Rebars

The arrangement and layout of a specific number of rebars in a column section substantially affects the moment capacity of section but does not significantly affect the axial load capacity. The layout also affects the location of Plastic Centroid, and hence indirectly affects the "net" eccentricity and the corresponding moment capacity. A symmetrical section with unsymmetrical rebar arrangement may be subjected to biaxial bending, even if the loads are concentric or are located along any of the principal axis as described in Chapter 3, Axial—Flexural Response of Cross-Sections. Generally, it is recommended to use the largest practical bar size because of ease of fabrication, checking and concreting. Small bar size produces congestion and requires closer spacing of lateral ties, which may complicate the construction. However, closer spacing of longitudinal and transverse bars can in fact be beneficial in increasing the ductility of the column for cyclic loads by providing greater confinement. Smaller bars also need smaller anchorage, embodiment, and splices. It is advisable to use at least one bar on each acute angle corner. The bars should be located with due consideration to the predominant direction and magnitude of moment. Often a few bars at the right location may be more effective than more bars at inappropriate location. Symmetrical rebars should be placed in symmetrical sections unless a clear unidirectional moment is present, as in the case of portal frames. Generally, large diameter bars are used in the corners and smaller diameters are used in the sides to maintain maximum spacing limits. Sometimes, highly complex rebar arrangements are used for bridge columns in seismic zones to enhance the confinement of concrete in the core. At least one bar should be placed on each corner to allow for the proper placement of transverse reinforcement.

Selecting Sizes and Layout of Transverse Rebars

The transverse rebars affects the shear capacity of column and confinement of concrete and hence, indirectly affects the ultimate capacity, the ductility and plastic strength of columns. There is no direct contribution of transverse rebars in the axial–flexural strength equations. However, the strength is enhanced indirectly by increased failure strength of confined concrete at higher strain levels. A well-confined concrete may have as much as twice the failure strength than that of an unconfined concrete. This aspect is discussed in detail in Chapter 6, Ductility of Cross-Sections. The effect of confinement is not reflected directly in the nominal maximum axial load capacity equation given in most design codes. The performance of columns subjected to cyclic loads, such as those produced by seismic effects can be enhanced significantly by using proper transverse rebars. The spacing and diameter of transverse bars also control the buckling of the longitudinal bars and hence, prevents premature failure. The transverse steel also resists the transverse stress in concrete due to axial load in hoop tension or in direct tension and hence, enhances the strength of column beyond the theoretically computed value given by axial–flexural equations. The effects of transverse steel are indirectly incorporated by ACI 318 in the strength equation by linking the capacity reduction factor to the type of transverse steel. It may, however, be noted that the effect of lateral reinforcement is more critical for compression-controlled sections rather than tension-controlled sections.

Following simple guidelines may be helpful in the selection of transverse rebars.

1. Start the transverse rebar design based on shear demand.
2. Select lateral ties based on the relative magnitude of loads and moments (tension or compression control).
3. Provide closer longitudinal and lateral spacing in the moment hinging regions for better ductility.
4. Spiral reinforcement is more effective than ties for enhancing the axial load capacity.
5. For large columns, it is not necessary to provide full-length intermediate ties. Embedment of ties should be just enough to anchor the tie in the compressed area.
6. Use smaller diameter ties for smaller bond and anchorage requirement and closer spacing.
7. Generally a transverse bar should tie alternate longitudinal bar.

Effect of Loading on Column Design

Theoretically a 3D frame member, specially a column is subjected to six actions at each end. Out of these, axial load is generally the most predominant load for high-rise buildings but moments may be more critical for columns in low-rise buildings using moment resisting frames. Moment may exist about any one or about two principal axes. These moments may act simultaneous (biaxial case) or independently about each axis (uniaxial case). If the cross-section is unsymmetrical about both the main axes, a uniaxial moment or pure axial load may in fact become a case of biaxial bending about the principal axis of the section.

The bending moments in columns may be produced due to several factors. Most common sources are:

1. Rotation of frame joints due to loads applied to beams
2. Rotation of frame joints due to lateral load on frames or direct application of force to the ends of cantilever column
3. Direct application of loads within the column height in a frame
4. Eccentricity of axial loads supported by columns
5. Unsymmetrical shape and reinforcement layout in column cross-sections subjected to concentric loads
6. Secondary $P-\Delta$ moment due to the slenderness effects
7. Eccentricity of load due to imperfections in column construction

In a particular column, the total moment may be a summation of some or all moment sources. Most design codes require that columns must be designed for some minimum moment even though there is no real computed moment acting. In fact, it is extremely difficult to apply purely concentric axial load to a column unless special load transfer mechanisms are used, such as electrometric bearing or pin bearings located on the plastic centroid axis. The exact load transfer mechanism between a monolithic column and beam joint is not known and depends on level of rotation, amount and distribution of reinforcement concrete strength, reinforcement strength, presence of external axial loads, moment shear force and even torsion.

In building and frame design for gravity loads only, shear force is often not critical for a column, partly due to relatively small value and partly due to the enhanced shear capacity of the column due to axial compression on the cross-section. Tension is generally not critical and often not considered in design, except for special tie or hanger columns. Torsion may be present in some columns, but because of the enhanced shear strength of concrete under compression, design for torsion is usually not carried out.

For frames resisting lateral loads, shear in columns could be significant and may need to be explicitly considered and designed for such column may also be subjected to tension in some cases.

Due to the direct axial–flexural interaction, the capacity of the column depends on the absolute values, relative magnitudes, and directions of the axial loads and the moments. The column shape, proportion, arrangement of rebars, etc., should be selected on the basis of predominant loading. However, this may not be easy, as the columns are required to resist several load cases and combinations, often in opposing direction and magnitude. It must be emphasized here that the columns cannot be designed or checked for "action envelops" choosing the highest values of load and moment. The design must be carried out for all load combinations using corresponding values of axial loads and moments, and the maximum rebars or critical section dimensions should be chosen.

Strength Versus Stability of Columns

Generally, it can be said that the strength relates to the material failure and stability relates to the overall member or structural failure due to lack of restraint. In fact, failure in stability ultimately leads to failure in material or failure in strength in most cases. However, failure in stability also entails excessive deformations or excessive vibration. Therefore, for the purpose at definition, we can say that when the deformations of a structural member or structural component become so large that it becomes unusable, then the stability failure has occurred. The failure in strength is, however, generally defined in terms of the load-carrying capacity of the member or cross-section and is computed from the material and cross-section parameters. We have discussed the determination of column capacity in Chapter 3, Axial–Flexural Response of Cross-Sections. The strength and stability become interdependent, however, in the case of columns in particular due to the $P-\Delta$ effects. A typical stability failure of column occurs by progressive increase of moment due to deflection and ultimately reaching the moment carrying capacity of the section.

Failure of RC columns due to sudden buckling is rare in practical situation as perfectly concentric loading on perfectly straight columns is difficult to achieve except may be in controlled laboratory test arrangements. The stability failure is also linked to the ductility of the cross-section as well as the member as a whole. Higher ductility allows larger deformations to be absorbed without significant loss of strength, thus improving stability indirectly. Another linkage between strength and stability is the stiffness. Generally the parameters that improve strength also improve the stiffness of the section and the member. Improved stiffness invariably improves the stability.

Designing for Strength

Designing the columns for strength basically means that the strength capacity ratio should be less than one, meaning that the section capacity for all action sets should be more than the applied actions. The definition of capacity ratio and methods to compute its value have been discussed in detail in Chapter 3, Axial–Flexural Response of Cross-Sections. The designing for strength entails the proportioning of the cross-section, selection of appropriate materials, and proper placement of reinforcement, etc. Proportioning for strength, however, does not automatically ensure proper serviceability, stability, ductility, and performance.

Designing for Stability

It is obvious that any parameter that improves strength, automatically improves stability as well. However, several factors that affect the stability directly may not have direct impact on the cross-section strength.

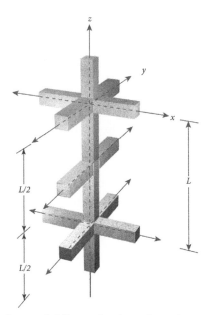

Figure 5.27 A practical case of different framings along the x- and y-axes.

Improving the column framing and bracing conditions directly improves the column stability, but may not improve the cross-section strength, as such. Similarly the column stability can be improved by reduction of load eccentricity and by improvement in the column alignment, straightness, verticality, etc. It is fairly obvious that the stability of a column is governed by the framing and bracing congestions on the weaker axis. If the framing and bracing conditions are similar along both principal directions, then the overall stability of the column can be improved by making the cross-section nearly symmetrical or, the cross-section proportions along and about the two principal directions should match with the framing, bracing and loading conditions or about that axis. For example, for a column with longer height in one plane than the other, it makes sense to make the section rectangular so that the kl/r ratio about the two axis, x and y are nearly the same if other loading and bracing are similar (see Fig. 5.27).

Column Stability and Structural Stability

As mentioned earlier, of all the structural members, the columns or piers are the most important members as far as the overall stability of the structure is concerned. The failure of a building column or a bridge pier can cause the collapse of a major portion of the structure or lead to progressive collapse of an entire structure. The stability failure of a column also causes significant rotation at the column ends, causing joint rotations, leading to propagating increase is beam end moments leading to secondary failures or complete collapses.

The structural stability is, therefore, directly linked to the column stability. Another equally vulnerable structural component is the flat slab to column connection and its failure in punching shear. This failure is basically a strength failure, but can be initiated or aggravated by the stability failure of the column. The importance of column strength and stability has prompted the well-established concept of strong column-weak beam design when it is ensured that on every joint, the combined strength of columns must be larger than the combine strength of beams. This requirement, especially applicable to structure in high seismic risk zones, is intended to limit the formation of joint hinges in the beams, rather than the columns.

SOME SPECIAL CASES AND CONSIDERATIONS
Design of Bridge Piers

The design of bridge piers needs certain special considerations, not relevant to the columns in the buildings These considerations can be divided into three main categories:

1. Loading consideration
2. Framing considerations
3. Cross-section consideration

The loading considerations can be further classified into two parts, the considerations based on type of loading and the ones based on number of loads. The bridge piers are generally subjected to several hundred or even several thousand load combinations or loading sets. The large number of loading sets can originate because of moving load analysis and or due to dynamic and nonlinear analysis. The large number of load sets can be however reduced somewhat by using the concept of "corresponding actions." The bridge piers are often subjected to well-defined uniaxial or nearly uniaxial moment directions due to use of special bridge framing and bearing conditions. For example, for bridge deck supported continuously on several columns or piers, most of the intermediate piers are not connected to the deck and hence are not subjected to moments in that direction. However, in the transverse direction, these piers act as cantilevers to resist the wind and the seismic forces, causing large moments. Due to this well-defined moment direction, the pier cross-section can be optimized both for shape and reinforcement distribution.

Designing Shear Walls as Columns

The shear walls can sometimes be designed as column cross-sections. The main consideration is the validity of the basic assumptions made in the derivation of capacity or stress resultant equations. One of the main assumptions is that the plane sections remains plane after deformation, or in other words the assumption of linear strain distribution across the entire cross-section. This assumption may not hold true in the case of shear walls, especially near the supports and across the openings. The strain distribution may also not remain linear in cellular walls, where the stiffening effect of the corners alters the strain. The effect of shear lag in large cellular walls also affects the linearity of the strain. The nonlinearity of the strain alters the stress destitution and hence the value of stress resultants. In general, the strain tends to be higher near the edges and corners. This produces higher stresses at those locations which alters the lever arm of the stress resultants. The effect of nonlinear strain can be offset by appropriate distribution of reinforcement. Generally higher concentration of rebar at the corners and edges produces more efficient and somewhat consistent solution to the shear wall design.

An alternate and probably more elaborate method of designing the shear walls and other large panel structures or deep beams with openings is to use the results of a finite element analysis considering various zones within the wall or panel structure. The detail of this approach can be found in the documentation of ETABS (CSI 2016). A brief summary of this approach is discussed here. In this method, the in-plane stresses in wall or panel are obtained, and are integrated within certain width or depth zones. These zones are either marked as piers or as spandrels. The pier portions are designed as column sections whereas the spandrel portions are designed as beam sections. The integration of stresses gives three stress resultants, the axial force, the shear force, and the beading moment. Any conventional design procedure can be used to design the cross-sections for these actions. The cross-sections used for the design is the same as that used to obtain the stress resultants. The actions obtained from a linear-elastic analysis will however not include the effect of cracking on the distribution of stresses and actions. It may be noted, that even though the local stress distribution given by a linear-elastic finite element analysis is not consistent with the actual stress distribution in concrete walls or panels, the overall equilibrium of stresses and the actions is still valid if proper reinforcement has been designed based on the elastic analysis results.

Design of Piles

The design of piles is generally concerned with two aspects: the geotechnical aspects and the structural design aspects. The geotechnical aspects are related to the interaction between the pile and soil and govern the length and cross-sectional size based on the load transfer from pile to soil. The geotechnical design of the pile depends on many factors including:

1. The type of soil layers throughout the embedded length of the pile
2. The end bearing conditions and the soil or rock properties at the tip
3. The construction or driving method used for installing the pile. The behavior of driver piles and bored piles is very different
4. The shape of pile cross-section
5. The inclination of pile from the vertical alignment
6. The spacing of adjacent pile in a group
7. The presence of lateral load in the pile, and the presence of negative friction due to soil subsidence or settlement
8. The presence of upward force due to heave
9. The scour depth and erosion zone in river piles

The structural design aspects relate to the sizing and design of pile cross-sections to resist the actions imposed on the pile. In this aspect, the pile cross-section design is similar to the column cross-section design. There are however some special considerations, as follows:

1. *Pile cross-section*: The piles often have special sections that may not be encountered in building columns.
2. *Prestressing*: Generally the building columns are not posttensioned or prestressed. The precast piles are however often prestressed to resist handling moments and loads.
3. *Framing conditions*: The piles are often constrained or supported along their length, unlike the building columns or bridge piers that are

generally connected at ends only. This complicates the determination of effective length and calculation of slenderness effects.

4. *Large variation in moment*: The moment along the length of a pile may vary abruptly. In fact, for most of the length, there is no moment and there is a high concentration of moment near the anchorage zone in soil. This variation of moment raises special considerations in detailing.

CONCLUDING REMARKS

The effective, economic, esthetic, and safe design of column cross-sections is one of the fundamental features of overall structural design and detailing process. In this chapter, an overview and issues related to behavior, response and design of column cross-sections (with emphasis on reinforced concrete) are discussed. Practical guidelines related to proportioning and detailing of RC columns are also included. However, it should be noted that these guidelines are in no way alternative to detailed analysis for determining actual design demands, and are intended to provide general design information based on past experience. Another important point is that the overall design process as well as the code-based procedures are intended to follow ultimate strength design philosophy, which is based on provision of adequate strength against factored and magnified moments. However, one of the key parameter for safe design of columns, subjected to lateral dynamic loads, is the provision of ductility. Therefore the discussion about column design may not be considered complete without addressing this key issue. The behavior of ductile sections, factors affecting their performance and ways to provide adequate confinement to reinforced concrete columns will be discussed in detail in Chapter 6, Ductility of Cross-Sections.

SOLVED EXAMPLES

Problem 1: Consider a column of 500 mm × 800 mm as shown in Fig. Ex. 5.1. The clear height of the column is 10 m. The column is braced in $X-X$-direction and unbraced in $Y-Y$-direction. Determine whether the column will be designed as slender column of short column.

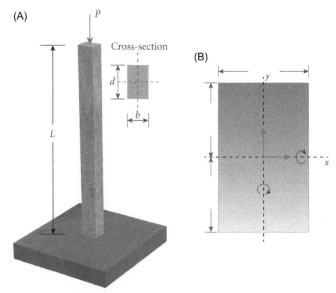

Figure Ex. 5.1 Slenderness check on a column. (A) An axial loaded column. (B) Cross-section of a column.

Assume the effective length factor $k_1 = 1.8$ for bending about x-axis and $k_2 = 1.8$ for bending about y-axis.

Given
$b = 500$ mm

$d = 800$ mm

$h = 10$ m

Solution
$l_{cx} = k_1 h$

$l_{cx} = 1.8 \times 10 = 18$ m

And

$l_{cx} = k_2 h$

$l_{cx} = 1.0 \times 10 = 10$ m

$\frac{l_{cx}}{h} = \frac{18}{0.8} = 22.5 > 10$

And

$\frac{l_{cy}}{b} = \frac{10}{0.5} = 20 > 15$

Result
Since $\frac{l_{cx}}{h} > 10$ for unbraced columns and $\frac{l_{cy}}{b} > 15$ for braced columns, the column should be designed as a slender column.

Problem 2: An aluminum alloy section $L\ 102 \times 76 \times 9.5$ is used as fixed end column. The length of column is 3 m. Determine maximum safe load for a column against buckling if factor of safety 2 is adopted. Use $E = 70$ GPa, $A = 1690$ mm^2, and $r_{min} = 16.4$ mm.

Given

$L = 3$ m

$F.O.S = 2$

$E = 70$ GPa

$A = 1690$ mm^2

$r_{min} = 16.4$ mm

Solution

For a fixed end, effective length is calculated as:

$L_e = 0.7\ L = 0.7 \times 3 = 2.1$ m $= 2100$ mm

$$\frac{L_e}{r_{min}} = \frac{2100}{16.4} = 128.09$$

Critical buckling load is calculated as:

$$P_{cr} = \frac{\pi^2 EA}{\left(L/r_{min}\right)}$$

$$P_{cr} = \frac{\pi^2 \times 70 \times 10^9 \times 1690 \times 10^{-6}}{128.048}$$

$P_{cr} = 71,209.65$ N $= 712.09$ kN

Now maximum safe load is calculated as:

$$P_{max} = \frac{P_{cr}}{F.O.S}$$

$P_{max} = \frac{712.09}{2} = 356.04$ kN

Result

Maximum safe load for a column against buckling is 356 kN.

Problem 3: Two channel sections $(C\ 229 \times 30)$ are shown in Fig. Ex. 5.2. The length of column is 10 m. Determine the buckling load under the following condition if $E = 200$ GPa.

1. If the sections act independently
2. If the sections are laced together at 125 mm

Given

$L = 10$ m

$E = 200$ GPa

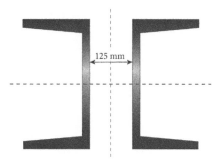

Figure Ex. 5.2 Cross-section of a column.

Solution

1. *When channels act independently*

   For channel section $C\ 229 \times 30$,

   $r_{min} = 16.4$ mm

   $A = 2845$ mm^2

   $\dfrac{L}{r_{min}} = \dfrac{10,000}{16.4} = 609.75$

   Hence the column is slender.

   $P_{cr} = \dfrac{\pi^2 EA}{\left(L/r_{min}\right)^2}$

   $P_{cr} = \dfrac{\pi^2 \times 200 \times 10^9 \times 2 \times 2845 \times 10^{-6}}{609.75^2}$

   $P_{cr} = 30,209.14$ N $= 30.209$ kN

2. *When channels are laced at 125 mm*

   For channel section $C\ 229 \times 30$,

   $I_{xc} = 21.2 \times 10^6$ mm^4

   $I_{yc} = 0.803 \times 10^6$ mm^4

   $I_x = 2I_{xc} = 2 \times 21.2 \times 10^6 = 42.4 \times 10^6$ mm^4

   $r_x = \sqrt{\dfrac{I_x}{A}} = \sqrt{\dfrac{21.2 \times 10^6}{2 \times 2845}} = 86.32$ mm

   $\dfrac{L}{r_{min}} = \dfrac{10,000}{16.3} = 613.4$

   $I_y = 2I_{yc} + Ad^2$

   $I_y = 2 \times 0.803 \times 10^6 + 2845 \times (75 + 14.9)^2$

   $I_y = 24.5 \times 10^6$ mm^4

   $r_y = \sqrt{\dfrac{I_y}{A}} = \sqrt{\dfrac{24.5 \times 10^6}{2 \times 2845}} = 65.61$ mm

   r_{min} is the lowest value of r_x and r_y; hence,

   $r_{min} = 65.61$ mm

$$\frac{L}{r_{min}} = \frac{10,000}{65.61} = 152.42$$

$$P_{cr} = \frac{\pi^2 EA}{\left(L/r_{min}\right)^2}$$

$$P_{cr} = \frac{\pi^2 \times 200 \times 10^9 \times 2 \times 2845 \times 10^{-6}}{152.42^2}$$

$$P_{cr} = 483,605 \text{ N} = 483.605 \text{ kN}$$

Result
1. Buckling load, when section acts independently is 30.209 kN.
2. Buckling load, when sections are laced at 125 mm is 483.605 kN.

Problem 4: Consider a short column of 500 mm × 500 mm as shown in Fig. Ex. 5.3. Compressive strength of concrete is 40 MPa and yield strength of reinforcement bars is 415 MPa. Design necessary reinforcements provided direct load in tension is 225 kN, ultimate bending moment is 225 kN−m, and ultimate shear force is 225 kN. Assume cover of 40 mm.

Given
$M_x = 225$ kN−m
$\quad P_t = 225$ kN
$\quad b = 500$ mm
$\quad d = 500$ mm
$\quad Cover = 40$ mm

Solution
$M_d = M_x - P_t\left(\frac{h}{2} - cover\right)$
$\quad M_d = 225 - 225 \times \left(\frac{0.5}{2} - 0.04\right)$

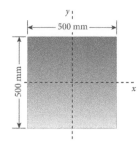

Figure Ex. 5.3 Cross-section of a column.

$M_d = 225 - 55.35 = 169.65$ kN$-$m

$k = 0.04 < 0.156$

Hence, no compressive reinforcement is required.

$$Z = d\left(0.5 + \sqrt{0.25 - \tfrac{k}{0.9}}\right)$$

$$Z = 460 \times \left(0.5 + \sqrt{0.25 - \tfrac{0.04}{0.9}}\right) = 440.79 \text{ mm}$$

$$A_s = \frac{M_d}{0.87 f_y z} + \frac{P}{0.87 f_y}$$

$$A_s = \frac{169.65 \times 10^6}{0.87 \times 415 \times 440.79} + \frac{225 \times 10^3}{0.87 \times 415}$$

$$A_s = 1689.17 \text{ mm}^2$$

Result

Use three bars of 29 mm diameter, having a total area of 1935 mm^2 on each side.

Problem 5: Consider a column of 500 mm $\times$ 700 mm as shown in Fig. Ex. 5.4. Compressive strength of concrete is 40 MPa. Clear height of column is 10 m. Direct load on column is 2600 kN. Bending moment in X- and Y-directions is 175 and 100 kN$-$m, respectively. Shear force in X- and Y-directions is 175 and 100 kN, respectively. Determine whether shear and crack width check are required for the column or not. Assume cover $= 40$ mm.

Given

$b = 500$ mm

$h = 700$ mm

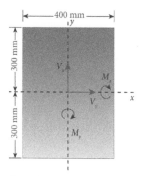

Figure Ex. 5.4 Cross-section of a column.

$$V_x = 175 \text{ kN}$$
$$V_y = 100 \text{ kN}$$
$$M_x = 175 \text{ kN−m}$$
$$M_y = 100 \text{ kN−m}$$
$$Clear\ height = 10 \text{ m}$$
$$f_c' = 40 \text{ MPa}$$
$$P = 2600 \text{ kN}$$

Solution

$$b' = b - cover = 500 - 40 = 460 \text{ mm}$$
$$h' = h - cover = 700 - 40 = 760 \text{ m}m$$

Shear Check

$$\frac{M_x}{P} = \frac{176}{2600} = 0.067$$
$$0.6h = 0.6 \times 700 = 0.42$$
$$\frac{M_x}{P} < 0.6h$$

And

$$\frac{M_y}{P} = \frac{100}{2600} = 0.038$$
$$0.6b = 0.6 \times 500 = 0.3$$
$$\frac{M_y}{P} < 0.6b$$
$$\frac{V_x}{bh'} = \frac{175 \times 10^3}{500 \times 660} = 0.53 \text{ N/mm}^2$$
$$\frac{V_y}{hb'} = \frac{100 \times 10^3}{700 \times 460} = 0.31 \text{ N/mm}^2$$
$$0.8\sqrt{f_c'} = 0.8 \times 40 = 5.05$$

Hence

$$\frac{V_x}{bh'} < 0.8\sqrt{f_c'}$$

Also

$$\frac{V_y}{hb'} < 0.8\sqrt{f_c'}$$

Crack Width Check

$$A_c = 0.95bh = 0.95 \times 500 \times 700 = 332,500 \text{ mm}^2$$
$$0.2f_c' A_c = 0.2 \times 40 \times 332,500 = 2,660,000 \text{ N} = 2660 \text{ kN}$$
$$0.2f_c' A_c < P$$

Result

1. From the above calculations of shear check, it can be concluded that no shear check is required for the given cross-section.
2. From the calculation of crack width check, it is concluded that crack width check is required for the given cross-section.

Problem 6: Consider a column of 300 mm × 375 mm as shown in Fig. Ex. 5.5. Compressive strength of concrete is 40 MPa whereas yield strength of reinforcement bars is 345 MPa. The column is subjected to load eccentricity of 250 mm. Area of steel is 1935 mm². Calculate safe normal load and nominal moment strength of column.

Given

$f_c' = 40$ MPa
$f_y = 345$ MPa
$e = 300$ mm
$b = 300$ mm
$h = 375$ mm
$A_s = 1935$ mm²
$E_s = 2 \times 10^5$ MPa
$d' = 50$ mm
$\beta_1 = 0.85$

Solution

Trial No. 1

Using trial and adjustment procedure, assume $c = 175$ mm
$a = \beta_1 c = 0.85 \times 175 = 148.75$ mm

Figure Ex. 5.5 Cross-section of a column.

$$f_s = 0.003E_s\left(\frac{c-d'}{c}\right)$$

$f_s = 0.003 \times 2 \times 10^5 \left(\frac{175-50}{175}\right) = 428.5$ MPa

$C_c = 0.85f'_c ab = 0.85 \times 40 \times 148.75 \times 300 = 1{,}517{,}250$ N

$C_s = A_s f_s = 1935 \times 428.5 = 829{,}147.5$ N

$T_s = A_s f_y = 1935 \times 345 = 667{,}575$ N

$P_n = C_c + C_s + T_s = 1{,}678{,}960.714$

$\bar{y} = \frac{h}{2} = \frac{375}{2} = 187.5$ mm

$d = 375 - 50 = 325$ mm

$M_n = C_c\left(\bar{y} - \frac{a}{2}\right) + C_s\left(\bar{y} - d'\right) + T_s(d - \bar{y})$

$M_n = 1{,}338{,}750 \times (187.5 - 74.375) + 667{,}575 \times$
$(187.5 - 50) + 1{,}678{,}960.714 \times (325 - 187.5)$

$M_n = 377{,}457{,}254.5$ N–m

$e = \dfrac{M_n}{P_n} = \dfrac{377{,}457{,}254.5}{1{,}678{,}960.714} = 224.81$ mm

Trial No. 2

Using trial and adjustment procedure, assume $c = 155$ mm

$a = \beta_1 c = 0.85 \times 155 = 131.75$ mm

$f_s = 0.003E_s\left(\frac{c-d'}{c}\right)$

$f_s = 0.003 \times 2 \times 10^5 \left(\dfrac{155-50}{155}\right) = 406.45$ MPa

$C_c = 0.85f'_c ab = 0.85 \times 40 \times 131.75 \times 300 = 1{,}343{,}850$ N

$C_s = A_s f_s = 1935 \times 406.45 = 786{,}483.871$ N

$T_s = A_s f_y = 1935 \times 345 = 667{,}575$ N

$P_n = C_c + C_s + T_s = 1{,}462{,}758.871$ N

$\bar{y} = \dfrac{h}{2} = \dfrac{375}{2} = 187.5$ mm

$d = 375 - 50 = 325$ mm

$M_n = C_c\left(\bar{y} - \dfrac{a}{2}\right) + C_s\left(\bar{y} - d'\right) + T_s(d - \bar{y})$

$M_n = 1{,}338{,}750 \times (187.5 - 65.875) + 667{,}575 \times ($
$187.5 - 50) + 1{,}678{,}960.714 \times (325 - 187.5)$

$M_n = 363{,}378{,}851$ N–m

$e = \dfrac{M_n}{P_n} = \dfrac{363{,}378{,}851}{1{,}462{,}758.871} = 248.42$ mm

Result

1. Nominal load capacity of column = 1462.758 kN.
2. Nominal moment capacity of column = 363,378.851 kN–m.

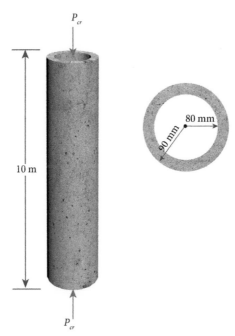

Figure Ex. 5.6 Cross-section of a column.

Problem 7: Consider a circular column having outer diameter of 90 mm and inner diameter of 80 mm. Length of the column is 10 m. Column is used as pin ended. Determine maximum allowable axial load that column can resist. Yield stress is 300 MPa. Also generate curve between slenderness ratio and critical stress for the given column cross-section. Use $E = 200$ GPa (see Fig. Ex. 5.6).

Given

$d_o = 900$ mm $= 0.09$ m

$d_i = 800$ mm $= 0.08$ m

$\sigma_y = 200$ MPa

$L = 10$ m

Solution

Critical Buckling Load

$A = \frac{\pi}{4}\left(d_o^2 - d_i^2\right) = \frac{\pi}{4} \times \left(0.09^2 - 0.08^2\right) = 0.00133 \text{ m}^2$

$I = \frac{\pi}{4}\left(d_o^4 - d_i^4\right) = \frac{\pi}{4} \times \left(0.09^4 - 0.08^4\right) = 1.93 \times 10^{-5} \text{ m}^4$

Now critical buckling load is calculated as:

$P_{cr} = \dfrac{\pi^2 EI}{L^2} = \dfrac{\pi^2 \times 200 \times 10^6 \times 1.93 \times 10^{-5}}{100} = 382.145 \text{ kN}$

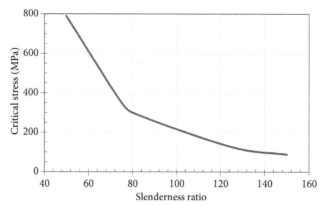

Figure Ex. 5.7 A curve between slenderness ratio and critical stress.

Now critical stress is calculated as:

$$\sigma_{cr} = \frac{P_{cr}}{A} = \frac{382.145}{0.00133 \times 1000} = 286.183 \text{ MPa}$$

Critical Stress and Slenderness Ratio Curve

$$r = \sqrt{\frac{I}{A}} = \sqrt{\frac{1.93 \times 10^{-5}}{0.00133}} = 0.1204 \text{ m}$$

$$\frac{kL}{r} = \frac{1 \times 10}{0.1204} = 83.05$$

This is the slenderness ratio at which critical buckling load is 382.145. Using the following equation and by varying value of slenderness ratio, we can plot a curve.

$$\sigma_{cr} = \frac{\pi^2 E}{\left(kL/r\right)^2}$$

Result

Critical buckling load for a given column cross-section is 382.15 kN.

The curve between slenderness ratio and critical stress is shown in Fig. Ex. 5.7.

Problem 8: W 150 × 24 column is shown in Fig. Ex. 5.8 is 10 m long and is fixed along X-axis. The load capacity is enhanced by bracing about Y-axis using struts which are pinned connected. Determine maximum load that column can resist without buckling. Use $A = 3060 \text{ mm}^2$ $I_x = 13.$ $4 \times 10^6 \text{ mm}^4$, $I_y = 1.83 \times 10^6 \text{ mm}^4$, $E = 200 \text{ GPa}$, and $\sigma_y = 300 \text{ MPa}$.

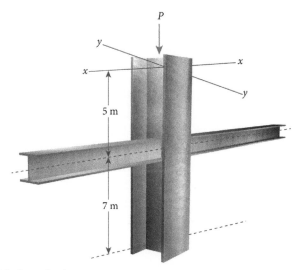

Figure Ex. 5.8 A steel column.

Given

$L = 10$ m
$\quad A = 3060$ mm^2
$\quad I_x = 13.4 \times 10^6$ mm^4
$\quad I_y = 1.83 \times 10^6$ mm^4
$\quad E = 200$ GPa
$\quad \sigma_y = 300$ MPa

Solution

Yield Stress
$$\sigma_y = \sigma_y A = \frac{300 \times 10^6 \times 3060 \times 10^{-6}}{1000} = 918 \text{ kN}$$
X–X Axis

The conditions for this axis are fixed bottom and top; hence,
$k_x = 0.5$
$r_x = 66.2$ mm
$$(P_{cr})_x = \frac{\pi^2 E I_x}{(k_x L)^2} = \frac{\pi^2 \times 200 \times 10^6 \times 13.4 \times 10^{-6}}{(0.7 \times 10)^2} = 1056.94 \text{ kN}$$
Y–Y Axis

The conditions for this axis are fixed bottom and pinned at top; hence,
$k_y = 0.7$
$r_y = 24.5$ mm

$$(P_{cr})_y = \frac{\pi^2 EI_y}{(k_y L)^2} = \frac{\pi^2 \times 200 \times 10^6 \times 1.83 \times 10^{-6}}{(0.7 \times 7)^2} = 150.448 \text{ kN}$$

Result

1. Maximum load along X-axis = 1056.94 kN.
2. Maximum load along Y-axis = 150.448 kN.

Problem 9: Column cross-section is shown in Fig. Ex. 5.9. Compressive strength of concrete is 40 MPa whereas yield strength of steel is 400 MPa. 10–25 mm bars are used as longitudinal reinforcements. Height of the column is 5 m. Use CSiCOL to model and determine capacity ratio provided $P_u = 1000$ kN, $M_{ux} = 200$ kN at the bottom, $M_{ux} = 120$ kN at the top, $M_{uy} = 250$ kN at the bottom, and $M_{uy} = 120$ kN at the top. Also generate MC curve, PM interaction curve, and stress view of the given column.

Given

$f_c' = 40$ MPa
$\quad f_y = 400$ MPa
$\quad P_u = 1000$ kN
$\quad M_{ux} = 200$ kN (at bottom)
$\quad M_{ux} = 120$ kN (at top)
$\quad M_{uy} = 250$ kN (at bottom)
$\quad M_{uy} = 120$ kN (at top)

Solution

The given column section will be modeled and analyze in the CSiCOL, commercial software for the analysis and design of column cross-sections. The step by step of procedure of column designing is illustrated below.

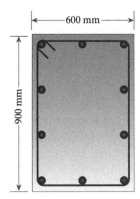

Figure Ex. 5.9 Given column section.

Material and Cross-Section Properties

The first step is to define material properties and section dimension in column design wizard. Fig. Ex. 5.10 shows window of material and cross-section properties in CSiCOL.

Column Framing and Effective Length Calculations

After defining material and cross-section properties, the next step is column facing and effective length calculation. CSiCOL 9 provides multiple options in terms of column framing and automatically calculates *K*-factor for braced and unbraced section. Fig. Ex. 5.11 shows windows of column framing conditions and effective length calculator in CSiCOL.

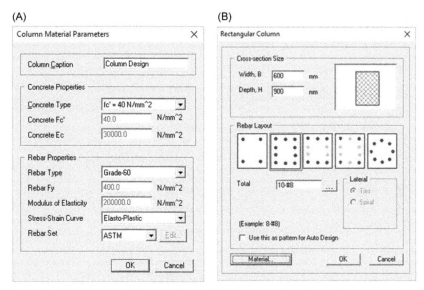

Figure Ex. 5.10 (A) Material properties and (B) cross-section dimensions.

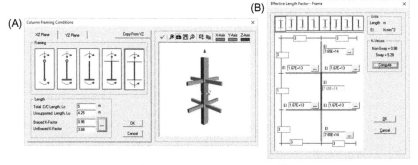

Figure Ex. 5.11 (A) Column framing conditions and (B) effective length calculator.

Loading Conditions

The next step is to assign loading conditions to the column that are given in the statement. Fig. Ex. 5.12 shows windows of loading conditions along *X*- and *Y*-axes in CSiCOL.

Calculation of Capacity Ratio

After running the design, CSiCOL gives us the capacity ratio which is defined as ratio of demand over the capacity of the cross-section. If this ratio is less than 1, then design is OK else the design needs to be revised. In our case, the capacity ratio of 0.56 which means that 56% of cross-section of the cross-section is being utilized under the given loading conditions. Fig. Ex. 5.13 shows window of capacity ratio results in CSiCOL.

Moment—Curvature Curve

Fig. Ex. 5.14 shows Moment-Curvature curve generated using CSiCOL.

PM Interaction Curve

Fig. Ex. 5.15 shows *PM* interaction curve generated using CSiCOL.

Stress View

Fig. Ex. 5.16 shows stress view generated using CSiCOL.

Figure Ex. 5.12 Loading conditions.

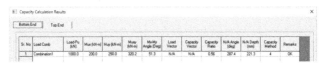

Figure Ex. 5.13 Capacity ratio of given column cross-section.

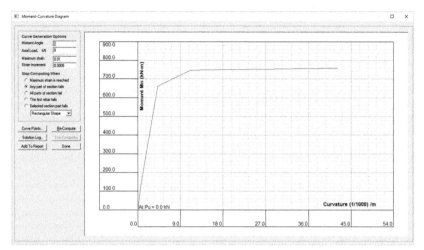

Figure Ex. 5.14 *MC* curve.

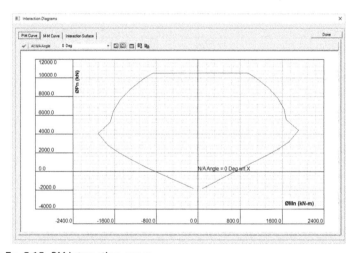

Figure Ex. 5.15 *PM* interaction curve.

UNSOLVED EXAMPLES

Problem 10: Consider a circular column having outer diameter of 100 mm and inner diameter of 85 mm. Length of the column is 8 m. Column is used as pin ended. Determine maximum allowable axial load, column can resist. Yield stress is 350 MPa. Also generate a curve between slenderness ratio and critical stress for the given column cross–section. Use $E = 200$ GPa.

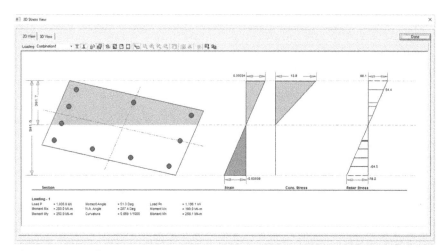

Figure Ex. 5.16 Stress view.

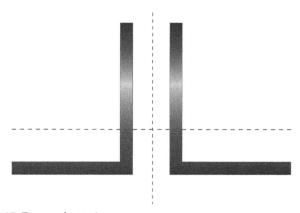

Figure Ex. 5.17 Two angle sections.

Problem 11: Two angle sections $(60 \times 40 \times 6)$ are shown in Fig. Ex. 5.17. The length of column is 8 m. Determine the buckling load under the following condition if $E = 200$ GPa.

1. If the sections act independently.

2. If the sections are laced together at 150 mm.

Problem 12: Consider a column of 400 mm $\times$ 575 mm cross-section. Compressive strength of concrete is 30 MPa whereas yield strength of reinforcement bars is 415 MPa. The column is subjected to load eccentricity of 275 mm. Area of steel is 2457 mm^2. Calculate safe normal load and nominal moment strength of column.

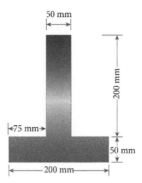

Figure Ex. 5.18 A column section.

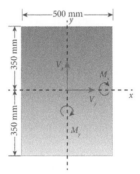

Figure Ex. 5.19 A rectangular column section.

Problem 13: A column cross-section is shown in Fig. Ex. 5.18. The length of the column is 5 m. The column comprised of two concrete blocks that acts together as unit. Determine the following:

1. Slenderness ratio
2. Buckling load
3. Axial stress in the column against buckling load

Problem 14: Consider a column of 500 mm × 700 mm as shown in Fig. Ex. 5.19. Compressive strength of concrete is 40 MPa. Clear height of column is 10 m. Direct load on column is 2500 kN. Bending moment in X- and Y-directions is 150 and 125 kN−m, respectively. Shear force in X- and Y-directions is 150 and 125 kN, respectively. Determine whether shear and crack width check are required for the column or not. Assume cover = 30 mm.

Problem 15: Consider a column of 400 mm × 600 mm as shown in Fig. Ex. 5.20. Compressive strength of concrete is 30 MPa and yield strength

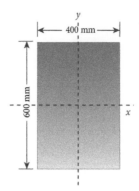

Figure Ex. 5.20 A rectangular column cross-section.

of reinforcement bars is 345 MPa. Design necessary reinforcements provided direct load in tension is 200 kN, ultimate bending moment is 175 kN−m, and ultimate shear force is 200 kN. Assume cover of 50 mm.

Problem 16: An 8 m high steel column is made of UC 203 × 203 × 60 steel section. The column is supported along *x*-axis and is fixed at both ends. Evaluate the buckling load. Use standard section properties.

REFERENCES

ACI 318. (2014). *Building code requirements for structural concrete (ACI 318-14) and commentary.* American Concrete Institute (ACI), and International Organization for Standardization.

Leet, K. (1982). *Reinforced concrete design.* New York: McGraw-Hill Science, Engineering & Mathematics.

MacGregor, J. G., Wight, J. K., Teng, S., & Irawan, P. (1997). *Reinforced concrete: Mechanics and design* (Vol. 3) Upper Saddle River, NJ: Prentice Hall.

McCormac, J. C., & Brown, R. H. (2015). *Design of reinforced concrete.* New York: John Wiley & Sons.

Nilson, A. (1997). *Design of concrete structures* (12th ed.). New York: McGraw-Hill.

FURTHER READING

ACI 318 (1963). *Commentary on building code requirements for reinforced concrete (ACI 318-11).* Detroit, MI: American Concrete Institute.

Green, R. (1997). *Reinforced concrete column design. Trends in structural mechanics* (pp. 281−288). Netherlands: Springer.

MacGregor, J. G., Breen, J. E., & Pfrang, E. O. (1970). Design of slender concrete columns. *ACI Structural Journal, 67*(1), 6−28.

Maekawa, K., Okamura, H., & Pimanmas, A. (2003). *Non-linear mechanics of reinforced concrete.* Boca Raton, FL: CRC Press.

Park, R., & Paulay, T. (1975). *Reinforced concrete structures.* New York, NY: John Wiley & Sons.

Tomlinson, M., & Woodward, J. (2014). *Pile design and construction practice.* Boca Raton, FL: CRC Press.

CHAPTER SIX

Ductility of Cross-Sections

Ductility is the Key to good (seismic) performance of Structures, confinement is the Key for improving Ductility in Reinforced Concrete Members, and Moment–Curvature Relationship is the Key for computing Cross-Section and Member Ductility.

WHAT IS DUCTILITY?
Introduction

Ductility is the ability of a material, cross-section, member or structure to sustain large deformation without fracture/failure. For most practical cases, it is defined in terms of the ratio of maximum deformation to the deformation level corresponding to a yield point. This ratio is often referred to as ductility ratio, μ in Fig. 6.1. The deformations can be strains, rotations, curvature, or deflections. Strain-based definition of ductility is used at material level, while rotation- or curvature-based definition also includes the effect of shape, size, and stiffness of cross-section. All seismic design codes around the world recognize the importance of ductility as it plays a vital role in structural performance against earthquakes. The greater the ductility, the greater is the capacity of the member to undergo large deformations without losing strength capacity and absorbing energy and its dissipation. Well-detailed steel and reinforced concrete (RC) structures, fulfilling the ductility requirements of codes, are expected to undergo large plastic deformations with little decrease in strength.

Limitations of Strength-Based Design

The strength design based on the internal stress resultants discussed in Chapter 3, Axial–Flexural Response of Cross-Sections, and Chapter 4,

Structural Cross Sections
DOI: http://dx.doi.org/10.1016/B978-0-12-804443-8.00006-3

391

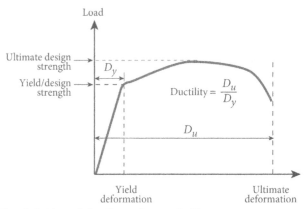

Figure 6.1 The definition of ductility ratio, $\mu = D_u/D_y$.

Response and Design for Shear and Torsion, primarily ensures that the members or rather cross-sections of the member are capable of resisting a certain value of actions based on assumed failure criterion. The actions are obtained often from linear elastic analysis, static or dynamic, and are factored to provide certain factor of safety. The strength design itself provides no information or control on the level of deformation produced at that factored load level. It also gives no information as to what will be the behavior of the member if loads or actions were to exceed the factored design load values. For example, in the design of reinforced concrete column sections, it is often assumed that concrete fails at a strain of 0.003–0.0035, so the design capacity is computed based on the assumption that the strain has reached this value and the corresponding stress resultants are computed and used as the section capacity. However, if the concrete is prevented from failing at 0.003 Strain level by improved detailing such as providing confinement, then the gain in strength due to this improvement is not reflected in conventional strength design procedures. Such explicit considerations of postyield characteristics and ductility resulted in latest performance-based design philosophies which basically look at the entire load versus deformation characteristics and push the strain levels on the cross-section until the materials in the entire section reach their maximum strain values. For example, in the case of reinforced concrete section, if significant amount of reinforcement is placed in the compression zone, then the maximum strain may well exceed the assumed 0.003 value without causing failure losing the capacity.

ACTION–DEFORMATION CURVES

The capacity surfaces and curves defined and developed in Chapter 3, Axial–Flexural Response of Cross-Sections, represent the interaction of stress resultants for a particular failure criterion such as maximum strain in concrete. However, during the deformation of an axial-flexural member, the maximum concrete strain varies as a function of the deformation. The deformation is related to the applied actions. The plot of relationship between action and deformation is often termed as Action–Deformation Curves.

The action–deformation curves provide very useful information about the overall behavior of the member. The entire response of structure or a member can be determined, in an integrated manner from these curves. The action–deformation curves can be obtained in different ways. The most common methods are:

1. By actual application of action on member and measurement of the corresponding deformation. This approach is used in laboratory testing, strength evaluation of existing structures, verification of theoretical models, and so on
2. By theoretical computations of stress resultants for assumed strain profiles corresponding to some deformation pattern
3. By the computation of deformations for various levels of action values using action–deformation relationships. This is useful in steel sections but is not very reliable for reinforced concrete members where the stiffness properties cannot be computed reliably for various levels of actions.

These relationships can be obtained at several levels as shown below:
- The Structural Level: Load–Deflection
- The Member Level: Moment–Rotation
- The Cross-Section Level: Moment–Curvature
- The Material Level: Stress–Strain.

Similar to stiffness, the overall ductility of a structure is derived from its members depending upon structural configuration, geometry, and member behavior. The ductility exhibited by an individual member is derived from cross-sectional response, which is ultimately dependent on its constituent material behavior. This concept is shown in Fig. 6.2.

As discussed in Chapter 2, Understanding Cross-Sections, the stress–strain curve is at the base of ductile behavior. The quantifications and

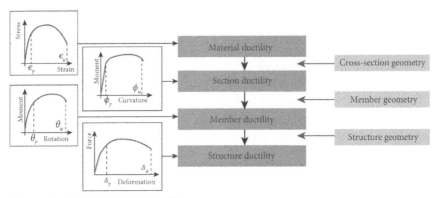

Figure 6.2 Various levels of ductility.

prediction of material ductility have been a topic of extensive research in last few decades. Various researchers have proposed a variety of theoretical and empirical relationships between stress and corresponding strain (often referred to as stress—strain models) for common engineering materials. Some common idealized forms of stress—strain behavior for concrete and steel were discussed in Chapter 2, Understanding Cross-Sections. In this chapter, specific expressions and models for reinforced concrete proposed by some important references will be focused. As an example, Fig. 6.3 shows four stress—strain models and their expressions. Depending upon the consideration of parameters, empirical or theoretical assumptions, and other controlling factors, the prediction from various stress—strain models can vary significantly. Therefore, it requires a careful investigation and understanding of all considered variables while selecting an adequate stress—strain relationship to model nonlinear behavior at material level.

Force—deformation relations at almost all levels exhibit some similar characteristics and can be represented as general force—deformation (F—D) relationship. The key features of a general F—D relationships will be discussed next.

General F—D Relationship

It is well known that a structural member when loaded or subjected to actions undergoes deformations. The force—deformation curve exhibits the entire behavior of the member right from initial loads to final loads or initial deformations to final deformations. The actual shape of the force—deformation curve and relative locations of various points depends on many factors including cross-section geometry, material properties, slenderness considerations, type of deformation, and corresponding force

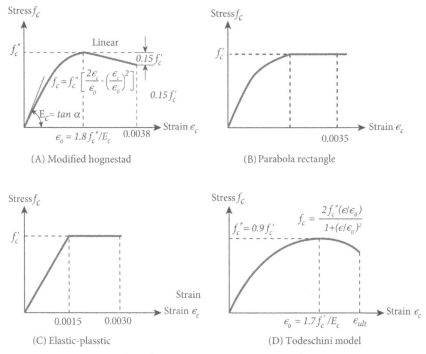

Figure 6.3 Some examples of stress–strain models for reinforced concrete. *Based on Wight, J. K., & MacGregor, J. G. (2011). Reinforced concrete: Mechanics and design, 5th Edition. Prentice Hall International, Inc.; Hognestad, E. (1951). A study of combined bending and axial load in reinforced concrete members.* Bulletin 399, University of Illinois Engineering Experiment Station, Urbana, IL; Todeschini, C. E., Bianchini, A. C., & Kesler, C. E. (1964). Behavior of concrete columns reinforced with high strength steels. ACI Journal Proceedings, 61(6).

being considered. A typical nonlinear force–deformation curve can be divided into at least three distinct parts or at least four distinct points as shown in Fig. 6.4. Broadly speaking, these points are:

1. Point "A" corresponds to the serviceability design considerations and working strength or allowable strength design concepts. Although the working design methods have not really been developed considering the load–deformation behaviors directly, it can indirectly be related to this linear, small deformation state.

2. Point "B" is the point up to which the relationship between load and deformation can be considered nearly linear and the deformations are relatively small. This is generally the start of nonlinearity and could indicate cracking of concrete or local buckling of parts of section, etc.

3. Point "C" roughly corresponds to the ultimate strength considerations or the design capacity consideration. It should be noted that ultimate

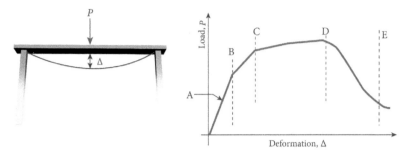

Figure 6.4 A general force—deformation relationship for reinforced concrete members. A, Serviceability range; B, initial cracking; C, yield point; D, maximum strength; E, residual strength.

strength design method is based on the material strength or material yielding criterion and is not based on the complete load versus deformation characteristics of members. This point "C" is often used as the start of a measure of ductility in many cases. At this point, the deformation starts to increase suddenly at more or less constant load value or with relatively small increase in the load.

4. Point "D" is the point at which the load value starts to drop with increasing deformations. It is a clear indication of capacities of deformation-based performance of the member but is rarely used in conventional design considerations due to a certain level of uncertainty near that point due to the failure of computational algorithms or the measurement of loads at sudden failure.

5. Point "E" is the point at which the load value is reduced to just a fraction of ultimate load (residual strength). This corresponds to complete failure or fracture of cross-section material. Further analysis or considerations will have no physical meaning, once this point is achieved.

Moment—Curvature ($M-\phi$) Curve

The load—deformation curves can be plotted between axial load and axial shortening, shear force and shear deformation, moment and curvature, and torsion and twist. Of all these curves, the moment—curvature relationship is probably the most important and useful action—deformation curve especially for flexural members such as beams, columns, and shear walls. It is also probably the least understood or utilized in normal design practice. Many of the design codes and design procedures or design handbooks do not provide sufficient information for the computation and use of $M-\phi$ relationships. This curve is dependent on several parameters including the cross-section stiffness (which itself is comprised on material

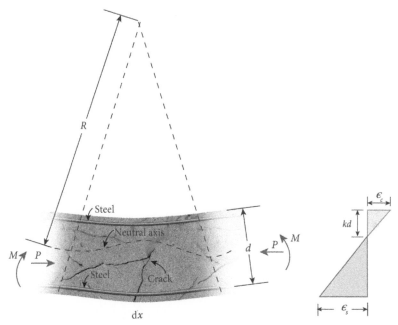

Figure 6.5 Definition of curvature for a reinforced concrete beam.

and geometric stiffnesses) and level of axial load on the member. The generalized stress resultant equations presented in Chapter 3, Axial–Flexural Response of Cross-Sections, can be used to obtain the actions corresponding to a certain deformation. For example, a certain strain profile on the cross-section corresponds to a particular curvature or axial shortening. The curvature can be defined in several contexts. In geometry, it is rate of change of rotation. In structural behavior, curvature is related to moment through stiffness. For a cross-section undergoing flexural deformation, it can be computed as the ratio of the strain to the depth of neutral axis. It is simply the inverse of the radius of curvature (R) that can be obtained from the slope of the strain profile (the ratio between the strain and the depth of neutral axis) and is measured in radians/length units. Fig. 6.5 shows the notations and definition of curvature for a reinforced concrete subjected to axial force P and moment M.

$$\frac{dx}{R} = \frac{\varepsilon_c dx}{kd} = \frac{\varepsilon_s dx}{d(1-k)} \tag{6.1}$$

$$\frac{1}{R} = \frac{\varepsilon_c}{kd} = \frac{\varepsilon_s}{d(1-k)} \tag{6.2}$$

$$\frac{1}{R} = \frac{\varepsilon_c}{kd} = \frac{\varepsilon_s}{d(1-k)} = \frac{\varepsilon_c + \varepsilon_s}{d} \qquad (6.3)$$

$$\text{Curvature } (\phi) = \frac{1}{R} = \frac{\varepsilon_c}{kd} = \frac{\varepsilon_s}{d(1-k)} = \frac{\varepsilon_c + \varepsilon_s}{d} \qquad (6.4)$$

There is no direct solution possible for reinforced concrete members and an iterative approach needs to be used for the determination of $M-\phi$ curve. The neutral axis depth is iteratively calculated for a fixed strain at the compression extreme for a given set of axial load. Varying the strain profile and computing the moment stress resultant corresponding to specified axial load gives various points on the curve. Once a moment—curvature set is obtained, the extreme fiber strain is changed and another solution is obtained to compute the curve (Fig. 6.6).

If the strain is varied to fairly large values (using a small increment), then the moment—curvature relationship can be used to indicate the cracking moment, yield moment, and ultimate or failure moment with corresponding deformation. The generation of moment—curvature curve can be terminated based on any number of specific conditions such as the maximum specified strain is reached, the first rebar reaches yield stress, or the concrete reaches a certain strain level. Also, during the generation of the moment—curvature curve, the failure or key response points can be recorded and displayed on the curve. This information is very useful for nonlinear analysis of a structure including the post-elastic behavior. These Moment—Curvature relationships are also the basis for the capacity-based or performance-based design methods. They are especially useful in the static pushover analysis as well as nonlinear time history analysis of structures, and in determining the rotational capacity (also a measure of

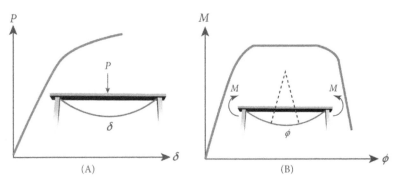

Figure 6.6 The load—deflection curve (A) and the moment—curvature curve (B) can be obtained either by physical testing of a member or by analytically solving the stress resultant equations for different strain profiles.

ductility) of plastic hinges formed during high seismic activity. From the $M-\phi$ curve, the ductility of the cross-section is defined by the ratio between the curvature at any given point to the curvature at yield of first rebar. In fact, several ductility ratios can be computed for various required or specified performance levels.

Moment—Rotation ($M-\theta$) Curve

The $M-\phi$ curve at the specified location can be integrated over the specified hinge length to obtain rotation from curvature resulting in the generation of moment—rotation ($M-\theta$) relationship. If a constant value of curvature is assumed over the hinge length, the simple multiplication of this length to the curvature scale would result in the moment—rotation curve. This curve is also useful in nonlinear analysis and modeling. A moment—rotation plastic hinge can also be used for nonlinear modeling of various members.

Axial Load—Deflection ($P-\delta$) Curve

The axial load—shortening curve can also be generated in a manner similar to that described for the generation of moment—curvature curve. However, in this case the iteration to determine the depth of neutral axis is not needed, as the neutral axis is assumed to be horizontal, in the absence at any moment. The strain is incremented and at each increment of strain, the corresponding axial load is determined using the appropriate material models.

Torque—Twist ($T-\theta$) Curve

Another important action—deformation relationship is the torque—twist ($T-\theta$) curve. It depicts the twisting response of structural members subjected to torsion. Similar to other action—deformation curves, various damage states or failure criteria (e.g., cracking, yielding, and crushing, etc.) can be indicated on this curve to get physical insight of member behavior. A typical torque—twist curve for reinforced concrete sections is shown in Fig. 6.7.

Fig. 6.8 shows the effect of total amount of reinforcement on postcracking torque—twist behavior of reinforced concrete beams. It can be seen that there is no effect of amount of reinforcement on cracking torque, which corresponds to the elastic torsional capacity of uncracked plain concrete. However, an increase in area of rebars is resulting in increased ultimate torsional capacity. More details can be found in various textbooks including Hsu (1992) and Wight and MacGregor (2011).

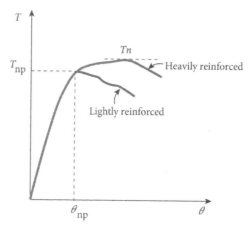

Figure 6.7 A typical torque–twist curve for reinforced concrete sections.

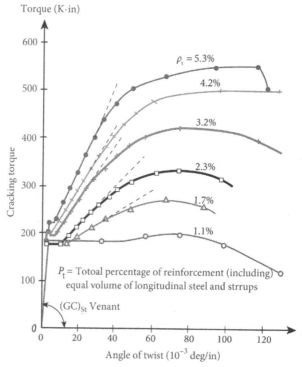

Figure 6.8 Torque–twist curves of reinforced concrete sections for various levels of reinforcement. *Based on Hsu, T.T.C. (1992). Unified theory of reinforced concrete. CRC Press, ISBN 9780849386138.*

SIGNIFICANCE OF THE MOMENT−CURVATURE (*M*−φ) CURVE

The information provided by the moment−curvature curve is very useful for nonlinear analysis of structures including the evaluation of post-elastic behavior. This relationship is also the basis for the capacity-based and performance-based design methods, and is especially useful in the analysis of structures using nonlinear static procedures as well as in determining the rotational capacity of plastic hinges formed during high seismic activity. It is interesting to note that a moment−curvature curves obtained from the stress resultants are independent of the member geometry or the bending moment diagram, and for a given axial load, they are a property of the cross-section, just like the capacity interaction surface.

Moment−Curvature (*M*−φ) Curve and Stiffness

The product of modulus of elasticity and moment of inertia (*EI*) is often used together in several expressions in structural analysis. This term generally called the cross-section stiffness can be determined in two ways. The first method is simply the product of the moment of inertia (*I*) and the elastic modulus or modulus of elasticity (*E*) based on section geometry and material properties. However, for reinforced concrete members, both the moment of inertia and its average modulus of elasticity depend on the amount of reinforcement, the extent of cracking and the level of strain; so the elastic stiffness determined using the product of basic *EI* is only valid in the linear elastic range and not useful in many applications.

Another way to indirectly determine the cross-section stiffness is from the slope of the *M*−φ curve. It is obvious that the stiffness of the cross-section depends on the level of axial load and bending moment. Therefore, a particular *M*−φ curve for a particular level of axial load and direction of bending can be used to determine the stiffness for a specified moment as shown in Fig. 6.9.

$$EI = \frac{M}{\phi} \tag{6.5}$$

It is important to note that the value of stiffness determined from slope of moment−curvature curve is the "effective stiffness." It includes the effect of various important phenomenona e.g., cracking and other material and section nonlinearities, the stiffness contribution due to

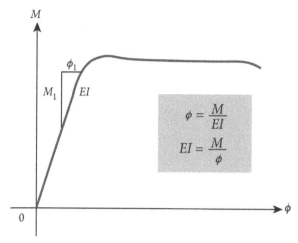

Figure 6.9 Determination of effective stiffness from the moment–curvature curve.

reinforcement, and the change in elastic modulus of concrete/steel due to a certain level of strain. Moreover, if the moment–curvature curve has been generated considering the short- and long-term effects, then the stiffness would automatically be adjusted for all those effects such as confinement, creep, shrinkage, and temperature change. The stiffness determined this way can be used directly in structural analysis such as stiffness matrix approach and finite element formulations. The stiffness can also be used to determine the deflections using the conventional deflection equations and formulae.

Important Outputs from the M–ϕ Curve

The moment–curvature curve is often termed as an open secret that is not utilized by structural engineers effectively in the routine design of structures, especially for earthquake effects and extreme events and even for the determination of several response quantities for normal loads. Following are some important outputs which can be obtained from the moment–curvature curves.

1. Cracking Point: This point corresponds to the onset of material cracking of a cross-section. It provides the moment and corresponding curvature for design considerations related to start of cracking.

2. Yield Point: This point corresponds to the onset of material yielding of a cross-section. It provides the moment capacity and corresponding curvature for strength design of section.

3. Failure Point: This point corresponds to the maximum curvature and defines the maximum deformation capacity of section.

4. Ductility: The ratio of ultimate curvature and yield curvature defines the section ductility.

$$\mu = \phi_u/\phi_y \tag{6.6}$$

5. Stiffness of the Section at Given M and ϕ: As mentioned in previous section, the slope of $M-\phi$ curve at any given point corresponds to the effective stiffness of the section, as shown below (Fig. 6.10).

$$\phi = \frac{M}{EI} \text{ and } EI = \frac{M}{\phi} \tag{6.7}$$

6. Slope of the Section at Given Moment: As discussed earlier, the $M-\phi$ curve can also be used to determine rotation at any point in a member, as follows.

$$\theta = \int_a^b \frac{M}{EI} dx \tag{6.8}$$

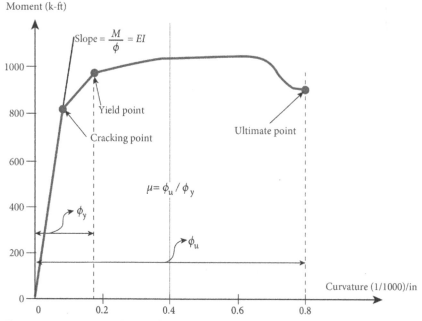

Figure 6.10 Curvature ductility from the $M-\phi$ curve (CSiCOL.).

7. **Deflection of the Section at Given Moment:** The deflection corresponding to a particular point in the $M-\phi$ curve is as follows.

$$\Delta = \int_a^b \left(\frac{M}{EI}\right) x \, dx \qquad (6.9)$$

8. **Strain at Given Moment:** Given the distance from the neutral axis to any point (c), the strain (ε) at that point can be determined as

$$\varepsilon = \phi c \qquad (6.10)$$

9. **Crack Width at Given Crack Spacing:** Using the strain value corresponding to a certain moment, the crack spacing in reinforced concrete beams can be determined as follows (Fig. 6.11):

$$W = \varepsilon_s X \qquad (6.11)$$

$$W = \phi_y X \qquad (6.12)$$

10. **Crack Spacing at Given Crack Width:** Alternatively, with known crack width, spacing between two consecutive cracks can be given as follows (Fig. 6.12).

$$X = W/\varepsilon_s \qquad (6.13)$$

$$X = W/\phi_y \qquad (6.14)$$

The procedure to determine the deflection of a point on a straight member using the $M-\phi$ curve will now be discussed. The deflection under bending in the direction perpendicular to the original straight axis of the member is equal to the moment of the M/EI diagram between those two points about the point where the deflection occurs. To

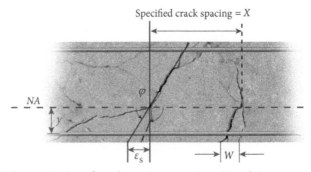

Figure 6.11 Determination of crack spacing at a given M and ϕ.

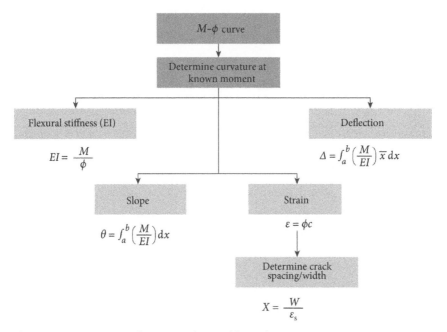

Figure 6.12 Important information obtained from the $M-\phi$ curve.

compute the deflections by using the Moment—Curvature curves, the following procedure can be used.

1. Design the cross-section for moment and axial load at various locations along the beam/column.
2. Generate the moment—curvature curves for the designed sections.
3. Plot the moment diagram and the axial load diagram. The axial load is generally constant along the length of the member so it can be used as the controlling variable while generating the Moment—Curvature curves.
4. For various locations along the member length, read the curvature for the applied moment using the appropriate moment—curvature curves. Use interpolation if the cross-section changes along the length.
5. Plot the M/EI diagram along the length of the member.
6. To compute the deflection at a specific location, calculate the area of this M/EI diagram up to that point, starting one end of the member.

It is important to note that the deflection determined this way includes the secondary effect of axial load on change is axial. The flexural stiffness of the moment—curvature curve has been computed for a specific axial load. Although the above procedure can be used manually with certain simplifications, it is best suited for computer applications where the generation of moment—curvature curves and the integration of M/EI diagram can be automatically accomplished.

Performance-Based Design of Cross-Sections

The performance-based design is basically a comparison between the capacity curve of the section, member or entire structure against the demand curve for the section, member or the structure as a whole. The capacity curve is a load—deformation curve and the demand curve or demand levels are expected deformation levels for certain loads or expected load capacity at specified deformation levels. A complete range of cross-sectional response quantities is required for its performance evaluation against anticipated loads. Fig. 6.13 shows an overview of cross-sectional responses required to ensure adequate design for both performance as well as strength.

As a simple example, let us consider a beam carrying a point load as shown in Fig. 6.14. The demand curves could simply be limited on deformation while carrying certain load. For example, it may be specified that the beam should not crack before reaching a load of 50 KN. The beam should not deflect more than say 20 mm for a load of 75 KN. The beam should not deflect more than 50 mm for a load of 100 KN. Now if

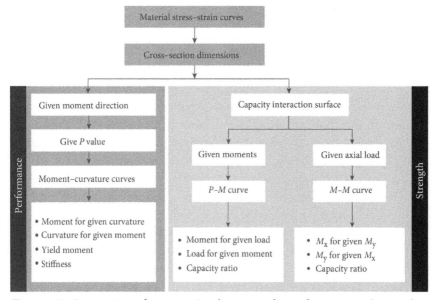

Figure 6.13 An overview of cross-sectional response for performance and strength.

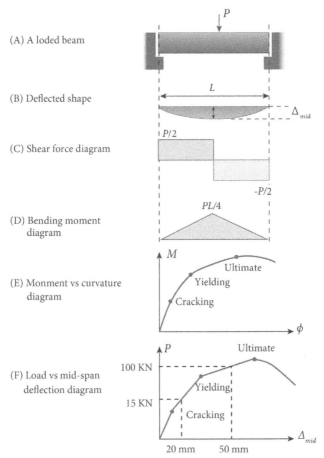

Figure 6.14 Performance evaluation of a simply supported point-loaded beam in terms of mid-span deflection.

a section for the beam is assumed, then a load—deformation curve can be generated by using a moment—curvature curve. These points can then be marked on this curve and the limits checked against the actual performance of the beam for carrying this load. If some of the demands are not satisfied, then the section can be revised and the load—deformation curve regenerated each time, until the desired performance levels are achieved. It is important to note that there is no need to specify a design load or ultimate load or load factors or capacity reduction factors. They are all automatically built into the expected demands and provided capacities. It covers everything from serviceability considerations such as cracking and service load deflection to maximum load-carrying capacity and ductility.

It is important to note that in this case the beam section is expected to meet all performance criteria. Simultaneously, it is also important to note that various modifications in the cross-section design will have different impact on performance levels. For example, if cracking criteria is not being met then section size may be increased, tension reinforcement may be increased, or concrete with greater modulus of rupture may be used. If the final ductility criterion is not being met, then concrete in compression may be confined or more compression reinforcement may be used.

A similar process can be done at structural level. The limiting failure criterion can be replaced by a set of detailed acceptance criteria for various damage states of complete structure. As the construction of tall buildings and complex structures is becoming more prevalent, the concern for the safety of the public from various natural and man-made hazards is becoming more relevant. So when clients and users of the buildings ask a structural engineer, an apparently simple question "Is my structure safe?", engineers are at odds to respond to this explicitly. Generally, the structural engineers follow the prescriptive provisions of the building and design codes, and probably the best answer they could offer is that "I am not sure, but I have designed this structure according to the building code!" Obviously, such a response may not be sufficient, or acceptable, but unless clearer and more refined design approaches and methodologies are used, this may be the only choice. The main reasons why an explicit answer is not possible are, first, the question itself is not well defined. It does not explain what a "safe design" is; and if it is safe against what effects, and what level of safety is required. It also does not allow the designer to carry out a clear-cut understanding and evaluation of the "strength" of the structure that is not bound by typical simplifications, arbitrary factors, and prescriptive limits. The development and recent advancements in the Performance-Based Design approach (PBD) provide an alternative, as well a progression to more explicit evaluation of the safety and reliability of structures for various hazards, specially earthquakes, and progressive collapse scenarios. PBD gives the opportunity to clearly define the levels of hazards to be designed against, with the corresponding performance to be achieved, and to evaluate the cost implications in the process. This essentially allows the clients, building owners, and team carrying out the PBD to evaluate the explicit risks at the site, consider the purpose and usage of the building, and set the design for appropriate performance levels, in line with international guidelines and practices.

The appropriate implementation of PBD process requires an in-depth understanding of ground-shaking hazards, structural materials behavior, and nonlinear dynamic structural response. These knowledge and skill sets are generally not imparted to structural engineers in a typical undergraduate civil engineering program. These are either acquired through a specialized master's degree programs or through extensive training and experience in PBD applications under the supervision of experts. Also, it is uncommon that a single person can possess all these abilities, thus performing PBD requires a specialized team of engineers to carry out the process. To increase reliability, the use of the PBD approach on a project often requires a peer review. This review is usually conducted by a third party which is also specialized in PBD. A thorough evaluation of the followed procedures, selection of hazard and performance criteria, and the final outcome is carried out. The willingness of building developers and clients to engage primary structural engineers, PBD specialists, and reviewers clearly indicates that the advantages of using PBD are becoming evident both in terms of enhanced performance and public safety and better cost-effectiveness. In a typical scenario, the client, architect, structural engineers using code-based design, specialized PBD consultants, and PBD peer reviewers jointly work to enhance the performance and reliability of the structure as shown in Fig. 6.15, while maintaining cost-effectiveness.

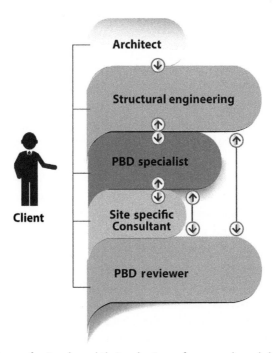

Figure 6.15 The professionals and their roles in performance-based design.

DUCTILITY AND CONFINEMENT OF REINFORCED CONCRETE SECTIONS

Ductility and Seismic Performance

The performance of columns is critical to the overall performance of structures during strong earthquake. It is well known that the performance of structures subjected to seismic forces does not depend so much on the design strength of the members as it depends on the ductility of the members and connections. The structures often fail due to the formation of hinges in primary components which usually form at the locations of high moments. The failure of a column can lead to the collapse of a complete structure or its major part, whereas the failure of a beam may lead to a local failure or excessive deflections. Most seismic design codes, including the ACI special provisions for seismic design, the SEAOC provisions, the UBC, etc., require that the columns should be stronger than the beams to control the mechanism. In a strong earthquake, the actions, especially the moments and shears in the primary lateral load resisting members exceed the design capacities. In such cases, the ductility is the primary source of additional performance.

Confinement of RC Sections

As discussed in Chapter 2, Understanding Cross-Sections, it is possible to improve a certain material property in a particular direction by altering a property in another direction. Imagine a small pile of sand which can barely support its own weight. If we put it in a closed container and now apply some load, it is almost impossible to compress it and will keep on carrying the applied load until the confining container breaks. It means that we can convert a weak material (in one direction) into a very strong one just by confining it with another suitable material. Same is the case with concrete also. If we wrap a concrete sample with a suitably strong material (e.g., carbon fiber), it is almost impossible for concrete sample to crush as it is not allowed to move out or expand in lateral direction (as governed by Poisson's effect shown in Fig. 6.16). If we confine the sample in a way that it is allowed to expand very slowly, we can control its ductility as well. For a material sample to crush, it must expand and the high ductility of confining material will be imparted to overall sample behavior. The only way for the whole wrapped sample to fail is when the confining material fails. In reinforced concrete members, concrete is confined using rectangular or circular steel

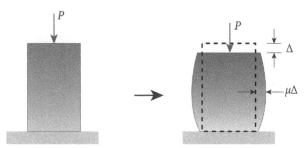

Figure 6.16 Poisson's effect for compressive force.

reinforcement hoops. So within one cross-section, we have concretes of two types, i.e., the confined concrete in the inner core and the cover concrete outside the core. Double confinement using multiple hoops is also quite common in bridges. The whole rebar cage of a member is very important in achieving optimum confinement. For reinforced concrete columns, usually a lot of attention is paid to vertical reinforcement and lateral reinforcement is not given much importance. In fact, most of the axial strength is contributed by the lateral reinforcement (perpendicular to the cross-section) due to confinement. Hence horizontal bars should be considered at least as important as axial reinforcement. A concrete with a ductility value of 1 (without confinement) can be improved to have a ductility of 10 with good confinement scheme. In short, confinement is the key to improve the ductility of a member which in turn is the key to improve the seismic performance of overall structure.

The above discussion follows that if the compression zone of a reinforced concrete member is confined by closely spaced transverse reinforcement in the form of stirrups, ties, hoops, or spirals, the ductility of the concrete may increase significantly. When compressive stress approaching the compressive strength of concrete, the transverse strains in the concrete increased rapidly and the concrete expands against the transverse reinforcement. The retaining pressure applied by the reinforcement to the concrete considerably improves the stress—strain behavior of the concrete at higher strain. Thus it helps to improve the ductility of the member. Studies have shown that the circular spirals confine the concrete more effectively than rectangular stirrups, ties or hoops because confining steel in the shape of circle applies a uniform radial pressure to the concrete, whereas a rectangle tends to confine the concrete mainly at the corner. The confining steel also acts as shear reinforcement, prevents buckling of longitudinal reinforcement, and prevents bond splitting

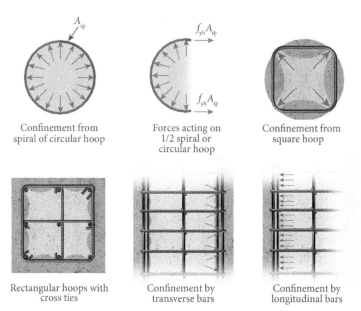

Figure 6.17 Various types and configurations of confinement of reinforced concrete members.

failures. Some of the common confinement configurations for circular and rectangular cross-sections are shown in Fig. 6.17.

Beside hoops and ties, spiral reinforcement is also one of the most efficient ways of providing confinement to reinforced concrete members. Fig. 6.18A shows the stresses developed in spiral rebars (f_{sp}) due to lateral expansion of concrete under axial loads. Fig. 6.18B shows the typical relationship between axial load and shortening (measured in terms of axial strain) for column members with ties and spiral reinforcement. It can be seen that spiral confinement results in a significant increase in ductility and ultimately improved performance.

Stress–Strain Models for Confined Concrete

During the last two decades, extensive experimental and analytical research is done in material modeling for concrete and steel. Different building codes have also provided guidelines about limits of yield and ultimate strains of both materials. BS 8110 recommends to use 0.0035 as ultimate strain of concrete while ACI 318 proposes 0.003. These models can be viewed as a set of guidelines to develop and predict an idealized behavior

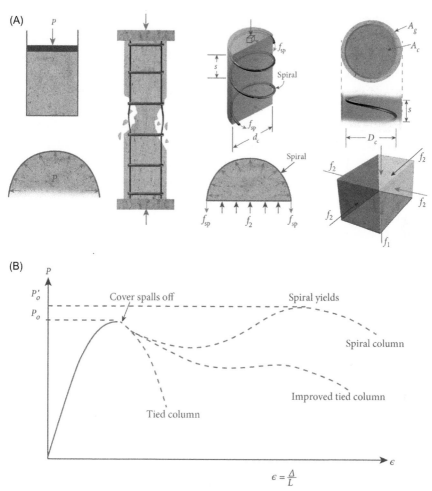

Figure 6.18 (A) Confinement provided by spiral reinforcement in RC members. (B) Comparison of axial force–deformation behaviors of reinforced concrete columns with various confinement configurations.

of material. Building codes on the other hand always use simplified formulations, expressions, and schemes. AC1 318 assumes a rectangular stress–strain behavior for concrete which is a rather crude idealization for convenience in calculations. However, some of the proposed models (e.g., Mander's Model, 1988, described in previous section) are widely accepted, gained popularity, and considered reasonably accurate among practicing engineers. Steel on the other hand is usually considered to be elastoplastic but in actual, it has the ability to exhibit additional strength after yielding. This strain hardening portion of steel's stress—strain curve (postyield part) can be utilized very intelligently to create many new types of steels having even more ductility and high strength. An application of this idea can be seen in TOR steel in which the steel bars are twisted in factories prior to use, making them deform under torsion. The yielding portion of stress—strain curve is removed by this process and the resulting "twisted steel" exhibits strain hardening (additional strength) directly after yield point. This TOR steel is very high-strength steel with no definite yield point. The ductility is also reduced as a necessary result which may prove fatal under severe ground motions. In fact, the use of TOR steel has already been disallowed in earthquake prone areas by various design guidelines and recommendations due to low ductility (nowadays mild steel is a general recommendation). Moreover, the use of high-strength steel (with yield strength >75 ksi) is also not recommended for most of the normal construction projects, owing to its low ductility compared to normal- and moderate-strength steels. If a reinforced concrete column member is tested in compression, to maintain the same strain rate, both concrete and steel bars have to deform together. Equal strains in both materials will produce unequal stresses because of their different stress—strain relationships. The overall member will keep on taking compression (without rebar buckling) until concrete is crushed. Similarly, if the sample is tested in tension, the situation is totally different. Concrete will soon crack and all tensile stress will be taken by vertical reinforcing steel. The overall member will keep on taking tension until yielding of vertical bars. This means both materials do not really behave independently. The behavior of overall member is far different as compared to individual behaviors of both steel and concrete. Engineers can exploit this interdependency to create economic and advanced composite materials exhibiting certain desirable properties.

Several researchers have proposed a variety of stress—stress models for confined concrete accounting for additional strength and ductility provided by confinement. One of the earliest models is the Kent and Park (1971) model which proposes not to use the extra strength coming from confinement but increased ductility (with increasing amount of confining steel) is considered. The one proposed by Mander, Priestley, and Park (1988) is widely accepted and used. A quick review of various parameters considered by some important stress—strain models and their expressions will be provided next.

Mander's Stress—Strain Model (1988)

This model has been developed for concrete confined by reinforcing bars (ties or spirals) of any shape but can be modified to be used for concrete-filled steel tubes and other sections with confined concrete. It is applicable for most shapes and sections, and is defined using the following equations:

$$f_c = \frac{f'_{cc} x r}{r - 1 + x^r} \tag{6.15a}$$

$$f'_{cc} = f'_c \left(2.254 \sqrt{1 + \frac{7.94 f'_1}{f'_c} - \frac{2 f'_1}{f'_c}} - 1.254 \right) \tag{6.15b}$$

$$x = \frac{\varepsilon_c}{\varepsilon_{cc}} \tag{6.15c}$$

$$\varepsilon_{cc} = 0.002 \left[1 + 5 \left(\frac{f'_{cc}}{f'_c} - 1 \right) \right] \tag{6.15d}$$

$$r = \frac{E_c}{E_c - E_{sec}} \tag{6.15e}$$

$$E_c = 60{,}000 \sqrt{f'_c} \ (\text{psi}) \ = \ 5000 \sqrt{f'_c} \ (\text{MPa}) \tag{6.15f}$$

$$E_{sec} = \frac{f'_{cc}}{\varepsilon_{cc}} \tag{6.15g}$$

The effective confining pressure or stress, f'_1 is expressed as

$$f'_1 = K_e f_1 \tag{6.15h}$$

where f_1 for circular sections can be determined using the condition of equilibrium as

$$f_1 = \frac{2 f_{yh} A_{sp}}{d_s s_h} \tag{6.15i}$$

$$K_e = \frac{A_e}{A_{cc}} \tag{6.15j}$$

ε_{cc} is the strain corresponding to the maximum compressive strength of confined concrete (f'_{cc}). E_c is the elastic modulus of concrete determined using Eq. (6.15f). K_e is also referred to as confinement coefficient (Fig. 6.19).

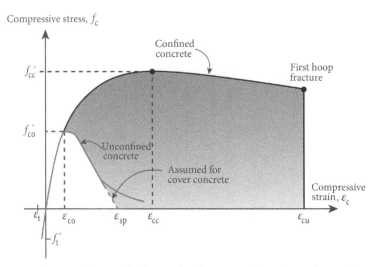

Compressive stress, f_c

Figure 6.19 The Mander's model for confined concrete (Mander et al., 1988).

Kent and Park Stress−Strain Model (1971)

The Kent and Park model is also one of the most widely used stress−strain behavior for confined concrete. It consists of two parts, that is, parabolic branch up to peak concrete stress and a linear falling tail up to a certain residual stress. Its governing expressions are as follows (Fig. 6.20):

$$f_c = f_c'[1 - Z(\varepsilon_c - \varepsilon_{co})] \tag{6.16a}$$

where

$$Z = \frac{0.5}{\varepsilon_{50h} + \varepsilon_{50u} - \varepsilon_o} \tag{6.16b}$$

and

$$\varepsilon_{50h} = \varepsilon_{50c} - \varepsilon_{50u} = \frac{3}{4}p''\sqrt{\frac{b''}{s}} \tag{6.16c}$$

$$\varepsilon_{50u} = \frac{3 + 0.29f_c'}{145f_c' - 1000} \,(f_c' \text{ is in MPa}) \tag{6.16d}$$

$$\varepsilon_{50u} = \frac{3 + 0.002f_c'}{f_c' - 1000} \,(f_c' \text{ is in psi}) \tag{6.16e}$$

$$p'' = \frac{2(b'' + d'')A_s''}{b''d''s} \tag{6.16f}$$

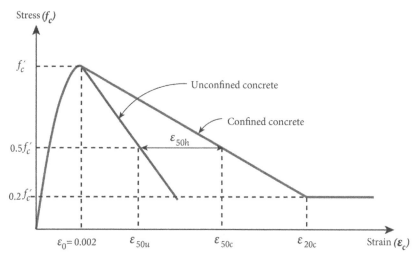

Figure 6.20 The Kent and Park (1971) model for confined and unconfined concretes.

The residual stress is $0.2f_c'$ corresponding to a strain of ε_{20c}.

f_c = Compressive strength of confined concrete

f_c' = Compressive strength of unconfined concrete

ε_{50u} = Strains corresponding to the stress equal to 50% of the maximum concrete strength for unconfined concrete

ε_{50c} = Strains corresponding to the stress equal to 50% of the maximum concrete strength for confined concrete

p'' = Volumetric ratio of confining hoops to volume of concrete core

b'' = Width of confined core

d'' = Depth of confined core

A_s'' = Cross-sectional area of hoop bar

s = Center-to-center spacing of hoops.

Scott et al. Stress–Strain Model (1982)

A modified version of the Kent and Park model is proposed by Scott et al. (1982), in which the maximum concrete stress attained is assumed to be Kf_c' and the strain at maximum concrete stress is 0.002K where "K" is defined in Eq. (6.18d). The governing equations are modified as follows (Fig. 6.21):

For $\varepsilon_c \leq 0.002K$

$$f_c = Kf_c' \left[\frac{2\varepsilon_c}{0.002K} - \left(\frac{\varepsilon_c}{0.002K} \right)^2 \right] \tag{6.18a}$$

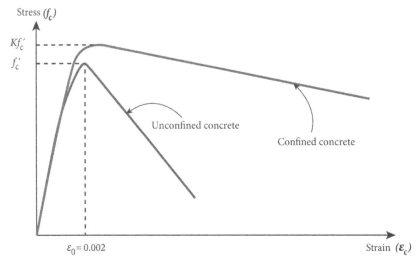

Figure 6.21 The Modified Kent and Park model for concrete confined by rectangular steel hoops proposed by Scott, Park, and Priestley (1982).

For $\varepsilon_c > 0.002K$

$$f_c = Kf_c'[1 - Z_m(\varepsilon_c - 0.002K)] \qquad (6.18b)$$

and

$$Z_m = \frac{0.5}{\dfrac{3 + 0.29f_c'}{145f_c' - 1000} + \dfrac{3}{4}\rho_s\sqrt{\dfrac{h''}{s_h}} - 0.002K} \text{(where } f_c' \text{ is in MPa)} \quad (6.18c)$$

$$K = 1 + \frac{\rho_s f_{yh}}{f_c'} \qquad (6.18d)$$

where
  f_{yh} = Yield strength of hoop reinforcement
  s_h = Center-to-center spacing of hoop sets
  ρ_s = Ratio of volume of rectangular steel hoops to volume of concrete core measured outside of the peripheral hoop
  h'' = Width of concrete core measured to the outside of peripheral hoop.

For high strain rates, the modified Kent and Park model can be used by multiplying the peak stress, the strain at the peak stress, and the slope of the falling branch, with 1.25. Thus for high strain rates the same expressions for f_c can be used, however, the values of K' and Z'_m are given by

$$K = 1.25 \left[1 + \frac{p'' f_{yh}}{f'_c} \right] \tag{6.18e}$$

$$Z_m = \frac{0.625}{\varepsilon_{50h} + \varepsilon_{50u} - 0.002K} \tag{6.18f}$$

Yong et al. Stress–Strain Model (1989)

Yong et al. (1989) proposed the following polynomial expressions for stress–strain relation of rectilinear confined high-strength concrete.

If $\varepsilon_c \leq \varepsilon_{cc}$

$$\frac{f_c}{f_{cc}} = \frac{A \left(\dfrac{\varepsilon_c}{\varepsilon_{cc}} \right) + B \left(\dfrac{\varepsilon_c}{\varepsilon_{cc}} \right)^2}{1 + (A - 2) \left(\dfrac{\varepsilon_c}{\varepsilon_{cc}} \right) + (B + 1) \left(\dfrac{\varepsilon_c}{\varepsilon_{cc}} \right)^2} \tag{6.19a}$$

If $\varepsilon_c > \varepsilon_{cc}$

$$\frac{f_c}{f_{cc}} = \frac{C \left(\dfrac{\varepsilon_c}{\varepsilon_{cc}} \right) + D \left(\dfrac{\varepsilon_c}{\varepsilon_{cc}} \right)^2}{1 + (C - 2) \left(\dfrac{\varepsilon_c}{\varepsilon_{cc}} \right) + (D + 1) \left(\dfrac{\varepsilon_c}{\varepsilon_{cc}} \right)^2} \tag{6.19b}$$

where

$$A = E_c \frac{\varepsilon_{cc}}{f'_{cc}} \tag{6.19c}$$

$$B = \frac{(A - 1)^2}{0.55} - 1 \tag{6.19d}$$

$$E_c = 27.55 w^{1.5} \sqrt{f'_c} \tag{6.19e}$$

$$C = \frac{(\varepsilon_{2i} - \varepsilon_i)}{\varepsilon_{co}} \left[\frac{\varepsilon_{2i} E_i}{f'_{cc} - f_i} - \frac{4\varepsilon_i E_{2i}}{f'_{cc} - f_{2i}} \right] \tag{6.19f}$$

$$D = (\varepsilon_i - \varepsilon_{2i}) \left[\frac{E_i}{f'_{cc} - f_i} - \frac{4 E_{2i}}{f'_{cc} - f_{2i}} \right] \tag{6.19g}$$

$$E_i = \frac{f_i}{\varepsilon_i} \tag{6.19h}$$

$$E_{2i} = \frac{f_{2i}}{\varepsilon_{2i}} \qquad (6.19i)$$

$$f_{cc}' = kf_c' = 0 + 0.0091\left(1 - \frac{0.245s}{h''}\right)\left(\rho'' + \frac{nd''}{8sd}\rho\right)\frac{f_y''}{\sqrt{f_c'}} \qquad (6.19j)$$

$$\varepsilon_{co} = 0.00265 + \frac{0.00035\left(1 - \frac{0.734s}{h''}\right)\left(\rho''f_y''\right)^{2/3}}{\sqrt{f_c'}} \qquad (6.19k)$$

$$f_i = f_{cc}'\left[0.25\left(\frac{f_c'}{f_{cc}'}\right) + 0.4\right] \qquad (6.19l)$$

$$\varepsilon_i = K\left[1.4\left(\frac{\varepsilon_o}{K}\right) + 0.003\right] \qquad (6.19m)$$

$$f_{2i} = f_{cc}'\left[0.025\left(\frac{f_{cc}'}{1000}\right) - 0.065\right] \qquad (6.19n)$$

$$\varepsilon_{2i} = 2\varepsilon_i - \varepsilon_{co} \qquad (6.19o)$$

Fig. 6.22 shows various symbols and notations used in above equations.

Bjerkeli et al. Stress–Strain Model (1990)

Based on detailed experimental results, Bjerkeli, Tomaszewicz and Jensen et al. (1990) proposed the following stress–strain equations for the confined concrete (Fig. 6.23).

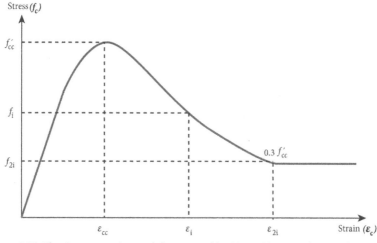

Figure 6.22 The Stress–strain model proposed by Yong, Nour, and Nawy (1989).

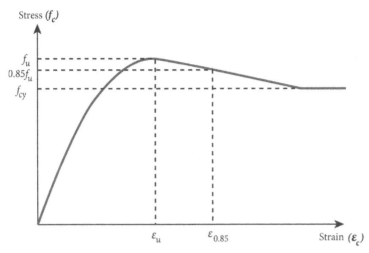

Figure 6.23 The Stress–strain model for confined concrete proposed by Bjerkeli et al. (1990).

For ascending branch

$$\sigma = \frac{E_c \varepsilon}{1 + \left(\dfrac{E_c}{E_o} - 2\right)\left(\dfrac{\varepsilon}{\varepsilon_u}\right) + \left(\dfrac{\varepsilon}{\varepsilon_u}\right)^2} \qquad (6.20\text{d})$$

For descending branch

$$\sigma = f_u - Z(\varepsilon - \varepsilon_u) \qquad (6.20\text{e})$$

For horizontal part

$$\sigma = f_{cy} = 4.87 \frac{d_{sp} A_{sh} f_{sy}}{s_p A_c} \qquad (6.20\text{f})$$

$$Z = \frac{0.15 f_u}{\varepsilon_{85} - f_u} \qquad (6.20\text{g})$$

$$E_o = \frac{f_u}{\varepsilon_u} \qquad (6.20\text{h})$$

$$E_c = 9500 \left(\frac{f_c}{2400}\right)^{1.5} \qquad (6.20\text{i})$$

Li et al. Stress—Strain Model (2000)

Li et al. (2000) proposed a stress—strain model consisting of three branches, for high-strength concrete confined by either normal or high-yield strength transverse reinforcement.

The equations are as follows (Fig. 6.24):

$$0 < \varepsilon_c \leq \varepsilon_{co} \quad f_c = E_c \varepsilon_c + \frac{(f_c' - E_c \varepsilon_c)}{\varepsilon_{co}^{\,2}} \varepsilon_c^{\,2} \tag{6.21a}$$

$$\varepsilon_{co} < \varepsilon_c \leq \varepsilon_{cc} \quad f_c = f_{cc}' - \frac{(f_{cc}' - f_c')}{(\varepsilon_{cc} - \varepsilon_{co})^2}(\varepsilon_c - \varepsilon_{cc})^2 \tag{6.21b}$$

$$\varepsilon_c > \varepsilon_{cc} \quad f_c = f_{cc}' - \beta \frac{f_{cc}'}{\varepsilon_{cc}}(\varepsilon_c - \varepsilon_{cc}) \geq 0.4 f_{cc}' \tag{6.21c}$$

The factor β controls the slope of the postpeak branch.

$$f_{cc}' = f_c' \left[-1.254 + 2.254 \sqrt{1 + \frac{7.94 f_1'}{f_c'}} - 2\alpha_s \frac{f_1'}{f_c'} \right] \tag{6.21d}$$

$$\alpha_s = (21.2 - 0.35 f_c') \frac{f_1'}{f_c'} \ (\text{If } f_c' \leq 52 \text{ MPa}) \tag{6.21e}$$

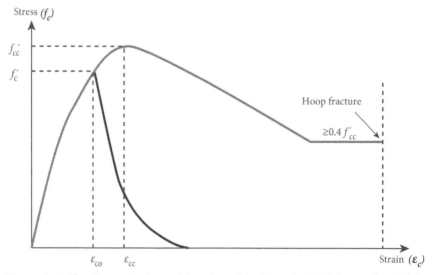

Figure 6.24 The Stress—strain model proposed by Li et al. (2000) for confined high-strength concrete.

$$\alpha_s = 3.1 \frac{f_1'}{f_c'} \text{(If } f_c' > 52 \text{ MPa)} \tag{6.21f}$$

$$f_1' = \frac{1}{2} K_e \cdot \rho_s \cdot f_{yh} \tag{6.21g}$$

For circular hoops,

$$K_e = \frac{\left(1 - \frac{s'}{2d_s}\right)^2}{1 - \rho_{cc}} \tag{6.21h}$$

For circular hoops,

$$K_e = \frac{1 - \frac{s'}{2d_s}}{1 - \rho_{cc}} \tag{6.21i}$$

where

ρ_s = Ratio of volume of transverse confining steel to volume of confined concrete core

f_{yh} = Yield strength of transverse reinforcement

f_1' = Effective lateral confining pressure

K_e = Confinement coefficient

ε_{cc} = Axial strain at maximum strength

ε_{cu} = Maximum concrete strain.

FACTORS AFFECTING MOMENT–CURVATURE RELATIONSHIP AND DUCTILITY OF RC SECTIONS

This section will discuss some of the important parameters which affect the ductility and moment–curvature relationship of a cross-section. Effect of individual factor will be discussed separately.

Effect of Compression Reinforcement

An important factor that improves the ductility is the presence of rebars in the compression zone. Fig. 6.25 shows the $M - \phi$ curves for a rectangular section with varying degree of compression reinforcement. A significant increase in ductility ratio is obvious with increase in amount of compression reinforcement. It is worth noting that increase in moment

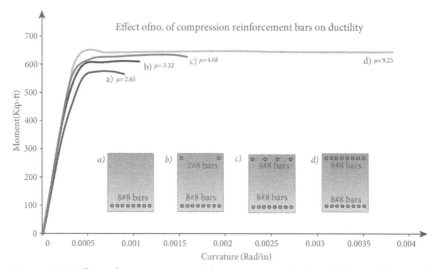

Figure 6.25 Effect of compression reinforcement on ductility of beams. Increased compression reinforcement increases ductility with marginal increase in the moment capacity (CSiCOL).

capacity is not significant. This is due to the fact that moment capacity is based on the tensile yield strength of the steel in rebars, whereas the ductility is affected by the ductile nature of the steel material.

Effect of Longitudinal Steel, Yield Strength, and Diameter of Rebars

Fig. 6.26 shows the effect of number of longitudinal bars on the $M - \phi$ curve. Rectangular sections (a), (b), and (c) corresponds to 2, 4, and 6 longitudinal bars, respectively. It can be seen that the section with least steel area in tension has the least ultimate moment capacity but highest ductility. As the amount of longitudinal steel in tension increases, the ultimate moment capacity increases, but ductility decreases. This invokes a careful consideration about increasing the steel area in cases with high ductility requirement.

Fig. 6.27 shows the effect of increasing the yield strength of longitudinal rebars, keeping their amount constant. The effect is similar as while increasing the steel amount. With increasing yield strength, the overall section capacity increases with decreasing ductility. This again provides a useful insight and guideline about the use of high-strength steel in high seismic risk areas.

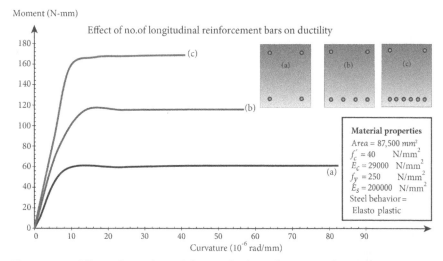

Figure 6.26 Effect of number of longitudinal reinforcement bars in tension on ductility. Increased tension reinforcement reduces ductility, increases moment capacity significantly (CSiCOL).

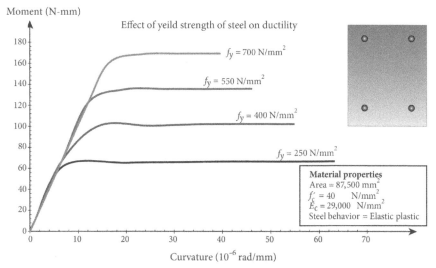

Figure 6.27 Effect of yield strength f_y of longitudinal tension reinforcement bars on ductility of beams. Increased f_y reduces ductility but increases the moment strength significantly (CSiCOL.).

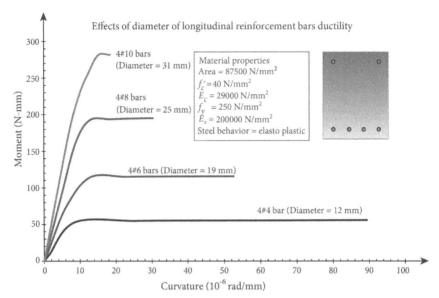

Figure 6.28 Effect of diameter of longitudinal reinforcement bars in tension on ductility of beams. Increased size reduces ductility but increases the moment capacity significantly (CSiCOL).

Fig. 6.28 shows the effect of increasing longitudinal reinforcement in tension, but in the form of increasing the diameter, while keeping the number of bars constant. Similar effect can be seen. Important considerations about the use of high-diameter bars increased are crack width and maximum aggregate size of concrete that can be placed.

It can be seen from the above examples that by increasing the tension force capacity (by increasing rebar area or rebar steel material strength), the failure shifts towards concrete, which has lower ductility at material level and hence reduces the overall cross-sectional ductility.

Fig. 6.29 shows the combined effect of increasing both compression and tensile reinforcement on ultimate moment capacity and ultimate curvature. The direct relation of moment capacity (as well as inverse relation of ductility) with increasing tensile reinforcement has been discussed earlier. Fig. 6.29 confirms that the effect of compression reinforcement in increasing moment capacity reduces with increasing amount of compression rebars. Therefore, increasing the compressive reinforcement is not a good idea to achieve high moment capacity. On the other hand, the ultimate curvature increases with increasing in compression reinforcement (increase in ductility) and its effect is further pronounced when used with higher amounts of longitudinal reinforcement.

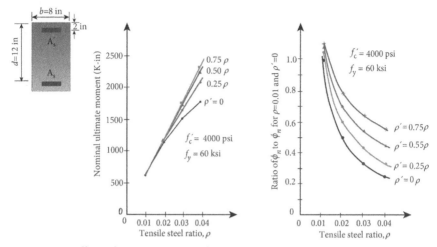

Figure 6.29 Effect of compression reinforcement on ultimate moment and ultimate curvature of beam sections. *Based on Kayani, K. R. (2011). Lectures on advanced concrete structures. Islamabad, Pakistan: National University of Sciences and Technology (NUST).*

Effect of Confinement Model for Concrete

As discussed in previous section, the confinement depends on several factors including the amount of longitudinal rebar and their distribution and spacing. This actually translates to confinement reinforcement volume ratio. The most effective confinement is the spiral reinforcement in circular cross-sections. The effect of concrete confinement on the ductility ratio is presented in following comparisons. It can be seen that there is a significant increase in ultimate curvature if a confined stress−strain model is used for concrete. Figs. 6.30 and 6.31 show the effects of various confinement models on the $M - \phi$ curve.

Effect of Cross-Sectional Shape

Fig. 6.32B compares the $M - \phi$ curves for five different cross-sections (with same amount of longitudinal reinforcement and cross-sectional areas as shown in Fig. 6.32A). Rectangular B section provides the highest moment capacity as well as ultimate curvature. Although the sectional shape is usually not governed by the considerations for ductility provision, it has an important role in configuration of confining steel and hence can play indirect role in overall structural performance.

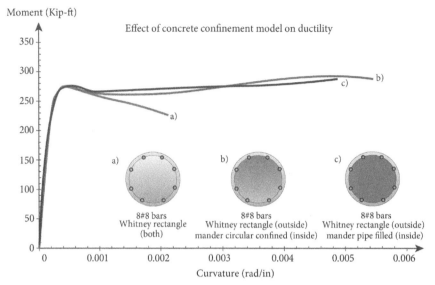

Figure 6.30 Effect of concrete confinement model on ductility (circular section).

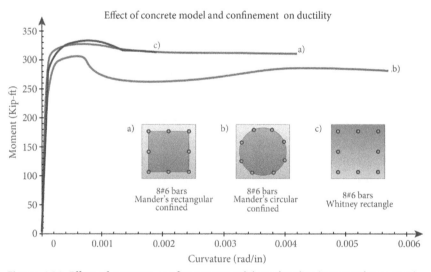

Figure 6.31 Effect of concrete confinement model on ductility (rectangular section).

Effect of Axial Load on Ductility

The curvature of the section is influenced greatly by the axial load, hence there is no unique $M-\phi$ relationship for a column section where axial load may change with time or for various load concentrations. However, it is possible to plot the combination of axial load P and Moment M which causes the section to reach the ultimate capacity (see chapter:

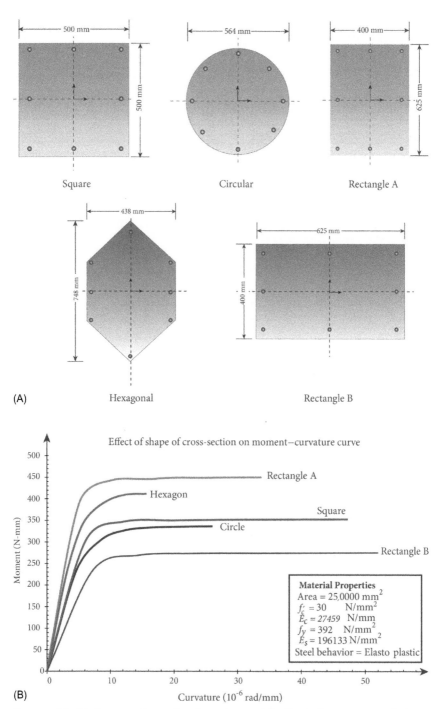

Figure 6.32 (A) Cross-sections of various shapes. (B) Effect of cross-sectional shape on ductility.

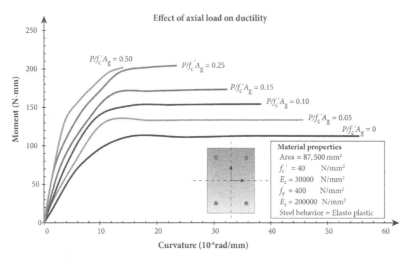

Figure 6.33 Effect of axial load on ductility.

Axial–Flexural Response of Cross-Sections). Because of the brittle failure of the unconfined columns at moderate axial load, ACI code recommends that the ends of the columns in ductile frame in earthquake areas be confined by closely spaced transverse reinforcement when axial load is greater than 0.4 times balanced load.

Fig. 6.33 shows the effect of axial load on the $M - \phi$ curve. It can be seen that the ductility of the column section is significantly reduced by the presence of axial load. A rectangular section behaving in a sufficiently ductile manner with a ductility ratio of 4–5 starts behaving as almost a brittle section at axial force ratio equal to half of its direct compression capacity. This observation is of particular interest as the dynamic earthquake forces can cause a significantly varying range of axial loads in individual members, resulting in different levels of exhibited ductility.

CONCRETE-FILLED STEEL TUBES

The most effective and efficient way of utilizing the strength of steel in a column under biaxial bending and load is to place the steel at the extremities of the cross-section. This is best achieved by placing concrete inside steel tubing. Composite columns consisting of concrete-filled steel tubes have become increasingly popular in structural applications around the world. By using composite columns consisting of concrete-filled steel tubes instead of traditional reinforced concrete columns, the problem of concrete cover spalling can be avoided. Furthermore, inward buckling of the steel tube is prevented by the concrete core, thus increasing the stability and strength of the column system. The shape of the composite may vary depending on the specific requirements for the particular case. Though effective, the use of concrete-filled steel tubes has been limited in the past, owing to the difficulty in analyzing the true behavior of the cross-section.

In a normal reinforced concrete section, the individual stress—strain relationships of concrete and steel reinforcement are generally considered to be independent of each other. However, in the case of steel tubes filled with concrete, the stress—strain curve for concrete is dependent on both the geometric and material properties of the steel tubes, due to the confining effect. The failure of concrete in compression is dependent on the failure of steel tubes in hoop tension due to the transverse strain produced in concrete. This is similar to the confinement effect of hoop reinforcement (ties or/and spiral), but is more significant, particularly when tubes of large thicknesses are used. It is obvious that circular steel tubing will have the greatest confining effect as compared to other shapes. This is because circular sections are placed in hoop tension as the concrete expands under uniaxial compression, which provides a continuous confining line load around the circumference of the enclosed concrete. On the other hand, rectangular tubes are only fully effective near the corners. Flexural deformation is prominent in rectangular steel tubes away from the corners as compared to axial deformation in circular tubes, resulting in lower stiffness and hence, less significant confinement.

Concrete filling has a significant effect on the axial load-carrying capacity of composite tubes. It does not, however, have a very considerable contribution towards the moment capacity of the section. Furthermore, inward buckling of the steel tube is prevented by the concrete core, which acts as a stiffener for the tube, thus increasing the stability and strength of the column. The presence of steel tube alters the performance of concrete significantly. In addition to increased earthquake resistant properties, such as high strength and high ductility, the ease of construction of such columns is appreciable. The steel tubing acts both as a reinforcement and formwork for the concrete core. In modern construction, the use of high-strength concrete has led to the emergence of tall buildings consisting of thin slender columns. This raises the question of brittle failure in such columns. To prevent any such failure, closely spaced strips are often used, increasing the risk of premature spalling of concrete cover. These problems are totally eradicated by the use of steel tubes as reinforcement, thus giving greater effective concrete area. Due to the presence of the confining steel tube, concrete is able to undergo large strains. The peak axial compressive stress that concrete can sustain is also increased with increase in confinement (and so is the strain at which this stress is reached). Experiments show that a small confining pressure of about 10% of the uniaxial cylinder compressive strength was sufficient to increase the load-bearing capacity of the specimen by as much as 50%. On the other hand, a small lateral tensile stress of about 5% of the uniaxial compressive strength was sufficient to reduce the capacity by the same amount. An efficient bonding between steel tubes and concrete core can be achieved by exploiting any of the following factors:

- Presence of mechanical connectors
- Interlocking at the interface of concrete and steel due to irregularities of surfaces
- Friction between the material due to normal forces
- Adhesion due to chemical reaction
- Creep in concrete.

The method of loading also has a very significant effect on the behavior and performance of concrete-filled steel tubes. These composites can be visualized as a system acting together and interacting with each other. As load is applied to the concrete uniaxially, it tends to expand outwards and is constrained due to the presence of steel tubing. This does not only produce confinement to concrete but also induces hoop stresses in the steel tube. Applying load only to the outer steel tubing will force it to behave like an empty tube except for the fact that the concrete filling restraints the tube from bucking inwards and thus prevents stability failure. It is, therefore, important to ensure that the load transfer to the column is effectively being distributed on/to the entire section and not only to the steel tube, in which case the capacity of the section will be underutilized.

The capacity of concrete-filled tubes can be further enhanced by the addition of reinforcement bars and/or additional steel shapes/sections (Fig. 6.34). Reinforced concrete-filled tubes can be used to carry various levels of axial load and moment while keeping the overall dimension and tube sizes constant (Fig. 6.35).

The determination of the axial—flexural stress resultants for the purpose of capacity estimation depends on the stress—strain relationships of both concrete and steel tube, including the effect of confinement. However, the maximum strain is generally limited to a predefined maximum strain value corresponding to the failure criteria. On the other hand, the determination of the moment—curvature or axial load—deformation relationship requires the evaluation of stress resultants far beyond the elastic stress level and

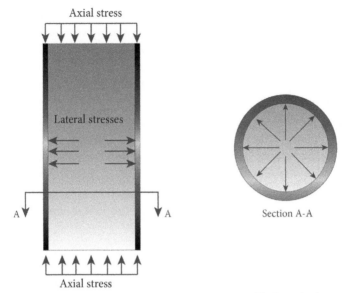

Figure 6.34 Development of lateral stresses in a concrete-filled steel tube.

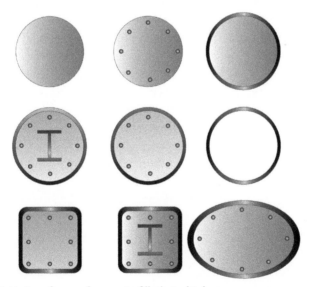

Figure 6.35 Various forms of concrete-filled steel tubes.

predefined strain limits. The entire stress—strain curve up to failure must be considered in this case. For ordinary reinforced or composite sections, the stress—strain curve for concrete is only considered up to the assumed maximum strain capacity (0.003 or 0.0035). For concrete-filled steel tubes, because of the confining effect of steel tube on concrete the concrete itself possesses significant ductility, which is modeled using the appropriate stress—strain model. The primary ductility is provided by the stress plateau and the strain hardening in steel. In the conventional determination of the capacity of a cross-section, where independent stress—strain curves are used for concrete as well as steel, the shape of the steel tube will not affect the total compressive load-carrying capacity as long as the area of concrete and steel are equal. In reality, the stress—strain curve of concrete is modified by the hoop tension capacity of the steel tube. It is obvious that the circular or nearly circular tubes will have the highest confinement effect and the least deformation in the transverse direction (Fig. 6.36).

A comparative study is carried out to analyze the effect of the type of stress—strain relationship used for concrete on the overall capacity of the section. For this purpose, two stress—strain relationships for concrete were considered, the ACI Whitney Rectangle and the Modified Mander Confined Stress—Strain Curve. A circular concrete section of diameter 300 mm, reinforced with 8#4 bars was considered as shown in Fig. 6.37. A 12.5-mm-thick steel tube is used to confine the circular concrete section (Fig. 6.37).

Fig. 6.38 presents the axial load—moment interaction curves for different cases of confinement model for above section. It can be observed that ACI Whitney's rectangular stress block is representing only the unconfined concrete and thus resulting in low axial—flexural capacity. The confinement effect included using the Mander's stress—strain model resulted in enhanced moment carrying capacity of the section. Similarly, the PM interaction diagrams for the section with steel tubing around the concrete resulted in a remarkable increase in both the moment and the axial load (both tensile and compressive) carrying capacities.

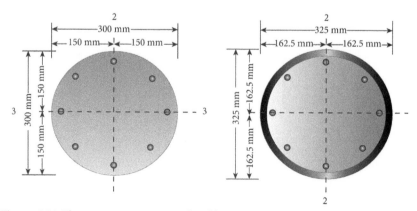

Figure 6.36 The concrete section confined by a steel tube.

Effect of stress–strain relationship and confinement of
concrete on capacity

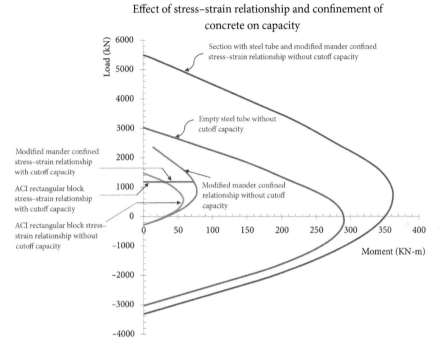

Figure 6.37 Effect of stress–strain relationship and confinement of concrete on the
capacity of concrete-filled tubes.

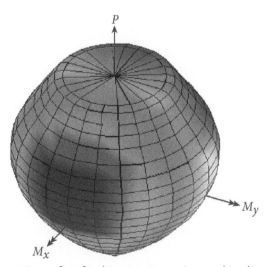

Figure 6.38 Interaction surface for the concrete section used in above example.

ROLE OF CROSS-SECTIONAL RESPONSE IN NONLINEAR ANALYSIS OF STRUCTURES

In recent years, nonlinear analysis has gained significant popularity and acceptance for evaluating the performance of structures, especially buildings, both for seismic and nonseismic loads. It provides improved understanding of building behavior and provides more accurate prediction of global displacement, more realistic prediction of earthquake demand on individual components and elements, and more reliable identification of "bad actors" in structural performance. The most common nonlinear analysis procedure is Pushover analysis which is a static procedure. In this analysis, the magnitude of the structural loading is incrementally increased in accordance with a certain predefined pattern. With the increase in the magnitude of the loading, weak links and failure modes of the structure are found. The procedure also helps to identify ductility requirements for various members. The loading is monotonic with the effects of the cyclic behavior and load reversals being estimated by using a modified mono-tonic force—deformation criteria and with damping approximations. Static pushover analysis is an attempt by the structural engineering profes-sion to evaluate the real strength of the structure and it promises to be a useful and effective tool for performance-based design. Pushover analysis is most suitable for determining the performance, especially for lateral loads such as earthquake or even wind. Since the structures do not respond as linearly elastic systems during strong ground shaking, the linear analysis cannot give true picture of structural system behavior.

The pushover analysis relies on the nonlinear cross-section response parameter as the basic input for determining the load—deformation and performance curves. Many softwares that carry out the pushover analysis have the capability to generate such parameter for simple section and material models. However, for complex cross-sections, material models, and behavior, computing these parameters often needs to be done sepa-rately and then provided as an input to the analysis programs.

Performance is generally of concern for lateral loads such as earth-quake and wind. The main factor that affects performance is the ductility of the members on the critical load path. In frame structures, the design of the joints between columns and beams is critical. The performance of shear walls is also of great importance for lateral load demands. The overall relationship of various response curves relevant to the

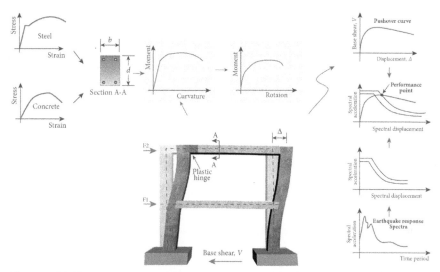

Figure 6.39 The moment–curvature curve is one of the key inputs in performance-based design using pushover analysis.

performance-based design or evaluation of structures using the Pushover analysis is shown in Fig. 6.39. This figure shows how the material stress–strain curves contribute to the Moment–Curvature curve, which in turn affects the moment–rotation behavior of plastic hinges, that in turn affect the structure performance curve, and finally relates to the seismic demand curve defined in Acceleration Response Spectra. It can be seen that the Moment–Curvature curve is the basic cross-section response that must be determined to carry on this type of analysis. The Moment–Curvature curve and the corresponding moment–rotation curve are generated after basic cross-section design is completed.

The capacity interaction PM curve and moment–curvature relationship can be used together to obtain useful information for the strength design of members. The strength of a reinforced concrete beam column is defined by its capacity interaction surface which is a plot of applied actions (P and M) for all orientations and rotations of the neutral axis plane for a fixed limiting strain. The actual applied actions (P and M) should be within this surface which is further represented by the P–M interaction curve for a particular bending direction and the M_x–M_y curve for a given axial load. The P–M interaction is in fact denoted by the yield or capacity point on the Moment–Curvature curve as shown in Fig. 6.40. It is, therefore, possible to generate the M–ϕ curve for a given

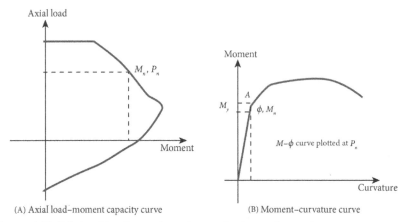

(A) Axial load–moment capacity curve   (B) Moment–curvature curve

Figure 6.40 Relationship of capacity point on the interaction and moment–curvature curves.

axial load and bending direction and then read off the capacity point to carry out strength design as well. The factors that affect strength include the basic material strength, the cross-section dimensions, the amount of rebars, and the overall structural framing conditions. The strength design is a prerequisite to the performance-based design and in fact required before the Pushover analysis can be carried out. This is because the cross-section dimension and reinforcement should be known for computation of the plastic hinge properties that are needed for Pushover analysis (Fig. 6.39).

The Magnitude and Location of Resultant Compressive Force from the Nonlinear Stress–Strain Model of Concrete

As discussed in Chapter 3, Axial–Flexural Response of Cross-Sections, in order to determine moment capacity of a reinforced concrete section, usually rectangular stress block is used, which makes it convenient to determine the total resultant compressive force on part of cross-section under compression (for a particular value of neutral axis). The total volume of stress block is the resultant compressive force acting at the center of block. The following simplified expressions are generally used for determining section moment capacity.

$$C_c = 0.85 f_c' a b$$

$$M_n = C_c(d - a/2)$$

(Continued)

The Magnitude and Location of Resultant Compressive Force from the Nonlinear Stress–Strain Model of Concrete (Continued)

where a is the depth of rectangular stress block. For any general nonlinear stress–strain model, if the actual ratio of average stress to maximum stress can be obtained for a given strain value, the resultant compressive force can still be obtained using the expression below.

$$C_c = k_1 f_c bkd$$

where k_1 is the ratio of average stress to maximum stress and kd is the depth of neutral axis. f_c is the stress in concrete such that $f_c \leq f_c'$. Similarly, if the distance from extreme fiber (compression) to the resultant of the compressive force is known for a nonlinear stress–strain model, we can determine the moment capacity from the following expression:

$$M_n = C_c(d - k_2 kd)$$

where k_2 is the ratio of distance from extreme fiber (compression) to the resultant of the compressive force to the distance to the neutral axis. It may be easy to derive the expressions for k_1 and k_2 for some stress–strain models. Here, two examples will be shown for the bilinear elastoplastic model and the Hognestad's concrete model with linear tail.

Bilinear Elastoplastic Model

Let's assume that ε_{cm} is the extreme fiber strain and ε_o represents the yielding strain, as shown in Fig. 6.41.

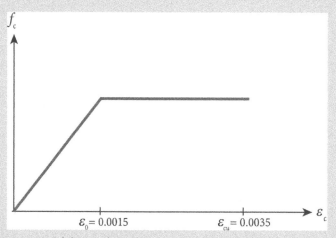

Figure 6.41 A bilinear elastoplastic stress–strain model.

(Continued)

The Magnitude and Location of Resultant Compressive Force from the Nonlinear Stress—Strain Model of Concrete (Continued)

For $\varepsilon_{cm} \leq 0.0015$

$$k_1 = \frac{0.5 f_c}{f_c'} = \frac{0.5 \varepsilon_{cm} E_c}{\varepsilon_\infty E_c} = \frac{0.5 \varepsilon_{cm}}{\varepsilon_\infty}$$

$$k_2 = \frac{1}{3} \frac{\varepsilon_{cm}}{\varepsilon_{cm}} = \frac{1}{3}$$

For $0.0015 \leq \varepsilon_{cm} \leq 0.0035$

$$k_1 = \frac{\sum_{i=1}^{2} A_i}{f_c'' \varepsilon_{cm}} = \frac{\frac{1}{2} f_c' \varepsilon_0 + f_c'(\varepsilon_{cm} - \varepsilon)}{f_c' \varepsilon_{cm}}$$

$$k_1 = \frac{\dfrac{\varepsilon_{co}}{2} + \varepsilon_{cm} - \varepsilon_0}{\varepsilon_{cm}} = \frac{\varepsilon_{cm} - \dfrac{\varepsilon_{co}}{2}}{\varepsilon_{cm}}$$

$$k_1 = \left[1 - \frac{\varepsilon_{co}}{2 \varepsilon_{cm}} \right]$$

$$k_2 \varepsilon_{cm} = \frac{\sum A_i y_i}{\sum A_i} = \frac{f_c'(\varepsilon_{cm} - \varepsilon_0)^2 + \dfrac{1}{2} \left(\dfrac{1}{3} \varepsilon_0 + \varepsilon_{cm} + \varepsilon_0 \right)}{f_c'(\varepsilon_{cm} - \varepsilon_0) + \dfrac{1}{2} f_c' \varepsilon_0}$$

$$k_2 = \frac{\varepsilon_{cm}^2 \varepsilon_0^2 2 \varepsilon_{cm} \varepsilon_0 + \varepsilon_0 \varepsilon_{cm} - \dfrac{2}{3} \varepsilon_0^2}{\varepsilon_{cm}(2 \varepsilon_{cm} - \varepsilon_{co})}$$

It can be noted that both k_1 and k_2 are the functions of extreme fiber strain and yield strain. Therefore, for a given ε_{cm}, the resultant compressive force, its location and moment capacity of the section can be determined.

Hognestad's Parabolic Model with Linear Tail

The Hognestad's stress—strain model consists of a parabolic function up to peak concrete stress and an descending tail, governed by the following equations.

For $0 \leq \varepsilon_{cm} \leq \varepsilon_0$

$$f_c = k_3 f_c' \left[2 \left(\frac{\varepsilon_{cm}}{\varepsilon_{co}} \right) - \left(\frac{\varepsilon_{cm}}{\varepsilon_{co}} \right)^2 \right]$$

For $\varepsilon_0 \leq \varepsilon_{cm}$

$$f_c = k_3 f_c'[1 - Z(\varepsilon_{cm} - \varepsilon_0)]$$

where Z is Slope constant for sloping portion. This model can be used for any value of maximum concrete strain (Fig. 6.42).

(Continued)

The Magnitude and Location of Resultant Compressive Force from the Nonlinear Stress–Strain Model of Concrete (Continued)

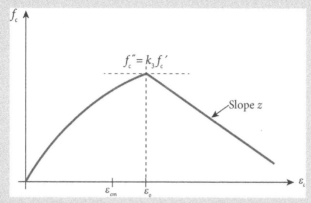

Figure 6.42 The Hognestad's parabolic model with linear tail.

Expressions for k_1 and k_2 as a function of ε_{cm} are derived as follows (Figs. 6.43–6.45).

For $\varepsilon_o \leq \varepsilon_{cm}$

$$k_1 = \frac{1}{f_c'}\left[\int_0^{\varepsilon_{cm}} f_c d\varepsilon_c\right]\frac{1}{\varepsilon_{cm}}$$

$$k_1 = \frac{1}{f_c'\varepsilon_{cm}}\int_0^{\varepsilon_{cm}} f_c'\left[2\left(\frac{\varepsilon_{cm}}{\varepsilon_{co}}\right) - \left(\frac{\varepsilon_{cm}}{\varepsilon_{co}}\right)^2\right]d\varepsilon_c$$

$$k_1 = \frac{1}{\varepsilon_{cm}}\left[\frac{\varepsilon_{cm}^2}{\varepsilon_o} - \frac{\varepsilon_{cm}^3}{3\varepsilon_o}\right] = \eta - \frac{1}{3}\eta^2$$

where

$$\eta = \frac{\varepsilon_{cm}}{\varepsilon_o}$$

$$k_2\varepsilon_{cm} = \varepsilon_{cm} - \frac{\int_0^{\varepsilon_{cm}} f_c\varepsilon_c d\varepsilon_c}{\int_0^{\varepsilon_{cm}} f_c d\varepsilon_c}$$

$$k_2 = 1 - f_c\int_0^{\varepsilon_{cm}} 2\left[\left(\frac{\varepsilon_c^2}{\varepsilon_o}\right) - \frac{\varepsilon_c^3}{\varepsilon_o^2}\right]d\varepsilon_c$$

$$k_2 = 1 - \frac{\left[\frac{2}{3}\left(\frac{\varepsilon_{cm}^3}{\varepsilon_o}\right) - \frac{1}{4}\left(\frac{\varepsilon_{cm}^4}{\varepsilon_o^2}\right)\right]}{\varepsilon_{cm}^2\left(\eta - \frac{1}{3}\eta^2\right)}$$

(Continued)

The Magnitude and Location of Resultant Compressive Force from the Nonlinear Stress–Strain Model of Concrete (Continued)

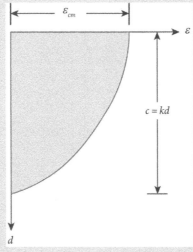

Figure 6.43 Neutral axis depth for $\varepsilon_o \leq \varepsilon_{cm}$.

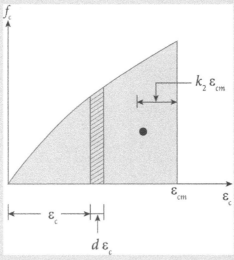

Figure 6.44 Centroidal distance for $\varepsilon_o \leq \varepsilon_{cm}$.

(Continued)

The Magnitude and Location of Resultant Compressive Force from the Nonlinear Stress–Strain Model of Concrete (Continued)

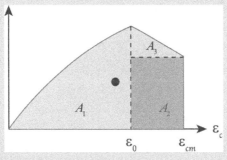

Figure 6.45 Centroidal distance for $\varepsilon_o \geq \varepsilon_{cm}$.

$$k_2 = 1 - \frac{\frac{2}{3}\eta - \frac{1}{4}\eta^2}{\eta - \frac{1}{3}\eta^2}$$

$$k_2 = 1 - \frac{\frac{1}{3} - \frac{1}{12}\eta}{1 - \frac{1}{3}\eta}$$

For $\varepsilon_o \leq \varepsilon_{cm}$

$$k_1 = \frac{\sum_{i=1}^{3} A_i}{f_c' \varepsilon_{cm}}$$

$$A_1 = \frac{2}{3} f_c' \varepsilon_o$$

$$A_2 = (\varepsilon_{cm} - \varepsilon_o)[1 - Z(\varepsilon_{cm} - \varepsilon_o)]f_c'$$

$$A_2 = f_c'[(\varepsilon_{cm} - \varepsilon_o) - Z(\varepsilon_{cm} - \varepsilon_o)^2]$$

$$A_3 = \frac{1}{2}(\varepsilon_{cm} - \varepsilon_o)Z(\varepsilon_{cm} - \varepsilon_o)f_c'$$

$$A_3 = f_c' \frac{Z}{2}(\varepsilon_{cm} - \varepsilon_o)^2$$

Now putting values of A_1, A_2, and A_3 in the above expression of k_1, we get

$$k_1 = \frac{1}{\varepsilon_{cm}}\left[\varepsilon_{cm} - \frac{\varepsilon_o}{3} - \frac{Z}{2}(\varepsilon_{cm} - \varepsilon_o)^2\right]$$

(Continued)

The Magnitude and Location of Resultant Compressive Force from the Nonlinear Stress–Strain Model of Concrete (Continued)

$$k_2 = 1 - \frac{1}{\varepsilon_{cm}} \frac{\sum A_i X_i}{\sum A_i}$$

$$k_2 = 1 - \frac{1}{\varepsilon_{cm} \sum A_i} \left[A_1 \left(\frac{5}{8} \varepsilon_o \right) + A_2 \left(\varepsilon_o + \frac{\varepsilon_{cm} - \varepsilon_o}{2} \right) + A_3 \left(\varepsilon_o + \frac{\varepsilon_{cm} - \varepsilon_o}{3} \right) \right]$$

Let

$$C_1 = \frac{\varepsilon_o}{\varepsilon_{cm}}, \ C_2 = \varepsilon_{cm} - \varepsilon_o, \ C_3 = \frac{1}{2} Z C_2^2, \ C_4 = C_2 - Z C_2^2, \ \text{and} \ C_5 = \varepsilon_{cm} - \frac{1}{3} \varepsilon_o - C_3$$

Then

$$k_2 = 1 - \frac{1}{C_5} \left[\frac{5}{12} C_1 \varepsilon_o + C_4 \left(C_1 + \frac{C_2}{2 \varepsilon_{cm}} \right) + C_3 \left(C_1 + \frac{C_2}{3 \varepsilon_{cm}} \right) \right]$$

SOLVED EXAMPLES

Outputs from the Moment–Curvature Curve

Problem 1: The Moment–curvature curve of beam shown in Fig. Ex. 6.1 is given in Fig. Ex. 6.2. Determine the following quantities given $f_c' = 40$ MPa, $f_y = 420$ MPa, and crack spacing of 450 mm.

1. Flexural stiffness
2. Deflection at the mid span and third point on the beam along the length
3. Strain in the bottom of steel

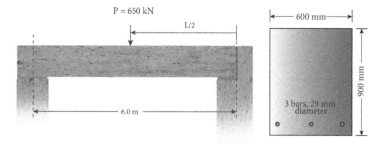

Figure Ex. 6.1 A simply supported beam and its cross-section.

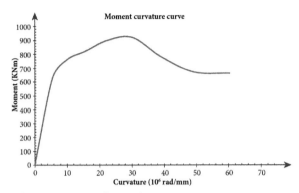

Figure Ex. 6.2 The $M-\phi$ curve of given beam cross-section.

4. Rotation of the beam at mid span
5. Crack width.

Given

$b = 600$ mm
$h = 900$ mm
Crack spacing $= 450$ mm
Cover $= 50$ mm
$L = 6.0$ m
$f'_c = 40$ MPa
$f_y = 420$ MPa

Solution

$$M = \frac{PL}{4} = \frac{650 \times 6}{4} = 975 \text{ kN-m}$$

From the $M-\phi$ curve, we can get the value of curvature for the above-mentioned M

$$\phi = 28 \times 10^{-6} \text{rad/mm}$$

Flexural Stiffness

$$EI = \frac{M}{\phi} = \frac{975 \times 10^6}{28 \times 10^{-6}} = 3.304 \times 10^{13} \text{N-mm}^2$$

Deflection at the Mid Span

$$\Delta_{\text{mid}} = \int_a^b \left(\frac{M}{EI}\right) \bar{x} dx$$

Using moment area theorem

$$\Delta_{mid} = \left(\frac{1}{2} \times \frac{925 \times 10^6}{3.304 \times 10^{13}} \times 3\right) \times \left(\frac{2}{3} \times 3\right) \times 10^6 = 83.989 \text{ mm} \cong 84 \text{ mm}$$

Deflection at the Third Point
   Again using moment area theorem

$$\Delta_{L/3} = \left(\frac{1}{2} \times \frac{616.67 \times 10^6}{3.304 \times 10^{13}} \times 2\right) \times \left(\frac{2}{3} \times 2\right) \times 10^6 = 24.886 \text{ mm} \cong 25 \text{ mm}$$

Strain in Bottom Steel at Mid Span
   Using rectangular stress block method, the location of neutral axis can be determined as

$$C = 0.85f'_c ab = 0.85 \times 40 \times 600 \times a = 20{,}400a$$

$$T = A_s f_y$$

For 3#9 bars the total area of steel is 1935 mm^2

$$T = 1935 \times 420 = 812{,}700 \text{ N}$$

$$C = T$$

$$a = \frac{812{,}700}{20{,}400} = 39.838 \text{ mm}$$

Now y can be calculated as

$$y = \frac{39.838}{\beta_1} = \frac{39.838}{0.85} = 46.86 \text{ mm}$$

Hence the neutral axis is at a distance of 46.86 mm from the top. Now c can be calculated as

$$c = h - (\text{cover}) - y = 900 - 50 - 46.86 = 803.14 \text{ mm}$$

$$\varepsilon_s = \phi c = 28 \times 10^{-6} \times 803.14 = 0.022$$

Slope at Mid Span

$$\theta = \int_a^b \frac{M}{EI} dx = \frac{1}{2} \times \frac{9.25 \times 10^6}{3.304 \times 10^{13}} \times 3 \times 1000 = 0.042 \text{ rad}$$

Crack Width

$$W = \varepsilon_s \times \text{crack spacing} = 0.022 \times 450 = 9.9 \text{ mm}$$

Results

1. Flexural Stiffness, $EI = 3.304 \times 10^{13} \text{N} - \text{mm}^2$
2. Deflection at Mid Span, $\Delta = 0.0459$ m
3. Strain in Steel, $\varepsilon_s = 0.022$
4. Slope at Mid Span, $\theta = 0.042$ rad
5. Crack Width, $W = 9.9$ mm

Kent and Park Confinement Model

Problem 2: RC cross-section is shown in Fig. Ex. 6.3. The unconfined compressive strength of concrete is $f'_c = 30$ MPa. The concrete is confined by hoops of 12 mm diameter and at 100 mm center-to-center spacing. Clear cover is assumed to be 35 mm. Generate stress—strain curve using the Kent and Park (1971) model provided $A_s = 113 \text{ mm}^2$, $\varepsilon_o = 0.002$ and $\varepsilon_c = 0.03$. Also carry out the parametric study by varying compressive strength of concrete, spacing. and diameters of hoop.

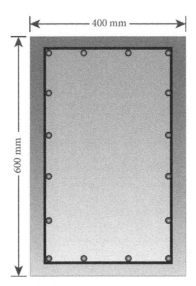

16-25 mm diameter longitudinal bars
Hoops of 12 mm diameter bars @100 mm

Figure Ex. 6.3 The given RC cross-section.

Given

$f_c' = 30$ MPa

$d = 12$ mm

Cover $= 35$ mm

$A_s = 113$ mm^2

$\varepsilon_o = 0.002$

$\varepsilon_c = 0.03$

Spacing $= 100$ mm

Solution

$$b'' = 400 - 2(35) = 330 \text{ mm}$$

$$d'' = 600 - 2(35) = 530 \text{ mm}$$

$$p'' = \frac{2(b'' + d'')A_s}{b''d''s} = \frac{2(330 + 530) \times 113}{330 \times 530 \times 100} = \frac{194,360}{17,490,000} = 0.011$$

$$\varepsilon_{50u} = \frac{3 + 0.29f_c'}{145f_c' - 1000} = \frac{3 + (0.29 \times 30)}{(145 \times 30) - 1000} = 0.0034$$

$$\varepsilon_{50h} = \frac{3}{4}p''\sqrt{\frac{b''}{s}} = \frac{3}{4} \times 0.011 \times \sqrt{\frac{330}{100}} = 0.0825 \times 1.817 = 0.0151$$

$$Z = \frac{0.5}{0.0034 + 0.0151 - 0.002} = \frac{0.5}{0.01779} = 30.06$$

Equations for the Stress–Strain Curve

For falling branch

$$f_c = f_c'[1 - Z(\varepsilon_c - \varepsilon_o)]$$

$$f_c = 30[1 - 30.06(\varepsilon_c - 0.002)$$

For ascending branch

$$f_c = f_c'\left[\frac{2\varepsilon_c}{\varepsilon_o} - \left(\frac{\varepsilon_c}{\varepsilon_o}\right)^2\right]$$

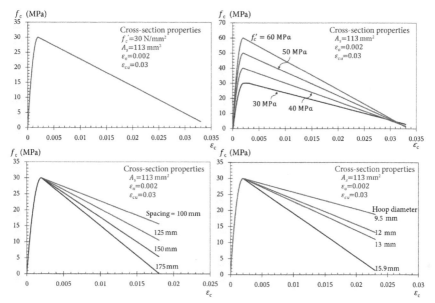

Figure Ex. 6.4 The stress—strain curves for various parameters, generated using the Kent and Park Model (1971). Calculations are shown in example for top left case.

$$f_c = 30 \left[\frac{2\varepsilon_c}{0.002} - \left(\frac{\varepsilon_c}{0.002} \right)^2 \right]$$

Results

Using these equations derived from the Kent and Park (1971) model, the stress—strain curve for confined concrete for the given section is shown in Fig. Ex. 6.4. The effect of concrete compressive strength, spacing, and diameter of hoops is also shown.

Modified Kent and Park (Scott et al.) Confinement Model

Problem 3: Determine the stress—strain behavior of the same cross-section as mentioned in Fig. Ex. 6.3 using the modified Kent and Park model (i.e., Scott et al., 1982 model). Also carry out the parametric study by varying compressive strength of concrete, spacing, and diameters of hoop.

Given

$f'_c = 30$ MPa

$d = 12$ mm

$f_{yh} = 275$ MPa
$\rho_s = 0.011$
Cover $= 35$ mm
$A_s = 113$ mm^2
$\varepsilon_o = 0.002$
$\varepsilon_c = 0.033$
$s_h = 100$ mm

Solution

For the modified Kent and Park model (Scott et al., 1982 model) expression for Z is different and another term k is introduced. The said parameters are calculated as

$$K = 1 + \frac{\rho_s f_{yh}}{f'_c} = 1 + \frac{0.011 \times 275}{30} = 1.101$$

$$Z = \frac{0.5}{\dfrac{3 + 0.29 f'_c}{145 f'_c - 1000} + \dfrac{3}{4}\rho_s\sqrt{\dfrac{h''}{s_h}} - 0.002k}$$

So Z will be

$$Z = \frac{0.5}{\dfrac{3 + (0.29 \times 30)}{(145 \times 30) - 1000} + \dfrac{3}{4} \times 0.011\sqrt{\dfrac{330}{100}} - (0.002 \times 1.101)} = 30.433$$

And for ascending branch

$$f_c = K f'_c \left[\frac{2\varepsilon_c}{0.002K} - \left(\frac{\varepsilon_c}{0.002K}\right)^2 \right]$$

$$f_c = 1.101 \times 30 \left[\frac{2 \times \varepsilon_c}{0.002 \times 1.101} - \left(\frac{\varepsilon_c}{0.002 \times 1.101}\right)^2 \right]$$

Now for falling branch is

$$f_c = K f'_c [1 - Z(\varepsilon_c - 0.002K)]$$

$$f_c = 1.101 \times 30 [1 - 30.433(\varepsilon_c - 0.002 \times 1.101)]$$

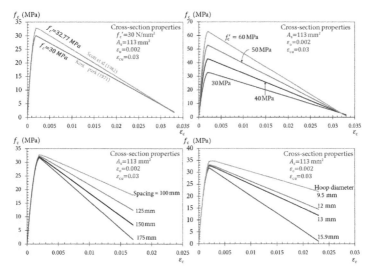

Figure Ex. 6.5 The stress—strain curves for various parameters, generated using the Scott et al. (1982) model. Calculations are shown in example for top left case.

Results

Using these equations derived from the Scott et al. (1982) model, the stress—strain curve for confined concrete for the given section is shown in Fig. Ex. 6.5. The effect of concrete compressive strength, spacing, and diameter of hoops are also shown in Fig. Ex. 6.5.

Mander et al. Confinement Model

Problem 4: RC cross-section is shown in Fig. Ex. 6.6. The unconfined compressive strength of concrete is $f_c' = 30$ MPa. The concrete is confined by hoops of 12 mm diameter and at 100 mm center-to-center spacing. Diameter of long bars is 25 mm. Yield strength of steel is 345 MPa. Clear cover is assumed to be 40 mm. Generate stress—strain curve using the Mander et al. confinement (1988) model provided $A_s = 113$ mm^2, $\varepsilon_o = 0.002$ and $\varepsilon_c = 0.03$. Also carry out the parametric study by varying compressive strength of concrete, yield strength of hoops, spacing, cover, and diameters of hoops.

Given

$f_c' = 30$ MPa
$f_{yh} = 345$ Mpa
$b = 500$ mm

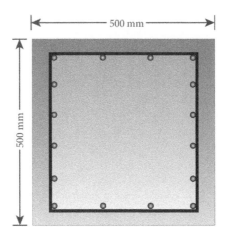

16–25 mm diameter longitudinal bars
Hoops of 12 mm diameter bars @100 mm

Figure Ex. 6.6 The given RC cross-section.

$d = 500$ mm
$A_s = 113$ mm^2
Diameter of transverse bars $= 12$ mm
Diameter of longitudinal bars $= 25$ mm
Spacing of transverse bars $= 100$ mm
No. of longitudinal bars $= 16$
Cover $= 40$ mm

Solution

$$b'' = 500 - 2(40) = 420 \text{ mm}$$

$$d'' = 500 - 2(40) = 420 \text{ mm}$$

$$f_{lx} = \rho_x f_{yh} = 1.858$$

$$f_{ly} = \rho_y f_{yh} = 1.858$$

where ρ_x and ρ_y are reinforcement ratio of transverse reinforcements.

$$\frac{f_{lx}}{f_c'} = \frac{1.858}{30} = 0.06$$

$$\frac{f_{ly}}{f_c'} = \frac{1.858}{30} = 0.06$$

Using Fig. 6.4 from Mander et al. (1988), we can determine ratio of f'_{cc}/f'_c which in this case is 1.6. Hence

$$\frac{f'_{cc}}{f'_c} = 1.6$$

$$f'_{cc} = 1.6 \times 30 = 48 \text{ MPa}$$

$$\varepsilon_{cc} = \varepsilon_{co}\left[1 + 5\left(\frac{f'_{cc}}{f'_c} - 1\right)\right] = 0.002\left[1 + 5\left(\frac{48}{30} - 1\right)\right] = 0.008$$

$$E_{sec} = \frac{f'_{cc}}{\varepsilon_{cc}} = \frac{48}{0.008} = 6000 \text{ MPa}$$

$$E_c = 5000\sqrt{30} = 27386.13 \text{ MPa}$$

$$n = \left[1 + 5\left(\frac{E_c}{E_c - E_{sec}}\right)\right] = \frac{27,386.13}{27,386.13 - 6000} = 1.28$$

Equation for the Stress−Strain Curve

$$\frac{f_c}{f'_{cc}} = \frac{n\left(\dfrac{\varepsilon_c}{\varepsilon_{cc}}\right)}{(n-1) + \left(\dfrac{\varepsilon_c}{\varepsilon_{cc}}\right)^n}$$

$$f_c = 48\left[\frac{1.28 \times \dfrac{\varepsilon_c}{0.008}}{(1.28 - 1) + \left(\dfrac{\varepsilon_c}{0.008}\right)^{1.28}}\right]$$

Result
Using equations derived from the Mander et al. (1988) model as shown above, the stress−strain curves for confined concrete for the given section are shown in Fig. Ex. 6.7.

Problem 5: Three RC cross-section of same gross area are shown in Fig. Ex 6.8A. The unconfined compressive strength of concrete is $f'_c = 40$ MPa. The concrete is confined by hoops of 12 mm diameter and at 100 mm center-to-center spacing. Diameter of long bars is 25 mm. Yield strength of steel is 415 MPa. Clear cover is assumed to be 40 mm. Generate stress−strain curve using the Mander et al. confinement (1988) model for each cross-section and carry out the comparison of the curves provided $A_s = 113$ mm^2, $\varepsilon_o = 0.002$ and $\varepsilon_c = 0.03$.

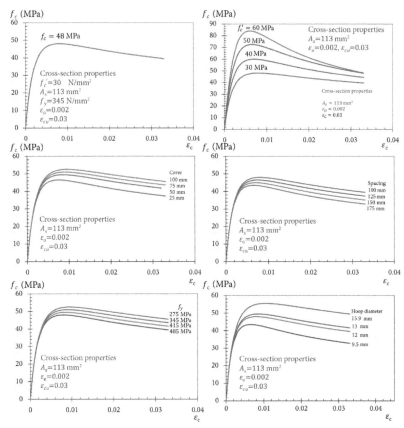

Figure Ex. 6.7 The stress—strain curves for various parameters, generated using the Mander et al. (1988) model. Calculations are shown in example for top left case.

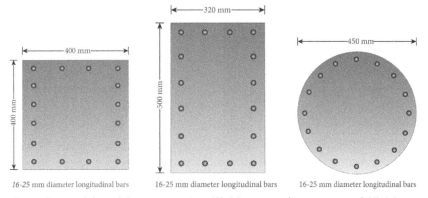

Figure Ex. 6.8 (A) An RC square section, (B) RC rectangular section, and (C) RC circular section.

Given

$f_c' = 40$ MPa
$f_{yh} = 345$ Mpa
$A_s = 113$ mm^2
Diameter of transverse bars = 12 mm
Diameter of longitudinal bars = 25 mm
Spacing of transverse bars = 100 mm
Cover = 40 mm

Solution

For Square Cross-Sections

$$b'' = 400 - 2(40) = 320 \text{ mm}$$

$$d'' = 400 - 2(40) = 320 \text{ mm}$$

$$f_{lx} = \rho_x f_{yh} = 3.3459$$

$$f_{ly} = \rho_y f_{yh} = 3.3459$$

where ρ_x and ρ_y are reinforcement ratio of transverse reinforcements.

$$\frac{f_{lx}}{f_c'} = \frac{1.858}{30} = 0.083$$

$$\frac{f_{ly}}{f_c'} = \frac{1.858}{30} = 0.083$$

Using Fig. 6.4 of Mander et al., 1988, we can determine ratio of f_{cc}'/f_c' which in our case is 1.6. Hence

$$\frac{f_{cc}'}{f_c'} = 1.75$$

$$f_{cc}' = 1.75 \times 40 = 70 \text{ MPa}$$

$$\varepsilon_{cc} = \varepsilon_{co}\left[1 + 5\left(\frac{f_{cc}'}{f_c'} - 1\right)\right] = 0.002\left[1 + 5\left(\frac{70}{40} - 1\right)\right] = 0.0095$$

$$E_{sec} = \frac{f_{cc}'}{\varepsilon_{cc}} = \frac{70}{0.0095} = 7386.42 \text{ MPa}$$

$$E_c = 5000 \times \sqrt{40} = 31,622.78 \text{ MPa}$$

$$n = \left[1 + 5\left(\frac{E_c}{E_c - E_{sec}}\right)\right] = \frac{31,622.78}{31,622.78 - 7368.421} = 1.303$$

Now equation for the stress—strain curve can be written as

$$\frac{f_c}{f_{cc}'} = \frac{n\left(\frac{\varepsilon_c}{\varepsilon_{cc}}\right)}{(n-1) + \left(\frac{\varepsilon_c}{\varepsilon_{cc}}\right)^n}$$

$$f_c = 70 \times \left[\frac{1.303 \times \frac{\varepsilon_c}{0.0095}}{(1.303 - 1) + \left(\frac{\varepsilon_c}{0.0095}\right)^{1.303}}\right]$$

For Rectangular Cross-Section

$$b'' = 320 - 2(40) = 240 \text{ mm}$$

$$d'' = 500 - 2(40) = 420 \text{ mm}$$

$$f_{lx} = \rho_x f_{yh} = 2.54$$

$$f_{ly} = \rho_y f_{yh} = 4.46$$

where ρ_x and ρ_y are reinforcement ratio of transverse reinforcements.

$$\frac{f_{lx}}{f_c'} = \frac{2.54}{40} = 0.063$$

$$\frac{f_{ly}}{f_c'} = \frac{4.46}{40} = 0.111$$

Using Fig. 6.4 of Mander et al., 1988, we can determine ratio of f_{cc}'/f_c' which in our case is 1.6. Hence

$$\frac{f_{cc}'}{f_c'} = 1.65$$

$$f_{cc}' = 1.65 \times 40 = 66 \text{ MPa}$$

$$\varepsilon_{cc} = \varepsilon_{co}\left[1 + 5\left(\frac{f_{cc}'}{f_c} - 1\right)\right] = 0.002\left[1 + 5\left(\frac{66}{40} - 1\right)\right] = 0.0085$$

$$E_{sec} = \frac{f_{cc}'}{\varepsilon_{cc}} = \frac{66}{0.0085} = 7764.706 \text{ MPa}$$

$$E_c = 5000 \times \sqrt{40} = 31{,}622.78 \text{ MPa}$$

$$n = \left[1 + 5\left(\frac{E_c}{E_c - E_{sec}}\right)\right] = \frac{31{,}622.78}{31{,}622.78 - 7764.706} = 1.325$$

Now equation for the stress–strain curve can be written as

$$\frac{f_c}{f_{cc}'} = \frac{n\left(\dfrac{\varepsilon_c}{\varepsilon_{cc}}\right)}{(n-1) + \left(\dfrac{\varepsilon_c}{\varepsilon_{cc}}\right)^n}$$

$$f_c = 66 \times \left[\frac{1.325 \times \dfrac{\varepsilon_c}{0.0085}}{(1.325 - 1) + \left(\dfrac{\varepsilon_c}{0.0085}\right)^{1.325}}\right]$$

For Circular Cross-Sections

$$d'' = 450 - 2(40) = 370 \text{ mm}$$

$$\rho_s = \frac{2\pi \dfrac{d''}{2} A_s}{\pi \dfrac{d''^2}{4} {''}_s} = \frac{2\pi \times \dfrac{370}{2} \times 129}{\pi \dfrac{370^2}{4} \times 100} = 0.0139$$

$$A_e = \pi \frac{d''^2}{4} = \pi \times \frac{370^2}{4} = 107{,}521 \text{ mm}^2$$

$$A_c = \pi \times \frac{(370 - 12.7)^2}{4} = 100{,}266.5 \text{ mm}^2$$

$$\rho_{cc} = \frac{16\pi d^2}{4} \times \frac{1}{A_e} = \frac{16\pi(25)^2}{4 \times 107{,}521} = 0.07305$$

Now

$$A_{cc} = A_c\left(1 - \rho_{cc}\right) = 100{,}266.5 \times (1 - 0.07305) = 92{,}942.4 \text{ mm}^2$$

$$k_e = \frac{A_e}{A_{cc}} = \frac{107{,}521}{92{,}942.4} = 1.156$$

$$f_t' = \frac{1}{2}k_e\rho_s f_{yh} = 0.5 \times 1.156 \times 0.07305 \times 415 = 3.347$$

$$f_{cc}' = f_c'\left[-1.254 + 2.254\sqrt{1 + \frac{7.94f_t'}{f_c'} - 2\frac{f_t'}{f_c'}}\right]$$

$$= 40 \times \left[-1.254 + 2.254\sqrt{1 + \frac{7.94 \times 3.347}{40}} - 2 \times \frac{3.347}{40}\right] = 59.46\,\text{MPa}$$

$$\varepsilon_{cc} = \varepsilon_{co}\left[1 + 5\left(\frac{f_{cc}'}{f_c'} - 1\right)\right] = 0.002\left[1 + 5\left(\frac{59.46}{40} - 1\right)\right] = 0.00687$$

$$E_{sec} = \frac{f_{cc}'}{\varepsilon_{cc}} = \frac{59.46}{0.00687} = 8660.38\,\text{MPa}$$

$$E_c = 5000 \times \sqrt{40} = 31{,}622.78\,\text{MPa}$$

$$n = \left[1 + 5\left(\frac{E_c}{E_c - E_{sec}}\right)\right] = \frac{31{,}622.78}{31{,}622.78 - 8228.588} = 1.377$$

Now equation for the stress–strain curve can be written as

$$\frac{f_c}{f_{cc}'} = \frac{n\left(\dfrac{\varepsilon_c}{\varepsilon_{cc}}\right)}{(n-1) + \left(\dfrac{\varepsilon_c}{\varepsilon_{cc}}\right)^n}$$

$$f_c = 59.46 \times \left[\frac{1.377 \times \dfrac{\varepsilon_c}{0.00687}}{(1.377 - 1) + \left(\dfrac{\varepsilon_c}{0.00687}\right)^{1.377}}\right]$$

Results

The equations derived for square, rectangular, and circular sections are written below, respectively.

$$f_c = 70 \times \left[\frac{1.303 \times \dfrac{\varepsilon_c}{0.0095}}{(1.303 - 1) + \left(\dfrac{\varepsilon_c}{0.0095}\right)^{1.303}}\right]$$

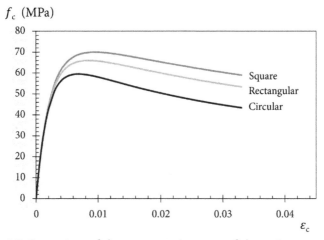

Figure Ex. 6.9 Comparison of the stress—strain curves of three given cross-sections generated using the Mander et al. (1988) model.

$$f_c = 66 \times \left[\frac{1.325 \times \dfrac{\varepsilon_c}{0.0085}}{(1.325 - 1) + \left(\dfrac{\varepsilon_c}{0.0085}\right)^{1.325}} \right]$$

$$f_c = 59.46 \times \left[\frac{1.377 \times \dfrac{\varepsilon_c}{0.00687}}{(1.377 - 1) + \left(\dfrac{\varepsilon_c}{0.00687}\right)^{1.377}} \right]$$

Using above equations derived from the Mander et al. (1988) model, the stress—strain curve for confined concrete for the given cross-sections is constructed by varying value of ε_c and is shown in Fig. Ex. 6.9.

Li et al. Confinement Model

Problem 6: RC cross-section is shown in Fig. Ex. 6.10. The unconfined compressive strength of concrete is $f_c' = 40$ MPa. The concrete is confined by hoops of 12 mm diameter and at 100 mm center-to-center spacing. Diameter of long bars is 25 mm. Yield strength of steel is 345 MPa. Clear cover is assumed to be 40 mm. Generate stress—strain curve using the Li et al. confinement (2000) model provided $A_s = 113$ mm^2, $\varepsilon_o = 0.002$ and $\varepsilon_c = 0.03$. Also carry out the parametric study by varying compressive strength of concrete, yield strength of hoops, spacing, and diameters of hoops.

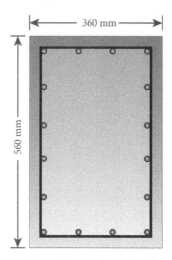

16–25 mm diameter longitudinal bars
Hoops of 12 mm diameter bars @100 mm

Figure Ex. 6.10 The given RC cross-section.

Given

$f_c' = 40$ MPa
$f_{yh} = 345$ Mpa
$b = 360$ mm
$d = 560$ mm
$A_s = 113$ mm^2
Diameter of transverse bars $= 12$ mm
Diameter of longitudinal bars $= 25$ mm
Spacing of transverse bars $= 100$ mm
Cover $= 40$ mm

Solution

$$b'' = 360 - 2(40) = 280 \text{ mm}$$

$$d'' = 560 - 2(40) = 480 \text{ mm}$$

$$\rho_s = \frac{2(b'' + d'')A_s}{b''d''s} = \frac{2(280 + 480) \times 113}{280 \times 480 \times 100} = \frac{171,760}{13,440,000} = 0.0127$$

$$A_e = 280 \times 480 = 134,400 \text{ mm}^2$$

$$A_c = 268 \times 468 = 125{,}424 \text{ mm}^2$$

$$\rho_{cc} = \frac{16\pi d^2}{4} \times \frac{1}{A_e} = \frac{16\pi(25)^2}{4 \times 134{,}400} = 0.0584$$

Now

$$A_{cc} = A_c\left(1 - \rho_{cc}\right) = 125{,}424(1 - 0.0584) = 118{,}094.6$$

$$k_e = \frac{A_e}{A_{cc}} = \frac{134{,}400}{118{,}094.6} = 1.138$$

$$f_1' = \frac{1}{2}k_e\rho_s f_{yh} = 0.5 \times 1.138 \times 0.0127 \times 345 = 2.50$$

$$\alpha_s = \left(21.2 - 0.35 f_c'\right)\frac{f_1'}{40} = [21.2 - (0.35 \times 40)]\frac{2.50}{40} = 0.4516$$

$$f_{cc}' = f_c'\left[-1.254 + 2.254\sqrt{1 + \frac{7.94 f_1'}{f_c'}} - 2\alpha_s \frac{f_1'}{f_c'}\right]$$

$$f_{cc}' = 40 \times \left[-1.254 + 2.254\sqrt{1 + \frac{7.94 \times 2.50}{40}} - 2 \times 0.4516 \times \frac{2.50}{40}\right]$$

$$= 57.92$$

$$\varepsilon_{cc} = \varepsilon_{co}\left[1 + 5\left(\frac{f_{cc}'}{f_c'} - 1\right)\right] = 0.002\left[1 + 5\left(\frac{57.92}{40} - 1\right)\right] = 0.006481$$

$$E_c = 5000\sqrt{f_c'} = 5000 \times \sqrt{40} = 31{,}622.77 \text{ MPa}$$

Equations for the Stress–Strain Curve (Fig. Ex. 6.11)

$$f_c = E_c\varepsilon_c + \frac{f_c' - E_c\varepsilon_{co}}{\varepsilon_{co}^2}\varepsilon_c^2 = 31{,}622.77\varepsilon_c + \frac{40 - 31{,}622.77(0.002)}{(0.002)^2}\varepsilon_c^2$$

$$\varepsilon_{co} < \varepsilon_c < \varepsilon_{cc}$$

$$f_c = f_{cc}' - \frac{\left(f_{cc}' - f_c'\right)}{\left(\varepsilon_{cc} - \varepsilon_{co}\right)^2}\left(\varepsilon_c - \varepsilon_{cc}\right)^2$$

$$f_c = 60.072 - \frac{(60.072 - 40)}{(0.00648 - 0.002)^2}\left(\varepsilon_c - 0.00648\right)^2$$

$$\varepsilon_c > \varepsilon_{cc}$$

$$f_c = f'_{cc} - \beta \frac{f'_{cc}}{\varepsilon_{cc}} (\varepsilon_c - \varepsilon_{cc})$$

$$\beta = \left(0.048 f'_c - 2.14\right) - (0.098 f'_c - 4.57) \left(\frac{f_1}{f'_c}\right)^{\frac{1}{3}}$$

$$\beta = [(0.048 \times 40) - 2.14] - [(0.098 \times 40) - 4.57] \left(\frac{2.50}{40}\right)^{\frac{1}{3}} = 0.038$$

$$f_c = 57.92 - 0.038 \left(\frac{57.92}{0.00648}\right) (\varepsilon_c - 0.00648)$$

Results

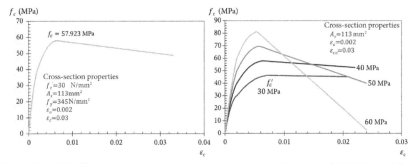

Figure Ex. 6.11 The stress–strain curve generated using the Li et al. (2000) model.

Yong et al. Confinement Model

Problem 7: RC cross-section is shown in Fig. Ex. 6.12. The unconfined compressive strength of concrete is $f'_c = 40$ MPa. The concrete is confined by hoops of 12 mm diameter and at 100 mm center-to-center spacing. Diameter of long bars is 25 mm. Yield strength of steel is 415 MPa. Clear cover is assumed to be 40 mm. Generate stress–strain curve using the Yong et al. confinement (1989) model provided $A_s = 113$ mm^2, $\varepsilon_o = 0.002$ and $\varepsilon_c = 0.03$. Also carry out the parametric study by varying compressive strength of concrete, yield strength of hoops and spacing of hoops.

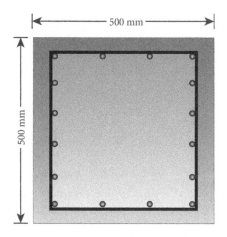

16–25 mm diameter longitudinal bars
Hoops of 12 mm diameter bars @100 mm

Figure Ex. 6.12 The given RC cross-section.

Given

$b = 500$ mm

$d = 500$ mm

$A_s = 113$ mm^2

Diameter of transverse bar $= 12$ mm

Diameter of longitudinal bar $= 25$ mm

$f_c' = 40$ MPa

$f_y = 415$ MPa

Cover $= 40$ mm

Spacing of transverse bars $= 100$ mm

Solution

$$b'' = 500 - 2(40) = 420 \text{ mm}$$

$$d'' = 500 - 2(40) = 420 \text{ mm}$$

$$p'' = \frac{2(b'' + d'')A_s}{b''d''s} = \frac{2(420 + 420) \times 113}{420 \times 420 \times 100} = 0.0107$$

$$A_e = 420 \times 420 = 176{,}400 \text{ mm}^2$$

$$A_c = 408 \times 408 = 166{,}464 \text{ mm}^2$$

$$\rho_{cc} = \frac{16\pi d^2}{4} \times \frac{1}{A_e} = \frac{16\pi(25)^2}{4 \times 176,400} = 0.0445$$

Now

$$A_{cc} = A_c\left(1 - \rho_{cc}\right) = 166,464(1 - 0.0445) = 159,052.405 \text{ mm}^2$$

$$k_e = \frac{A_e}{A_{cc}} = \frac{176,400}{159,052.405} = 1.109$$

$$f_t' = \frac{1}{2}k_e\rho_s f_{yh} = 0.5 \times 1.109 \times 0.0107 \times 415 = 2.47$$

$$f_{cc}' = k_e f_c' = 1.109 \times 40 = 44.36 \text{ MPa}$$

$$\varepsilon_{cc} = \varepsilon_o\left[1 + 5\left(\frac{f_{cc}'}{f_c'} - 1\right)\right] = 0.002\left[1 + 5\left(\frac{44.36}{40} - 1\right)\right] = 0.00309$$

$$E_c = 0.0309 \times (2400)^{1.5} \times \sqrt{40} = 26,770.06 \text{ MPa}$$

$$\varepsilon_{co} = \frac{0.0265 + 0.0035\left(1 - \dfrac{0.734s}{d''}\right)\left(p''f_y''\right)^{\frac{2}{3}}}{\sqrt{f_c'}}$$

$$\varepsilon_{co} = \frac{0.0265 + 0.0035\left(1 - \dfrac{0.734 \times 100}{420}\right)(0.0107 \times 415)^{\frac{2}{3}}}{\sqrt{40}} = 0.005429$$

$$f_i = f_{cc}'\left[0.25\left(\frac{f_c'}{f_{cc}'}\right) + 0.4\right] = 44.36\left[0.25\left(\frac{40}{44.36}\right) + 0.4\right] = 27.744 \text{ MPa}$$

$$\varepsilon_i = k_e\left[1.4\left(\frac{\varepsilon_o}{k_e}\right) + 0.003\right] = 1.109\left[1.4\left(\frac{0.002}{1.109}\right) + 0.003\right] = 0.006127$$

$$f_{2i} = f_{cc}'\left[0.0025\left(\frac{f_{cc}'}{1000}\right) - 0.065\right] \geq 0.3f_{cc}'$$

$$f_{2i} = 44.36\left[0.0025\left(\frac{44.36}{1000}\right) - 0.065\right] \geq 0.3f_{cc}' = -2.878 \ngeq 13.308 \text{ MPa}$$

Hence

$$f_{2i} = 13.308 \text{ MPa}$$

$$\varepsilon_{2i} = 2\varepsilon_i - \varepsilon_{co} = 2(0.006127) - 0.005692 = 0.006825$$

$$E_i = \left(\frac{f_i}{\varepsilon_i}\right) = \frac{27.744}{0.006127} = 4528.154 \text{ MPa}$$

$$E_{2i} = \left(\frac{f_{2i}}{\varepsilon_{2i}}\right) = \frac{13.308}{0.006825} = 1949.94$$

Now the factors are calculated as

$$A = \left(\frac{E_c \varepsilon_{cc}}{f'_{cc}}\right) = \frac{26{,}770.06 \times 0.00309}{44.36} = 1.864$$

$$B = \left[\frac{(A-1)^2}{0.55}\right] - 1 = 0.36$$

$$C = \left(\frac{\varepsilon_{2i} - \varepsilon_i}{\varepsilon_{co}}\right)\left[\frac{\varepsilon_{2i} E_i}{f'_{cc} - f_i} - 4\frac{\varepsilon_i E_{2i}}{f'_{cc} - f_{2i}}\right]$$

$$C = \left(\frac{0.006825 - 0.006127}{0.005692}\right)$$

$$\times \left[\left(\frac{0.006825 \times 4528.154}{44.36 - 27.744}\right) - \left(\frac{4 \times 0.006127 \times 1949.94}{44.36 - 13.308}\right)\right]$$

$$C = 0.04125$$

$$D = (\varepsilon_i - \varepsilon_{2i})\left[\frac{E_i}{f'_{cc} - f_i} - 4\frac{E_{2i}}{f'_{cc} - f_{2i}}\right]$$

$$D = (0.006127 - 0.006825) - \left[\frac{4528.154}{44.36 - 27.744} - \frac{4 \times 1949.94}{44.36 - 13.308}\right]$$

$$D = -0.0148$$

Equations for the Stress—Strain Curve (Fig. Ex. 6.13)

$$\varepsilon_c \leq \varepsilon_{cc}$$

$$\frac{f_c}{f'_{cc}} = \frac{A\left(\dfrac{\varepsilon_c}{\varepsilon_{cc}}\right) + B\left(\dfrac{\varepsilon_c}{\varepsilon_{cc}}\right)^2}{1 + (A-2)\left(\dfrac{\varepsilon_c}{\varepsilon_{cc}}\right) + (B+1)\left(\dfrac{\varepsilon_c}{\dot{\varepsilon}_{cc}}\right)^2}$$

$$f_c = 44.36\left[\frac{1.865\left(\dfrac{\varepsilon_c}{0.003091}\right) + 0.3605\left(\dfrac{\varepsilon_c}{0.003091}\right)^2}{1 - 0.135\left(\dfrac{\varepsilon_c}{0.003091}\right) + \left[1.3605\left(\dfrac{\varepsilon_c}{0.003091}\right)^2\right]}\right]$$

And for

$$\varepsilon_c > \varepsilon_{cc}$$

$$\frac{f_c}{f_{cc}'} = \frac{C\left(\dfrac{\varepsilon_c}{\varepsilon_{cc}}\right) + D\left(\dfrac{\varepsilon_c}{\varepsilon_{cc}}\right)^2}{1 + (C-2)\left(\dfrac{\varepsilon_c}{\varepsilon_{cc}}\right) + (D+1)\left(\dfrac{\varepsilon_c}{\varepsilon_{cc}}\right)^2}$$

$$f_c = 44.36\left[\frac{0.04125\left(\dfrac{\varepsilon_c}{0.003091}\right) + (-0.01488)\left(\dfrac{\varepsilon_c}{0.003091}\right)^2}{1 + \left[(0.04125 - 2)\times\left(\dfrac{\varepsilon_c}{0.003091}\right)\right] + \left[(-0.01488 + 1)\times\left(\dfrac{\varepsilon_c}{0.003091}\right)^2\right]}\right]$$

Results

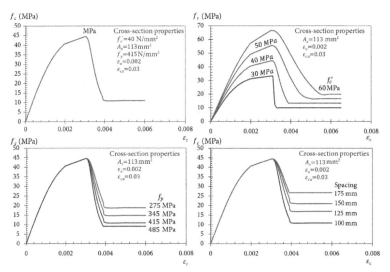

Figure Ex. 6.13 The stress–strain curves for various parameters, generated using the Yong et al. (1989) confinement model. Calculations are shown in example for top left case.

Determination of the Moment–Curvature Curve

Problem 8: RC cross-section is shown in Fig. Ex. 6.14. The compressive strength of concrete is $f_c' = 30$ MPa. Yield strength of steel is 420 MPa. Generate moment–curvature curve for the cross-section

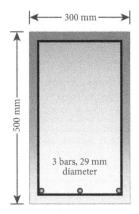

16–25 mm diameter longitudinal bars

Figure Ex. 6.14 The given RC cross-section.

provided $\varepsilon_o = 0.002, \varepsilon_{sh} = 0.006, E_c = 30,000$ MPa, $E_s = 2.4 \times 10^5$ MPa,
$Z = 150, f_r = 3.25$ MPa, $E_{sh} = 3750$ MPa, $f_{su} = 600$ MPa.

Given

$b = 300$ mm
$d = 450$ mm
$h = 500$ mm
$f_c' = 30$ MPa
$f_y = 420$ MPa
$f_r = 3.25$ MPa
$f_{su} = 600$ MPa
$Z = 150$
$\varepsilon_o = 0.002$
$\varepsilon_{sh} = 0.006$
$E_c = 30,000$ MPa
$E_{sh} = 3750$ MPa
$E_s = 2.4 \times 10^5$ MPa

Solution

$$n = \frac{E_s}{E_c} = \frac{2.4 \times 10^5}{30,000} = 8$$

$$A_g = 500 \times 300 = 150,000 \text{ mm}^2$$

Area of one bar is 645 mm^2, so

$$A_s = 3 \times 645 = 1935 \text{ mm}^2$$

$$C_t = \frac{A_g\left(\dfrac{h}{2}\right) - nA_sd}{A_g + A_s} = \frac{150{,}000\left(\dfrac{500}{2}\right) - 8 \times 1935 \times 450}{150{,}000 + 1935}$$

$$= 292.66 \text{ mm} \approx 290 \text{ mm}$$

$$I_{tr} = \frac{bh^3}{12} + bh\left(\frac{h}{2} - C_t\right) + (n-1)A_s(d-C_t)^2$$

$$I_{cr} = \frac{300 \times 500^3}{12} + 300 \times 500\left(\frac{500}{2} - 290\right) + (8-1)1935(450-290)^2$$

$$= 3.465 \times 10^9 \text{ mm}^4$$

Now the first point of $M-\phi$ curve can be calculated as

$$M_1 = \frac{f_t I_{tr}}{C_b} = \frac{3.25 \times 3.465 \times 10^9}{500 - 290} = 5.36 \times 10^7 \text{N--mm}$$

$$\phi_1 = \frac{M_1}{I_{cr} E_c} = \frac{5.36 \times 10^7}{3.465 \times 10^9 \times 30{,}000} = 5.156 \times 10^{-7} \, 1/\text{mm}$$

Point Just Before Cracking

$$\varepsilon_c = \frac{M_1 C_t}{I_{cr} E_c} = \frac{5.36 \times 10^7 \times 290}{3.465 \times 10^9 \times 30{,}000} = 0.000149$$

Set top $\varepsilon_{cm} = 0.0002$
Try $kd = 192.5$ mm

$$\varepsilon_s = \left(\frac{d - kd}{kd}\right)\varepsilon_{cm} = \left(\frac{450 - 192.5}{192.5}\right) \times 0.0002 = 267 \times 10^{-6}$$

$$\varepsilon_{cr} = \frac{f_r}{E_c} = \frac{3.25}{30{,}000} = 108 \times 10^{-6}$$

$$X = \left(\frac{\varepsilon_{cr}}{\varepsilon_{cm}}\right)kd = \left(\frac{108 \times 10^{-6}}{0.0002}\right) \times 192.5 = 103.95$$

$$f_c = \varepsilon_{cm} E_c = 0.0002 \times 30{,}000 = 6 \text{ MPa}$$

$$f_s = \varepsilon_s E_s = 267 \times 10^{-6} \times 2.4 \times 10^5 = 64.08 \text{ MPa}$$

$$C_c = 0.5 f_c bkd = 0.5 \times 6 \times 300 \times 192.5 = 173,250 \text{ N}$$

$$T_c = 0.5 f_r bX = 0.5 \times 3.25 \times 300 \times 103.95 = 50,675.625 \text{ N}$$

$$T_s = A_s f_s = 1935 \times 64.08 = 123,994.8 \text{N}$$

As

$$C_c \cong T_c + T_s$$

Hence the second point of $M-\phi$ curve can be calculated as

$$M_2 = C_c \left(d - \frac{kd}{3} \right) - T_c \left(d - kd - \frac{2X}{3} \right)$$

$$M_2 = 173,250 \left(450 - \frac{192.5}{3} \right) - 50,675.625 \left(450 - 192.5 - \frac{2 \times 103.95}{3} \right)$$

$$= 5.73 \times 10^7 \text{N}-\text{mm}$$

$$\phi_2 = \frac{\varepsilon_{cm}}{kd} = \frac{0.0002}{192.5} = 1.03 \times 10^{-8} 1/\text{mm}$$

Point No 3 ($\varepsilon_{cm} = 0.001$)
From the Hognestad's model,

$$k_1 = 0.417 \text{ and } k_2 = 0.35$$

Try $kd = 182.5$ mm

$$\varepsilon_s = \left(\frac{d - kd}{kd} \right) \varepsilon_{cm} = \left(\frac{450 - 182.5}{182.5} \right) \times 0.001 = 146 \times 10^{-5}$$

$$\varepsilon_{cr} = \frac{f_r}{E_c} = \frac{3.25}{30,000} = 108 \times 10^{-6}$$

$$X = \left(\frac{\varepsilon_{cr}}{\varepsilon_{cm}} \right) kd = \left(\frac{108 \times 10^{-6}}{0.001} \right) \times 182.5 = 19.71 \text{ mm}$$

$$f_s = \varepsilon_s E_s = 146 \times 10^{-5} \times 2.4 \times 10^5 = 350.4 \text{ MPa}$$

$$C_c = k_1 f_c bkd = 0.417 \times 30 \times 300 \times 182.5 = 684,922.5 \text{ N}$$

$$T_c = 0.5 f_r bX = 0.5 \times 3.25 \times 300 \times 19.71 = 9608.625 \text{ N}$$

$$T_s = A_s f_s = 1935 \times 350.4 = 678,024 \text{N}$$

As

$$C_c \cong T_c + T_s$$

Hence the third point of $M-\phi$ curve can be calculated as

$$M_3 = C_c(d - k_2 kd) - T_c\left(d - kd - \frac{2X}{3}\right)$$

$$M_3 = 684{,}922.5(450 - 0.35 \times 182.5)$$

$$- 9608.625\left(450 - 182.5 - \frac{2 \times 19.71}{3}\right)$$

$$= 2.62 \times 10^8 \text{N-mm}$$

$$\phi_3 = \frac{\varepsilon_{cm}}{kd} = \frac{0.001}{182.5} = 5.47 \times 10^{-6} 1/\text{mm}$$

At Yield Point

$$\varepsilon_s = \frac{f_y}{E_s} = \frac{420}{2.4 \times 10^5} = 1.75 \times 10^{-3}$$

Try $kd = 170$ mm, from Fig. Ex. 6.15 and using law of similar triangles,

$$\frac{\varepsilon_{cm}}{kd} = \frac{\varepsilon_s}{d - kd}$$

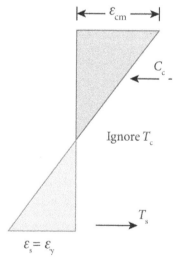

Figure Ex. 6.15 Strain profile at yield point.

$$\varepsilon_{cm} = \frac{\varepsilon_s}{d - kd} \times kd = \frac{1.75 \times 10^{-3}}{450 - 170} \times 170 = 1.0625 \times 10^{-3}$$

Now using the Hognestad's model, k_1 and k_2 can be calculated as

$$k_1 = 0.53$$

$$k_2 = 0.384$$

$$C_c = k_1 f_c b k d = 0.53 \times 30 \times 300 \times 170 = 812,700 \text{ N}$$

$$T_s = A_s f_s = 1935 \times 420 = 812,430 \text{ N}$$

As

$$C_c \cong T_s$$

Hence the fourth point of $M-\phi$ curve can be calculated as

$$M_4 = C_c(d - k_2 kd)$$

$$M_4 = 812,700 \times (450 - 0.384 \times 170) = 3.12 \times 10^8 \text{N}-\text{mm}$$

$$\phi_4 = \frac{\varepsilon_{cm}}{kd} = \frac{1.0625 \times 10^{-3}}{170} = 6.24 \times 10^{-6} 1/\text{mm}$$

Point beyond Yield Point ($\varepsilon_{cm} = 0.003$)

$$\frac{\varepsilon_{cm}}{\varepsilon_o} = \frac{0.003}{0.002} = 1.5$$

Now using the Hognestad's model, k_1 and k_2 can be calculated as

$$k_1 = 0.7527$$

$$k_2 = 0.4145$$

$$\rho = \frac{A_s}{bd} = \frac{1935}{300 \times 450} = 0.0143$$

$$f_s = \frac{k_1 f_c'}{\rho}\left(\frac{\varepsilon_{cm}}{\varepsilon_{cm} + \varepsilon_s}\right) = \frac{0.7527 \times 30}{0.0143}\left(\frac{0.003}{0.003 + \varepsilon_s}\right)$$

Using the above equation, we can plot a curve which intersects the stress—strain curve of steel which is shown in Fig. Ex. 6.16. The stress corresponding to intersecting point is used in calculations.

From Fig. Ex. 6.16, for $\varepsilon_{cm} = 0.003$, the corresponding $f_s = 425$ MPa

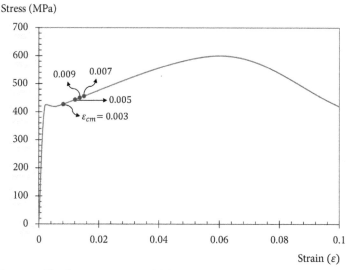

Figure Ex. 6.16 The Stress–strain model for steel including strain hardening.

Now kd can be calculated as

$$kd = \frac{A_s f_s}{k_1 f_c' b} = \frac{1935 \times 425}{0.7527 \times 30 \times 300} = 121.39 \text{ mm}$$

$$C_c = k_1 f_c bkd = 0.7527 \times 30 \times 300 \times 121.39 = 822{,}332.277 \text{N}$$

Hence the fifth point of $M-\phi$ curve can be calculated as

$$M_5 = C_c(d - k_2 kd)$$

$$M_5 = 822{,}332.277 \times (450 - 0.415 \times 121.39) = 3.28 \times 10^8 \text{N}-\text{mm}$$

$$\phi_5 = \frac{\varepsilon_{cm}}{kd} = \frac{0.003}{121.39} = 24.7 \times 10^{-6} 1/\text{mm}$$

Point beyond Yield Point ($\varepsilon_{cm} = 0.005$)

$$\frac{\varepsilon_{cm}}{\varepsilon_o} = \frac{0.005}{0.002} = 2.5$$

Now using the Hognestad's model, k_1 and k_2 can be calculated as

$$k_1 = 0.7316$$

$$k_2 = 0.4824$$

$$\rho = \frac{A_s}{bd} = \frac{1935}{300 \times 450} = 0.0143$$

$$f_s = \frac{k_1 f_c'}{\rho}\left(\frac{\varepsilon_{cm}}{\varepsilon_{cm} + \varepsilon_s}\right) = \frac{0.7316 \times 30}{0.0143}\left(\frac{0.005}{0.005 + \varepsilon_s}\right)$$

From Fig. Ex. 6.16, for $\varepsilon_{cm} = 0.005$, the corresponding $f_s = 450$ MPa
Now kd can be calculated as

$$kd = \frac{A_s f_s}{k_1 f_c' b} = \frac{1935 \times 450}{0.7316 \times 30 \times 300} = 128.53 \text{ mm}$$

$$C_c = k_1 f_c b kd = 0.7316 \times 30 \times 300 \times 128.53 = 870{,}700.779 \text{ N}$$

Hence the sixth point of $M - \phi$ curve can be calculated as

$$M_6 = C_c(d - k_2 kd)$$

$$M_6 = 870{,}700.779 \times (450 - 0.4824 \times 128.53) = 3.45 \times 10^8 \text{ N--mm}$$

$$\phi_6 = \frac{\varepsilon_{cm}}{kd} = \frac{0.005}{128.53} = 38.9 \times 10^{-6} 1/\text{mm}$$

Point Beyond Yield Point $(\varepsilon_{cm} = 0.007)$

$$\frac{\varepsilon_{cm}}{\varepsilon_o} = \frac{0.007}{0.002} = 3.5$$

Now using the Hognestad's model, k_1 and k_2 can be calculated as

$$k_1 = 0.6369$$

$$k_2 = 0.5460$$

$$\rho = \frac{A_s}{bd} = \frac{1935}{300 \times 450} = 0.0143$$

$$f_s = \frac{k_1 f_c'}{\rho}\left(\frac{\varepsilon_{cm}}{\varepsilon_{cm} + \varepsilon_s}\right) = \frac{0.636 \times 30}{0.0143}\left(\frac{0.007}{0.007 + \varepsilon_s}\right)$$

Again from Fig. Ex. 6.16, for $\varepsilon_{cm} = 0.007$, the corresponding $f_s = 460$ MPa
Now kd can be calculated as

$$kd = \frac{A_s f_s}{k_1 f_c' b} = \frac{1935 \times 460}{0.6369 \times 30 \times 300} = 155.28 \text{ mm}$$

$$C_c = k_1 f_c b k d = 0.6369 \times 30 \times 300 \times 155.28 = 890{,}080.488 \text{ N}$$

Hence the seventh point of $M-\phi$ curve can be calculated as

$$M_7 = C_c(d - k_2 kd)$$

$$M_7 = 890{,}080.488 \times (450 - 0.5460 \times 155.28) = 3.25 \times 10^8 \text{N/mm}$$

$$\phi_7 = \frac{\varepsilon_{cm}}{kd} = \frac{0.007}{155.28} = 45.07 \times 10^{-6} 1/\text{mm}$$

Point Beyond Yield Point $(\varepsilon_{cm} = 0.009)$

$$\frac{\varepsilon_{cm}}{\varepsilon_o} = \frac{0.009}{0.002} = 4.5$$

Now using the Hognestad's model, k_1 and k_2 can be calculated as

$$k_1 = 0.5175$$

$$k_2 = 0.6263$$

$$\rho = \frac{A_s}{bd} = \frac{1935}{300 \times 450} = 0.0143$$

$$f_s = \frac{k_1 f_c'}{\rho}\left(\frac{\varepsilon_{cm}}{\varepsilon_{cm} + \varepsilon_s}\right) = \frac{0.5175 \times 30}{0.0143}\left(\frac{0.007}{0.007 + \varepsilon_s}\right)$$

Again from Fig. Ex. 6.16, for $\varepsilon_{cm} = 0.007$, the corresponding $f_s = 445$ MPa

Now kd can be calculated as

$$kd = \frac{A_s f_s}{k_1 f_c' b} = \frac{1935 \times 445}{0.5175 \times 30 \times 300} = 184.879 \text{ mm}$$

$$C_c = k_1 f_c b k d = 0.5175 \times 30 \times 300 \times 184.879 = 861{,}073.9425 \text{ N}$$

Hence the eighth point of $M-\phi$ curve can be calculated as (Fig. Ex. 6.17)

$$M_7 = C_c(d - k_2 kd)$$

$$M_8 = 861{,}073.9425 \times (450 - 0.6263 \times 184.879) = 3.25 \times 10^8 \text{N} - \text{mm}$$

$$\phi_8 = \frac{\varepsilon_{cm}}{kd} = \frac{0.009}{155.28} = 48.6 \times 10^{-6} 1/\text{mm}$$

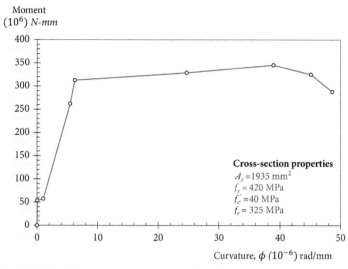

Figure Ex. 6.17 The Moment–curvature curve for the given cross-section.

Results

Point No.	ε_{cm}	kd(mm)	$M(10^6$ N−mm)	$\phi(10^{-6}1/\text{mm})$
0	—	—	0	0
1	—	—	53.625	0.05
2	0.0002	192.5	57.3084	1.03
3	0.001	182.5	262.021	5.47
4	0.00113	175	312.558	6.24
5	0.003	47.06	328.672	24.7
6	0.005	33.79	345.428	38.9
7	0.007	25.35	325.072	45.07
8	0.009	25.35	287.77	48.6

UNSOLVED PROBLEMS

Problem 9: The moment–curvature curve of beam shown in Fig. Ex. 6.18 is given in Fig. Ex. 6.19. Determine the following quantities given $f_c' = 40$ MPa, $f_y = 420$ MPa, and crack width of 0.50 mm.
1. Flexural Stiffness
2. Deflection at the mid span and third point on the beam along the length

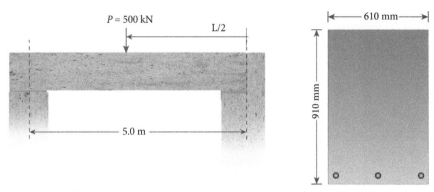

Figure Ex. 6.18 A simply supported beam and its cross-section.

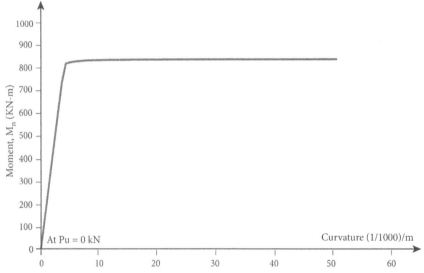

Figure Ex. 6.19 The $M-\phi$ curve of the given cross-section.

3. Strain in the bottom of steel
4. Rotation of the beam at mid span
5. Crack Spacing

Problem 10: RC cross-section is shown in Fig. Ex. 6.20. The unconfined compressive strength of concrete is $f_c' = 40$ MPa. The concrete is confined by hoops of 12 mm diameter and at 100 mm center-to-center spacing. Clear cover is assumed to be 30 mm. Generate stress—strain

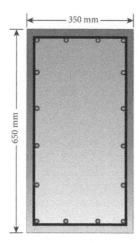

16–25 mm diameter longitudinal bars
Hoops of 12 mm diameter bars @ 100 mm

Figure Ex. 6.20 The given RC cross-section.

curve using the Kent and Park (1971) model provided $A_s = 113$ mm^2, $\varepsilon_o = 0.002$ and $\varepsilon_c = 0.03$.

Problem 11: Determine the stress–strain behavior of the same cross-section as mentioned in Fig. Ex. 6.20 using the modified Kent and Park model (i.e., Scott et al., 1982 model). Also carry out the comparison of the curves generated using the above-mentioned models. Use same given data as mentioned in problem 9.

Problem 12: RC cross-section is shown in Fig. Ex. 6.21. The unconfined compressive strength of concrete is $f_c' = 40$ MPa. The concrete is confined by hoops of 12 mm diameter and at 100 mm center-to-center spacing. Diameter of long bars is 25 mm. Yield strength of steel is 420 MPa. Clear cover is assumed to be 30 mm. Generate stress–strain curve using the Mander et al. confinement (1988) model provided $A_s = 113$mm^2, $\varepsilon_o = 0.002$ and $\varepsilon_c = 0.03$.

Problem 13: RC cross-section is shown in Fig. Ex. 6.22. The unconfined compressive strength of concrete is $f_c' = 30$ MPa. The concrete is confined by hoops of 12 mm diameter and at 100 mm center-to-center spacing. Diameter of long bars is 25 mm. Yield strength of steel is 420 MPa. Clear cover is assumed to be 30 mm. Generate stress–strain curve using the Li et al. confinement (2000) model provided $A_s = 113$ mm^2, $\varepsilon_o = 0.002$ and $\varepsilon_c = 0.03$.

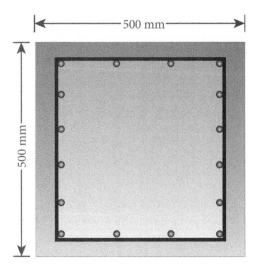

16–25 mm diameter longitudinal bars
Hoops of 12 mm diameter bars @ 100 mm

Figure Ex. 6.21 The given RC cross-section.

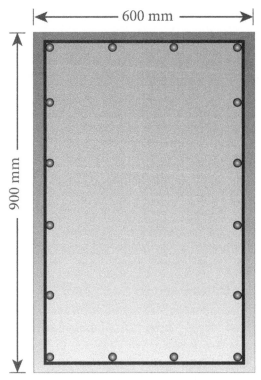

16-25 mm diameter longitudinal bars
Hoops of 12 mm diameter bars @ 100 mm

Figure Ex. 6.22 The given RC cross-section.

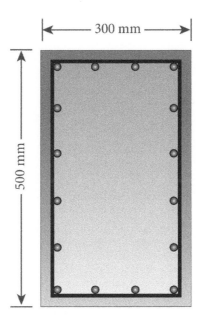

16–25 mm diameter longitudinal bars
Hoops of 12 mm diameter bars @ 100 mm

Figure Ex. 6.23 The given RC cross-section.

Problem 14: RC cross-section is shown in Fig. Ex. 6.23. The unconfined compressive strength of concrete is $f_c' = 27.5$ MPa. The concrete is confined by hoops of 12 mm diameter and at 100 mm center-to-center spacing. Diameter of long bars is 25 mm. Yield strength of steel is 345 MPa. Clear cover is assumed to be 40 mm. Generate stress–strain curve using the Yong et al. confinement (1989) model provided $A_s = 113 \text{mm}^2, \varepsilon_o = 0.002$ and $\varepsilon_c = 0.03$.

Problem 15: RC cross-section is shown in Fig. Ex. 6.24. The compressive strength of concrete is $f_c' = 40$ MPa. Yield strength of steel is 420 MPa. Generate moment–curvature curve for the cross-section provided $\varepsilon_o = 0.002, \varepsilon_{sh} = 0.006, E_c = 30,000$ MPa, $E_s = 2.0 \times 10^5$ MPa, $Z = 150, f_r = 3.50$ MPa, $E_{sh} = 7000$ MPa, $f_{su} = 600$ MPa.

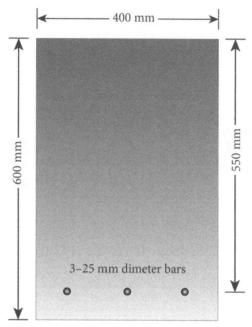

Figure Ex. 6.24 The given RC cross-section.

REFERENCES

Bjerkeli, L., Tomaszewicz, A., & Jensen, J. J. (1990). Deformation properties and ductility of high-strength concrete. In *High-strength concrete: Second international symposium*, (pp. 215–238). Detroit, MI: American Concrete Institute.

Hognestad, E. (1951). *A study of combined bending and axial load in reinforced concrete members. Bulletin* (399). Urbana, IL: University of Illinois Engineering Experiment Station.

Hsu, T. T. C. (1992). *Unified theory of reinforced concrete.* Boca Raton: CRC Press. ISBN 9780849386138.

Kayani, K. R. (2011). *Lectures on advanced concrete structures.* Islamabad, Pakistan: National University of Sciences and Technology (NUST).

Kent, D. C., & Park, R. (1971). Flexural members with confined concrete. *Journal of the Structural Division, Proceedings of the American Society of Civil Engineers, 97*(ST7), 1969–1990.

Li, B., Park, R., & Tanaka, H. (2000). Constitutive behavior of high-strength concrete under dynamic loads. *ACI Structural Journal, 97*(4), 619–629.

Mander, J. B., Priestley, M. J. N., & Park, R. (1988). Theoretical stress–strain model of confined concrete. *Journal of Structural Engineering, 114*(8), 1804–1826.

Scott, B. D., Park, R., & Priestley, M. J. N. (1982). Stress-strain behavior of concrete confined by overlapping hoops at low and high strain rates. *Journal of the American Concrete Institute, 79*, 13–27.

Todeschini, C. E., Bianchini, A. C., & Kesler, C. E. (1964). Behavior of concrete columns reinforced with high strength steels. *ACI Journal Proceedings, 61*(6).

Wight, J. K., & MacGregor, J. G. (2011). *Reinforced concrete: Mechanics and design* (5th ed.). Englewood Cliffs, NJ: Prentice Hall International, Inc.

Yong, Y. K., Nour, M. G., & Nawy, E. G. (1989). Behavior of laterally confined high-strength concrete under axial loads. *Journal of Structural Engineering, 114*(2), 332–351.

FURTHER READING

Park, R., & Paulay, T. (1975). *Reinforced concrete structures*. New York: John Wiley & Sons.

ACECOMS (1997). *General Engineering Assistant and Reference, GEAR. Released by Asian Center for Engineering Computations and Software, ACECOMS.* Thailand: Asian Institute of Technology.

CSI (2001). *Section Builder. Released by Computers and Structures Inc.* Berkeley, CA: CSI.

CSI (2008). *CSICOL 8.4.0. Released by Computers and Structures Inc.* Berkeley, CA: CSI.

CSI (2012). *Section Designer ETABS 2013. Released by Computers and Structures Inc.* Berkeley, CA: CSI.

McGregor, J. G. (1992). *Reinforced concrete: Mechanics and design* (2nd ed.). Upper Saddle River, NJ: Prentice Hall International, Inc.

Nilson, A., Darwin, D., & Dolan, C. (2004). *Design of concrete structures* (14th ed.). New York, NY: McGraw-Hill Higher Education.

CHAPTER SEVEN

Retrofitting of Cross-Sections

Retrofit can be global or local. Local retrofit enhances the capacity and performance of individual components, which in turn enhances the performance of the overall structure. Global retrofit tends to enhance the performance by modifying the structural system itself, either by altering, removing, or adding one or more components or subsystems.

OVERVIEW

Retrofitting is the process of modifying the existing structures in order to achieve desirable performance. The need for strengthening and retrofitting of existing members may arise due to several reasons. The most common objectives of retrofitting are to overcome the design, construction, or material deficiency; recover lost load-carrying capacity caused by major damages or to efficiently carry additional loads that may be applied as a result of the change in the usage of the structure. The process of retrofitting involves the determination of the existing and new demands for the structure and the identification of the member or system to be strengthened. The techniques used to regain the original strength and performance are referred to as "Rehabilitation," "Repair," or "Restoration." On the other hand, the terms "Preservation" and "Conservation" are used for procedures which ensure that the original performance of the structure is maintained over time.

Repair, strengthening, and retrofit of reinforced and prestressed concrete members have become increasingly important issues as the built infrastructure deteriorates with time due to environmental effects. Existing structures often require rehabilitation or retrofit measures in order to meet any changes in occupancy or scope of building.

Structural Cross Sections
DOI: http://dx.doi.org/10.1016/B978-0-12-804443-8.00007-5

483

Inappropriate design and detailing, or need to conform to new building codes may also lead to retrofit.

Strength evaluation may be required if there are any indications of deficiencies in construction process, or a structure is deteriorated, a structure is required to serve a new function, or a structure or a portion does not comply with the requirements of code. The strength evaluation can be done analytically or experimentally. The analytical evaluation is carried out if strength deficiency is well understood, material properties and dimensions required for analysis are known or can be obtained. If it is not feasible to establish these parameters, then load test may need to be carried out.

Once the strength evaluation has been carried out and the needs for retrofitting are established, the designer needs to ensure that some or all of the following criteria are met after retrofitting, such as:

• New performance needs imposed on the members are satisfied.
• New loads are transferred to the new system and/or existing members are relieved from existing loads, as the case may be.
• The old and new systems act together or the old and new systems do not act together, as intended by the designer.

The type and extent of any repair or strengthening work can be decided after ascertaining both the cause as well as the nature of structural inadequacy. The selected technique must enhance the performance of members under the most probable failure mode. Material aging, loading history, modified stress—strain properties, and long-term phenomena such as creep and shrinkage are also need to be considered.

With our increasing understanding of applied actions and resulting responses, as well as with increasing availabilities of data from actual damage under excessive loadings or earthquakes, various procedures have evolved over last few decades to enhance structural performance and safety of existing buildings. These procedures cover a wide range of approaches including the provision of additional structural components or energy-dissipating devices as well as strengthening techniques for existing components. The local response of individual cross-sections to applied loading is the key consideration in later procedures. This chapter deals with some of the important considerations for the selection of retrofit strategy, rehabilitation methodologies, retrofit approaches for cross-sections, and analysis of retrofitted sections.

Refurbishment is a process carried out to improve the existing structure or its facilities by altering internal areas, provision of decorating components and materials, and new facilities and equipment, etc.

Repair is the process of restoring the performance of a damaged building component to its original condition by fixing it.

Renovation refers to the process carried out to upgrade an existing structure to improve performance by either altering the scope of structure, providing additional facilities or improving existing facilities.

Restoration refers to the process carried out to bring a deteriorating structure or its component back to its original design plan. Usually, this is required for old existing structures of any historical importance.

Retrofitting refers to the measures or process carried out to restore or enhance the load carrying capacity and/or performance of a structure or its components. These measures may vary depending upon intended purpose, type of structural component and extent of damage, and are not conceived or foreseen in the original design of structure.

Rebuilding *refers to demolishing and reconstruction of a structure or its components and is usually considered if retrofitting is not an adequate option due to economic or other reasons.*

Strengthening *refers to the measures or process carried out to increase the strength of a structural component. It is considered in cases where expected loading may exceed the capacity of member.*

Rehabilitation *refers to the process carried out to extend the life of a structure and may include options from strengthening and retrofitting. The purpose of rehabilitation is to restore the intended function of a building or structure which is rendered uninhabitable due to some man-made or natural disaster.*

Conservation *refers to the measures taken to reduce structural deterioration due to environmental conditions and is usually associated with historical structures and monuments.*

OVERALL RETROFIT PROCESS AND STRATEGIES FOR STRUCTURES

A retrofit strategy is a systematic approach adopted to improve the probable performance of any existing structure (or its component) or to reduce the existing risk to an acceptable level. Both technical and management strategies can be employed to achieve these goals. The technical strategies may include consideration and approaches to modifying the basic demand and response parameters of the structure and may require the following considerations:

- Measures resulting in overall strengthening of system
- Actions for enhancing the stiffness of system
- Enhancing deformational capacity of load-resisting system
- Enhancing energy dissipation capacity of structure
- Measures resulting in reduced demand on structure.

Under extreme conditions, e.g., lateral dynamic forces resulting from earthquakes and strong winds, the structures may experience lateral displacements causing high levels of inelasticity and damage in various components. For reliable seismic performance, a building must have a complete lateral force resisting system, capable of limiting the lateral displacements. As an example, for earthquake forces, some basic parameters governing this desirable performance are the structure's mass, stiffness, damping and energy-dissipating capacity, structural configuration, deformation capacity of its individual components, and ground motion characteristics. The retrofit and rehabilitation strategies against earthquakes are required to explicitly (or implicitly) consider all of these factors. We will discuss some general considerations which should be evaluated and play a key role in the selection of retrofit strategy. Table 7.1 lists some of the international standards providing explicit guidelines for the evaluation and retrofit of structures, specially for seismic effects.

Provision of an Efficient and Integrated Load Path

Sometimes the existing structures may lack important linking components which are necessary for an integrated and continuous load transfer from its elements to ground. This may result in reduced performance, local failures, and enhanced vulnerability of structures. Common deficiencies include, for example, lack of suitable chord and collector elements for diaphragms, insufficient bearing length at element supports, inadequate connection detailing, and improper bracing of various components. This problem can be avoided by the provision of missing links, by ensuring efficient element connectivity, and by suitable alteration to existing load paths.

Strengthening and Stiffening of Structures

The most common consideration while deciding a retrofit strategy (for the performance improvement of existing structures) is system strengthening and

Table 7.1 Various International Standards Dealing with Seismic Evaluation
and Retrofitting of Existing Building Structures

Organization/Document	Year	Title
Applied Technology Council, report ATC-40	1996	Seismic evaluation and retrofit of concrete buildings
FEMA 356 Federal Emergency Management Agency (FEMA)	2000	Prestandard and Commentary for the Seismic Rehabilitation of Buildings
ASCE/SEI 31-03 American Society of Civil Engineers (ASCE)	2003	Seismic Evaluation of Existing Buildings
Eurocode 8: Part 3	2005	Design of Structures for Earthquake Resistance—Part 3: Assessment and Retrofitting of Buildings
ASCE/SEI 41-06 American Society of Civil Engineers (ASCE)	2007	Seismic Rehabilitation of Existing Buildings
ASCE/SEI 41-06 American Society of Civil Engineers (ASCE)	2007	Supplement No. 1, Seismic Rehabilitation of Existing Buildings

stiffening. These two parameters are different but closely related to each other. The term strengthening refers to procedures adopted to increase the level of external forces required to initiate damage in a structure. Stiffening, on the other hand, results in the reduction of displacements caused by same external forces. Both system strengthening and stiffening techniques are usually performed in parallel to achieve performance objectives. Typical procedures employed for both stiffening and strengthening include the addition of new load-resisting elements, such as columns, shear walls, and infill walls, as need increasing the strength and stiffness of critical structural components. Additional elements may also include external braced framing and buttresses. The provision of special devices (e.g., buckling restrained braces and damping systems) can also be considered. These smart systems result in increased stiffness of overall framing, as well as increased energy dissipation capacity and ductility of structure.

Identification of Possible Failure Modes

It is important to identify all the relevant failure modes possible for the system/member under consideration and ensure that failure and performance needs are met. These failure modes may be divided into two categories: material failures and stability failures, as are listed in Table 7.2.

Table 7.2 Possible Failure Modes in System/Member/Section

Material Failures	Stability Failures
Direct tension failure	Axial—flexural buckling
Direct compression failure	Lateral buckling
Direct flexure failure	Lateral torsional buckling
Direct shear failure	Local buckling
Failure in shear due to tension failure	Crippling
Failure in shear due to compression failure	$P-\Delta$ effects
	Sliding
	Overturning
	Fatigue
	Bond, splice, anchorage failure
	Loss of prestress

Once the relevant failure modes as listed above have been identified, the designer must ensure that these are prevented. The basic methods adopted to prevent failure are:

1. Limit the modes of failure (prevent buckling, sliding, etc.)
2. Limit allowable stresses (adequate strength design)
3. Limit combined stresses and ratios (consider interaction of actions)
4. Redistribute forces to other components.

Increase in Deformational Capacity of Sections and Members

Stiffness attracts the load. One way to reduce undesirable damage is to introduce local stiffness reductions or modifications resulting in controlled damage only to selected unimportant components of structure. Sometimes extraordinarily stiff elements may attract unproportionally high force demands producing additional torsional moments and discontinuities in load paths. This can also be avoided by adequately reducing the unnecessary stiffness in various members which also results in significant cost reductions.

An efficient method of improving the overall structural performance (especially against strong lateral forces, e.g., resulting from ground motions) is to improve the deformational capacity of individual components of a structure. Some methods to achieve this goal include the following measures:

1. *Provision of Adequate Confinement*: Confinement is the key to enhance the deformational capacity of any structural member. The provision of exterior confinement (e.g., through jacketing of steel, reinforced

concrete, or fiber-reinforced plastic fabrics) to beams and columns results in enhanced deformational capacity, ductility, and improved performance.

2. *Strong Columns—Weak Beams Principle*: A common design discrepancy is strong beam—weak column formation of most gravity load-designed structures. Such configuration is vulnerable to the development of soft-story mechanism (under strong lateral forces) in which drift demands concentrate in single story resulting in significant amounts of local inelastic deformations. The columns and primary lateral load-resisting components require proper stiffness, strength, and deformational capacity to avoid such mechanisms.

Reducing Force Demands on Sections and Members

The conventional approach towards structural design is to provide them with sufficient strength to withstand loads and with the ability to deform in a ductile manner. Structures designed following this approach may have several limitations. With constant load-carrying and energy-dissipating capacities, conventionally designed structures may not be able to adapt to changing environmental conditions and excitations. They have to rely on their inherent damping (which may not be sufficient) to dissipate energy imposed by ideally unknown external excitations. With the developments in structural control measures, various techniques based on the modification of force demands on structures have been developed. The focus is towards controlling the demand instead of capacity or response. As an example, for lateral dynamic forces resulting from strong ground motions, this can be achieved either by modifying the structure's mass or by using any energy-dissipating device or base isolation technique. The modification in structure's mass results in change in natural time period which can ultimately be used to control the seismic forces. Various control devices (mass dampers/viscous dampers, tendons, braces, etc.) are also available which serve as a source of additional energy-dissipating capacity. Base isolation systems are intended to cut off or reduce the energy transmission of earthquake ground motions to the structure. Systems that apply a control force or alter vibration characteristics resulting in increased difference between predominant period of ground motion and natural period of structure can also be used to control seismic demand. The use of such special systems also results in significantly higher cost compared to conventional approaches for rehabilitation strategy. Therefore, they are often recommended for only important structures and critical occupancies (Anwar, Aung, & Najam, 2016).

Esthetic Considerations

Esthetics is an important factor while selecting a retrofit strategy. It is important that retrofitting does not significantly alter the usability and visual appearance of the structure. Often, special attention to the minor details and explicit architectural considerations can improve the aesthetic value and acceptability of retrofit options, without affecting the overall cost and constructability. Techniques that result in the reduction of ceiling heights or corridor widths are usually undesirable.

Final Selection of Retrofit Scheme

It is important to set the basic performance objectives and the related structural discrepancies prior to the start of process. All relevant options and approaches for rehabilitation strategy should be evaluated. After the detailed comparison, the most suitable option practically and economically capable of addressing the identified discrepancies is selected. Various analysis procedures are available to quantitatively identify most suitable measure resulting in best performance in comparatively less cost. The feasibility and requirements for available options can be compared in terms of reduced demand-to-capacity ratios. Cost estimates are made considering each strategy and benefit versus effectiveness relationship for all options is evaluated against the cost of new construction. The choice between the retrofit and rehabilitation versus the construction of new facility may depend on a predecided criteria based on the experience and budget limitations. Typically, if the retrofit cost is more than 20−30% of new construction, then new construction may be preferred. Researchers in different countries have suggested that this percentage should be linked to the age of the existing buildings. This could be in the range 30−50% for older buildings (older than 50% of expected life) and may be as high as 60−70% for new construction. While comparing the cost of retrofit versus the new construction, it is important to consider the "total project cost" of new construction including all allied works, and not only the structural cost. In general, it is preferable to rebuild or replace an older building, and retrofit or rehabilitate a newer building. The justification is that replacing an older building not only improves its structural resilience but also improves its overall functionality and restart its usable life.

Retrofitting an older building does not necessarily extend its expected life and may not provide the functional and architectural enhancement. However, the unavailability of funds and other financial limitations may override this consideration and lead to the selection of retrofit option.

Preliminary Retrofit Design

After the selection of final retrofit strategy, the next stage is to develop a preliminary retrofit design which includes the determination of approximate member sizes and the selection of optimum locations for all important elements involved in retrofit process. The preliminary design may either be used for the development of initial cost estimates for comparing a certain retrofit strategy with available options or for the coordination between various stakeholders. The first step is to determine demands and corresponding capacity parameters for the unretrofitted structure. The comparison of these indicators provides the expected performance of structure without any retrofitting. If structure is incapable of meeting the desired performance objectives, the deficiencies and corresponding rehabilitation measures are identified. For most building and systems especially the multistory frame or shear wall systems, lateral deflection is the most important demand parameters for structural performance against earthquake ground motions, whereas gravity load-carrying capacity is a consideration for other cases.

Basic Retrofitting Approaches

The retrofit or rehabilitation approaches can be broadly classified as
- Local retrofit
- Global retrofit
- Combination of local and global retrofits.

Local retrofit refers to the enhancement of the capacity and performance of individual components, which in turn enhances the performance of the overall structure or system. Global retrofit tends to enhance the overall performance of the system by modification of the structural system itself, either by altering, removing, or adding one or more components or subsystems. Global retrofit can be used to divert or reduce demands on the components that may be too difficult or too expensive to retrofit. Often a combination of global and local retrofits results in the optimum solution. Fig. 7.1 shows various global and local retrofit technologies. A variety of both local and global techniques have been implemented worldwide (especially for seismic actions). Retrofit for gravity loads

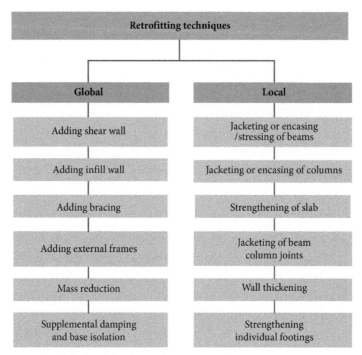

Figure 7.1 Basic global and local retrofitting techniques.

often relies on local strengthening measures for beams, slabs, columns, etc. Fig. 7.2 shows the lateral force—displacement response of a portal frame retrofitted using different methods. The relative effect of each rehabilitation measure can be easily evaluated based on such comparisons.

A flowchart explaining the overview of all retrofitting techniques for masonry and concrete buildings is given in Fig. 7.3. It covers both local and global techniques being used nowadays. Subsequent sections present their details and important factors affecting their selection as a suitable remedy to rehabilitate a damaged building or component.

Lateral Thinking Approach for Retrofitting

In formulating the strengthening or retrofitting solution, it often helps to consider unusual ideas or approaches by concentrating on the source of the problem rather than the problem itself. Some of the examples are:

1. Rather than strengthening a weak member which may be difficult due to construction, operational constraints, alternate load paths may be provided to reduce the action on the members, or strengthen/stiffening nearby members to redistribute the loads.

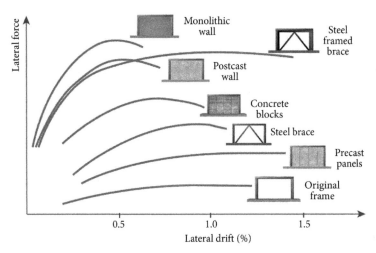

Figure 7.2 Force–displacement response of a portal frame with different retrofitting techniques. *Sugano, 1996.*

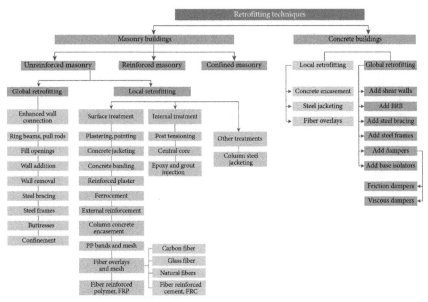

Figure 7.3 Overview of some retrofitting techniques for masonry and concrete buildings.

2. Avoid strengthening methods that increase dead load. For example, if a flat slab is weak in punching shear, consider adding column capitals rather than an attempt to increase thickness.

3. For continuous and indeterminate members, strengthen location that is easily accessible even if it is not at weak location as the new loads

will be redistributed to section of higher stiffness and old loads can be resisted with the help of plastic deformations.

4. If many members in a structure are found to be defective or need strengthening, consider modifying the basic structural system. For example, a moment resisting frame may be converted to braced frame by adding external/internal bracing/walls.

5. Utilize nonstructural components as structural components. For example, the partition walls, door and window frames, floor finishes, etc., can be modified to assist in carrying loads.

6. In case of foundation strengthening, consider reevaluating or investigating the soil properties by retesting. Often modifying the factor of safety from 3 to 2.5 or by computing the allowable bearing capacity for each footing individually can eliminate or reduce the need to strengthen the footing.

RETROFITTING TECHNIQUES FOR CROSS-SECTIONS

As the main focus of this book is the analysis and design of cross-sections, we will not focus on various approaches and considerations for retrofitting of cross-sections. These approaches tend to enhance the capacity and performance at cross-section level, which in turn enhance the performance of the overall member, structure, or system. These techniques include treatment on surface as well as internal treatment. Often a combination of both is also used in order to get optimum results.

Surface Treatment, Plastering, and Pointing

This is the most common approach to rehabilitate or improve an existing reinforced concrete or masonry section. These techniques involve the application of some reinforcing and/or confining material on surface only and are relatively easy to practice. These techniques are largely developed through experience and directly affect the architectural or historical appearance of the structure. Traditionally, Portland cement plaster is applied to brick masonry which results in durable, attractive, and reasonably weather-resistant walls. On the other hand, unplastered masonry is vulnerable to all kinds of environmental deteriorations and long-term effects. In some cases, another efficient way referred to as Pointing is applied, in which the original mortar in bed joints is leveled and replaced with a better quality mortar. This results in better structural performance and enhanced integrity against applied actions. In some cases, the use of reinforced plaster may also be a suitable option to enhance capacity of masonry wall sections. In this technique, a layer of cement plaster strengthened with steel reinforcement is applied to walls which results in improved appearance especially under seismic actions. Surface treatment is often not considered as structural retrofit and may not significantly enhanced strength or performance.

Jacketing and Encasement

The covering of a lower strength material with a higher strength material is referred to as jacketing. Usually steel plates or sections are jacketed over concrete sections for additional strength. Encasing is the covering of a material with another material of equal or lower strength. Examples of encasing are concrete over concrete or steel sections. Jacketing is the most common method for retrofitting and strengthening of cross-sections. Fig. 7.4 shows some of the application of the jacketing and encasing techniques. Primary objective of jacketing or encasing is to enhance capacity of cross-sections. Main purposes fulfilled by jacketing are:

- To increase axial—flexural capacity of cross-section by adding new reinforcements
- To enhance concrete confinement by the introduction of transverse reinforcements
- To increase the shear and tension capacity of cross-section by additional thickness or reinforcements.

Jacketing can be done in numerous ways but the most common approaches are discussed below.

Concrete Jacketing or Encasing

Concrete jacketing is the most commonly used method to rehabilitate severely damaged structural components (e.g., shear walls) under strong ground motions. This method is also the retrofit measure for masonry walls where cement or concrete coatings can be applied to one or both faces of walls. It must be noted that repairing of damaged regions of cross-section should be carried out prior to jacketing. In cases where the masonry is plastered, plaster should be removed first. The cracks in the wall are grouted, surface is cleaned and moistened with water. In cases where reinforced concrete is used, the reinforcing bars are threaded through the holes which are then grouted with either cement paste or

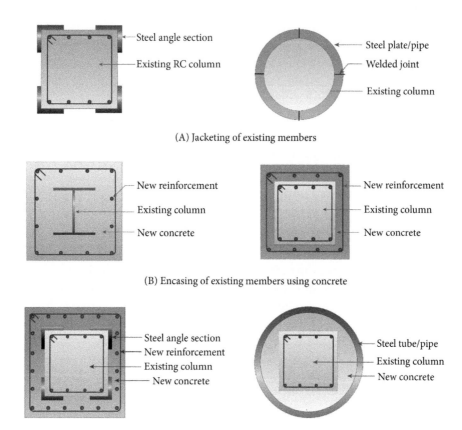

(A) Jacketing of existing members

(B) Encasing of existing members using concrete

(C) Combined jacketing and encasing

Figure 7.4 Some basic jacketing and encasing techniques.

suitable epoxy material. If reinforcing mesh is used, special attention should be paid towards its connection with anchors. Lastly, the finishing layer consisting of cement paste is applied. Concrete used for jacketing should have same properties with the one used in the construction. It is recommended that compressive strength should be more than used for the construction. Monolithic behavior of composite column must be assured and four sided jacketing is recommended. Minimum area of transverse and longitudinal reinforcements must be evaluated on the basis of standards and codes used for design.

Steel Jacketing

Confining RC columns in steel jackets is also an effective method to increase basic strength capacity. Steel jacketing not only provides enough confinement but also prevents deterioration of shell concrete, which is the main reason of bond failure and buckling of longitudinal bars. Various studies related to cyclic load testing of columns have shown that otherwise brittle columns can be retrofitted to perform extremely well up to as much as the displacement ductility ratio of 7. For rectangular columns, rectangular steel jacket may not be as effective as circular jackets and therefore, the use of stiffeners can also be considered.

FRP Jacketing

Fiber-reinforced polymer (FRP) jacketing is a relatively new technique of jacketing in which strengthening is carried out by using composite jackets made up of FRPs. Recent studies are investigating feasibility of using FRP to improve seismic capacity of cross-sections by wrapping them with high-strength carbon fibers around the cross-section. FRP jacketing can also provide very efficient confinement due to high-strength carbon fiber and high modulus of elasticity compared to both steel and reinforced concretes. Another advantage of using FRP is that it is light weight and rust free. However, the FRP may be more expensive than steel or concrete jacketing.

Use of FRP Laminates, Fiber-Reinforced Cement, and Mesh

The use of fiber-reinforced plastic is a relatively new technique for improving strength of cross-section members of the structures. FRP composite materials can be used for jacketing and wrapping of cross-section as mentioned in the previous section. The main advantage associated with using such materials is its high specific stiffness and specific weight with addition to high durability. Advanced composite materials such as carbon fiber-reinforced plastic (CFRP) are gaining popularity nowadays and are much stronger and lighter than steel. Various recent studies based on extensive experiments have shown that wrapping structural members with carbon fiber-reinforced plastic sheets results in improved strength and ductility without adding stiffness to the elements. Improved installation methods are making the use of CFRP sheets a very economical and efficient alternative for the retrofitting of existing structures. Reinforcing fibers can be classified into four basic categories as follows:

- Carbon Fibers
- Glass Fibers
- Natural Fibers
- Polymer Grids.

The primary advantage of strengthening cross-sections with FRP composite materials (enhancement of deformation capacity) is limited only to members with relatively low deficiencies in terms of resistance against applied actions. In most of practical cases, more than one retrofitting techniques need to be adopted at cross-section level as part of overall rehabilitation strategy. Another related technique is the use of high-strength fiber glass grid or mesh used in conjunction with thin layer of fiber-reinforced cement (FRC). Applying a FRC overlay to masonry walls results in both enhanced strength and ductility performance. It can also transform unreinforced masonry into load-bearing reinforced masonry wall with improved structural performance under strong ground motions.

RETROFITTING OF REINFORCED CONCRETE MEMBERS

In reinforced concrete columns and bridge piers, the requirement to satisfy the need for better esthetics, higher efficiency, greater performance, smaller size is continuously leading to highly complex cross-sections made of several materials and complicated shapes, with arbitrary distribution of rebars and holes. Retrofitting design for such complex members may sometimes become an onerous task. Moreover, the retrofit scheme for strengthening a structural member against a particular applied action may not be adequate against the other applied action. Here, some common techniques applied to reinforced concrete beams and columns will be discussed.

Strengthening of Beams and Slabs in Flexure

Two basic areas of concern for beams and slabs in flexure are excessive deflection and flexural strength. Flexural strength for beams/slabs may be increased by providing steel plate or steel section on the tension side of the beam as shown in Fig. 7.5. The newly added steel section or plates may be protected by encasing the cross-section with ferrocement or mortar. Wire mesh may be added to hold the new material in place. Providing posttensioning in beams is another effective way to increase the flexural performance of the member. Posttensioning reduces bending moment, increases the strength of the system, reduces cracking- and tension-related problems, and provides direct uplift forces to balance the loads as shown in Fig. 7.6.

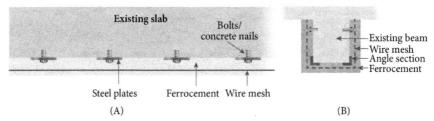

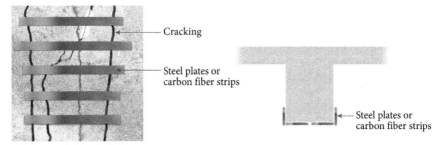

Figure 7.5 Strengthening of (A) slab and (B) beam using steel plates and ferrocement.

Figure 7.6 Strengthening of slab using steel plates and/or carbon fiber strips.

Hanging is another suitable method to reduce deflection and to increase the flexural capacity of beams. In this method, an I-beam (steel) is provided over the beam in question and is supported at the ends (over the support of the beam) on bearing pads. U bolts or a set of bolts with a steel plates are passed under the beam (to be strengthened) and is supported on the I-beam. This is done on several locations along the beam. The bolts are tightened producing an uplift effect on the beam.

Carbon fiber-reinforced polymer (CFRP) strips may be used at the bottom side of the slab and beams for reinforcement as shown in Fig. 7.6. This technique is effective in reducing deflection and crack width and in providing additional reinforcement. The plates or strips are glued or anchored to the slab and are covered with a layer of ferrocement increasing the flexural strength of the slab.

Strengthening of Beams in Shear

Steel plates may be bolted to the beam with shear cracks or shear deficiency. These may be inclined or in vertical direction. Carbon fiber strips may also be used instead of steel plates. Another method is to provide extra external post stressing in the form of U bolts that will uplift the portion in shear (Fig. 7.7).

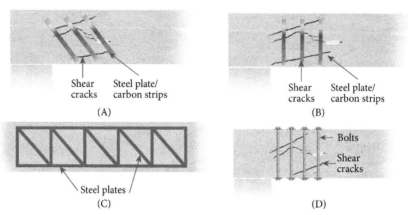

Figure 7.7 Strengthening of beams in shear (A) and (B), minor deficiency (C), major deficiency (D), using transverse posttensioning.

External prestressing is the dynamic strengthening technique as it leads to the reduction of deformations, stresses, and cracks in cross-sections. Prestressing is applied in such a manner that internal stresses produced as a result of prestress forces tend to cancel out or reduce the effect of stresses developed due to applied actions. This concept is generally adopted in following cases:
- Change in structural member configurations
- To enhance flexural strength of structural members
- Excessive deformation of structural members.

A common example of this retrofitting technique is external posttensioning of concrete beams using steel tendons. High-strength steel tendons with specific mechanical properties (e.g., higher elastic modulus) are used for this purpose. This results in elastic actions that can produce high opposing or secondary forces to reduce both the demand and net stresses. There are three main types of tendons that are used for prestressing, i.e., wires, strands, and bars. Prestressing wires are made with cold working process and have several types of cross-sectional shapes and surface conditions such as round, oval, twisted, or crimped. Strands are made by the combination of several wires. Prestressing bars are made of high-strength steel and concrete with smooth or ribbed surfaces.

The complete system consists of prestressing tendons, anchorage devices, and corrosion protection systems. CFRP tendons are also recently developed which provide high corrosion resistance and resistance

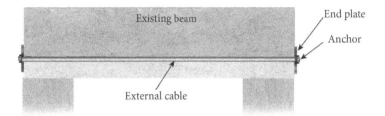

Figure 7.8 Post-tensioning a weak beam by external cable.

to fatigue but are costly and hence not frequently used. Relevant standards must be followed while selecting strength of prestressing tendons. Anchorage devices must be designed to provide effective profile of prestressing cables and to transfer forces to the member ends or anchors locations without damage. Appropriate corrosion system is also an important prerequisite for successful retrofitting by external prestressing. It must also be noted that temperature has significance effect on stiffness and strength of steel. Higher temperature results in significantly reduced stiffness and strength of steel and thus protection from fire is important. Furthermore, prestressed steel has shorter lifetime because steel is stressed up to significant percentage of its ultimate strength and provisions to replace the tensions should be made. Fig. 7.8 shows the conceptual applications of Post-tensioning for retrofitting of beams, weak in flexure or shear.

Strengthening of Columns

Reinforced concrete columns are among the primary structural components of a building. Under severe loading conditions, especially the case of high intensity seismic action, often the need for rehabilitation or enhancing the section capacity of RC column arises. Encasing and Jacketing are the two simplest and widely used methods for increasing the axial and moment capacity of columns. Concrete column section may be simply jacketed with steel plates or channel sections, which will increase the load-carrying capacity of the section significantly. Another effective way of strengthening concrete columns is to provide angle sections and steel strips at the corners and then encase the new section with reinforced concrete. Concrete columns may also be encased by new reinforced concrete to account for the new loads. In case of deficiency in the moment carrying capacity of columns, it may be best to provide extra reinforcement in the column as well. Full jacketing or combination of jacketing

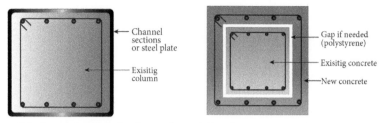

Figure 7.9 Strengthening of columns for axial loads.

and encasing can increase the confinement and ductility capacity as well. Jacketing by transverse FRP strips can greatly enhance the confinement and ductility but may not increase the bending capacity. Figs. 7.4 and 7.9 show some of the possible solutions for strengthening of columns.

ANALYSIS OF CROSS-SECTIONS EXPOSED TO FIRE

Damage to existing buildings due to fire is a common scenario and may require evaluation and retrofit. This section discusses the axial—flexural behavioral aspects of beam—column cross-sections under axial—flexural, exposed to high temperatures. The performance of structures when exposed to fire for a considerable duration of time is one of the concerns of the structural engineering industry. Much emphasis is being placed on increasing the fire performance of buildings/structures. This section presents a computational method for the determination of cross-section capacity of reinforced concrete beam—column sections. A time progressive analysis may be performed, reporting the loss of capacity versus sustained temperature, or as the intensity of fire is altered. A modified stress—strain relationship can be assigned to the portion of cross-section affected by fire, which can be determined using the finite element method for steady-state heat transfer using general procedures presented in Chapter 3, Axial—Flexural Response of Cross-Sections.

Effect of High Temperature on Structural Cross-Sections

High temperatures typically produced by exposure to fire have several effects on the behavior and response of structural members. Some of these effects are global, affecting the entire members or parts of the structure.

Some of them are local, affecting the member cross-sections, whereas some effects are related to microstructure or the cross-section materials.

The temperature change causes strains in the cross-section, which in turn tends to cause deformations in the members. If the members are completely unrestrained, these deformations do not change the stress conditions or the external actions on the cross-sections. However, most structural members are restrained by other members or by boundary conditions. This changes the deformation pattern and also induces additional actions that ultimately modify the strain and stress distribution within the cross-section. The additional actions produced in structural members due to temperature variation can be determined by temperature analysis using appropriate Finite Element Analysis programs. This analysis basically requires the input of temperature change and the coefficient of thermal expansion for all the materials of the members. Generally, this analysis assumes the elastic properties of the cross-section and materials at normal temperature (say $20°C$). The stiffness of cross-section is also affected by high temperature especially in the range generally encountered during exposure to fire. Therefore, a detailed analysis would require that the change in stiffness properties due to temperature is considered for determining modified response. The stiffness properties affected by fire include the elastic constants E, G, and V as well as the effective cross-section. The determination of modified stiffness is one of the issues associated with the analysis of cross-sections exposed to elevated temperatures. Another effect of temperature change is the change in mechanical properties of material due to high temperature and hence the change in internal stress resultants or action capacity. This section presents a procedure to determine the cross-section capacity, including the effects on various material properties that change due to exposure to high temperatures. This capacity can then be used to check the adequacy or safety of structural members against the original as well as modified actions due to temperature change.

Determination of Section Capacity

When exposed to fire, the properties of the reinforced concrete material change drastically. When calculating the capacity, load—deflection curves, or other kinds of relationships of fire impaired cross-sections, these modified material properties need to be considered to determine the actual behavior of materials.

The capacity of a general cross-section made up of several shapes and materials can be determined by applying basic concepts of the fiber section model. In this approach, the stress at each point (or fiber) in the cross-section is determined and then integrated to find the internal stress resultants. Although the concept is simple, its application requires the handling of several aspects that are not as simple. For example, how to discretize the cross-section into appropriate fibers/mesh, how to determine stress attach point, and how to integrate the stresses accurately across the areas. Several approaches and methods have been developed and proposed by researches to handle these issues. Some are applicable to simple or specific sections and materials whereas others are quite general in application.

In Chapter 3, Axial—Flexural Response of Cross-Sections, a fairly general procedure for automatic discretization of a general cross-section made up of several materials, including the effect of nonlinear strain distribution, nonlinear stress—strain relationship, and time-dependent effects such as creep, shrinkage, and relaxation, was presented. This section presents a further specialization and extension of that scheme to include the effects of high temperatures produced due to exposure to fire. The issues to handle in this special case are:

1. Determine temperature at the surface of the members, which translates to the temperature at the boundary of the cross-section.
2. Determine the temperature distribution across and within the cross-section for a specific sustained temperature on the boundary.
3. Determine the temperature-dependent properties of the cross-section materials at various locations.
4. Discretize/mesh the cross-section in such a way that the effects of cross-section shape, material variation, temperature variation, and stress variation are captured appropriately.
5. Once these issues have been tackled, stress distribution can be determined to compute internal stress resultant. The proposed solution to each one of these aspects is presented next.

Determination of Surface Temperatures

The surface temperature (or ambient temperature) during a fire is generally determined by time—temperature curves or relationships. Various guidelines and standards (e.g., BS 426 and ASTM EK9) prescribe generalized forms of such relationships. It is, therefore, assumed for the purpose of analysis that the surface temperature is known at a given time. It is generally assumed that a temperature of 800°C is reached within half an hour of exposure to fire whereas after about 2 hours, the temperature becomes almost constant at about 1000°C. The temperature on the surface of the member or the boundary of the cross-section will depend on the location of fire with respect to location of the member. Fig. 7.10 shows some of the practical cases of fire exposure to various configurations of structural cross-sections.

It can be seen that the surface temperature distribution will be different in each of these practical cases. Most references, dealing with fire resisting design, assume that the slabs and beams are exposed to fire from below completely and that columns are exposed from all sides. This situation may not be so realistic where several masonry walls are used as partition or semiload-bearing elements.

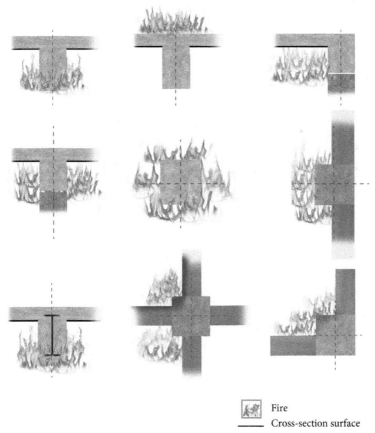

🔥 Fire
——— Cross-section surface

Figure 7.10 Practical cases of fire exposure to various configurations of structural cross-sections.

Determination of Temperature Distribution within the Cross-Section

The determination of temperature distribution within a cross-section for a given surface temperature variation is a complex problem. Several factors, including the thermal conductivity of material, shape of the cross-section, temperature variation on the boundary, change of temperature with time, radiation heat loss, material density, specific heat, etc., affect the temperature distribution within the cross-section. For reinforced concrete cross-sections, additional factors such as the type of aggregate, moisture content, and moisture vapor movement also become important. Various references provide (Kong, Evans, Cohen, & Roll, 1983; Lie & Celikkol, 1991; Wu, Su, Li, & Yuan, 2002) the typical temperature distribution for standard cross-section shape exposed to fire. These distributions are typically presented as graphs between penetration depth and temperature for slabs and estimated temperature contours for other cross-sections. The analytical solution of the temperature distribution problem is governed by a partial differential equation of the form:

$$\frac{K_1}{C\gamma}\left[\frac{\partial^2 T}{\partial^2 x} + \frac{\partial^2 T}{\partial^2 y}\right] = \frac{\partial T}{\partial t} \qquad (7.1)$$

where
   K_1 = Thermal conductivity
   $C\gamma$ = Volumetric heat capacity
   T = Temperature
   t = Time.

The solution of the above partial differential equations can be obtained numerically using the finite element method. The steady-state heat transfer problem can be considered, neglecting the effects of radiation heat loss and rapid temperature variation with time. For the application of the finite element solution, the following procedure is generally used.

1. The cross-section is first meshed automatically in such a way that all variations in cross-shape and its constituent materials are divided into simple, convex polygons. These polygons are further meshed or subdivided into smaller quadrilateral and triangular elements to enforce a specified mesh size or refinement level.
2. The surface temperature is then applied at boundary nodes corresponding to the exposure and insulation conditions.
3. The cross-section material thermal properties such as conductivity and specific heat are assigned to each element depending upon its location in the cross-section and the material at that point.

4. The finite element solution is carried out to give the nodal temperature on the exterior, as well as the interior of the cross-section. This nodal temperature is then used to determine temperature of rebars in the case of reinforced concrete and of steel shapes in the case of composite and encased sections.

This analysis is carried out in the plane of the cross-section at a particular location along the member length.

Determination of Capacity Interaction Surface

Once the temperature distribution within the cross-section has been determined for specified surface temperature, at a given time, the next step involves the determination of internal stress resultants. For beams and columns, the axial—flexural stress resultants can be best represented by the capacity interaction surface. This capacity interaction surface is generated by using the following procedures:

- A fracture criteria for concrete in compression and steel in tension is established, and a corresponding strain profile is determined for a particular depth of neutral axis.
- For concrete in normal conditions, the limit on compressive strain of 0.003—0.0035 is generally used. For concrete at high temperature, higher values of failure strain have to be specified and values as high as 0.006 may be used for the extreme fiber.
- Based on this reference strain profile, strain is determined at every node of the discretized mesh elements. The strain at these nodes is then used to determine corresponding stress at normal temperature condition using the appropriate stress—strain curve.
- The stress obtained from this curve is then modified using the temperature at that node and the temperature—stress curve for the material at node.
- This modified stress at each node is then used to determine the overall stress distribution in each element. The stress within the element is then integrated using the appropriate shape function and assumed variation. If the mesh is refined enough a linear variation of stress can be assumed within the element. Similar procedure is followed for rebars to determine the total stress resultant for a particular strain profile. The strain profile is varied across and around the cross-section to generate complete capacity interaction surface. The process is represented by the flow diagram in Fig. 7.11.

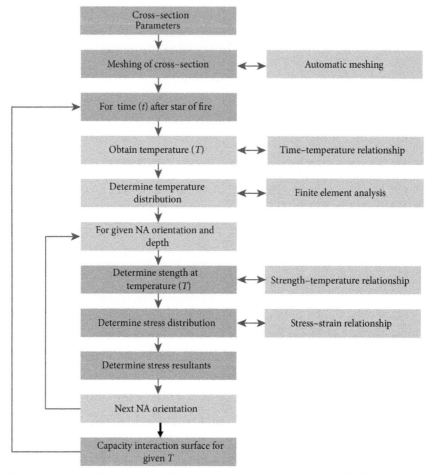

Figure 7.11 Overall process for determining the section capacity of a fire impaired cross-section.

Fig. 7.12 shows typical relationship among moment capacity and temperature, capacity ratio and temperature, temperature and time, and capacity ratio and time. The Capacity Ratio versus Temperature plot gives a clear idea of the failure of a section at a specified temperature. Since the Temperature—Time relationship is known, it can be superimposed on the Capacity Ratio versus Temperature plot to obtain the Capacity Ratio versus Time graph. This graph marks the time at which a cross-section is expected to fail.

An example application of this methodology is applied to four reinforced concrete cross-sections (rectangular, circular, hollow rectangular,

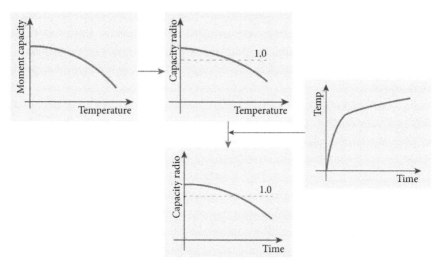

Figure 7.12 Relationship among moment capacity and temperature, capacity ratio and temperature, temperature and time, and capacity ratio and time.

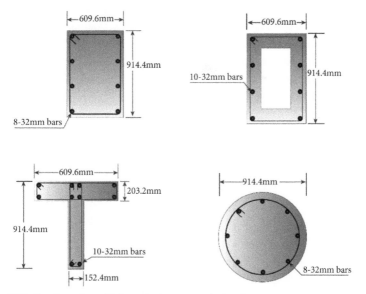

Figure 7.13 Sections for which analysis was carried out.

and tee shapes as shown in Fig. 7.13) to predict their moment capacity when exposed to high temperatures. Using a computer application developed for this model, the temperature distribution within a cross-section for steady-state heat exposure is computed. The corresponding cross-section properties, axial stress distributions, and temperature-modified material properties are also determined. The effect of cover on the capacity of a fire impaired rectangular section is also evaluated. In addition to the basic stress distribution, the load—deflection curves, capacity interaction surfaces, and capacity interaction curves can also be determined using this approach.

Effect of Concrete Cover

P—M curves are plotted for a rectangular section (916 × 610 mm or 24 × 36 in. reinforced with 8 #10 bars) for covers varying from 25 to 150 mm and are shown in Fig. 7.14B. As is evident that the moment capacity of cross-sections at elevated temperatures is not much affected with the increase in cover. All the curves in Fig. 7.14B almost have the same capacity except for a little increase in the tensile capacity of the section with very large cover (150 mm). This little increase may be due to the fact that when the cover is increased significantly the rebars with larger covers may still contribute to overall moment capacity owing to a relatively lower level of temperature. The moment capacity of the same section (at room temperature) is not affected by the change in cover in the high tension and compression zones as shown in Fig. 7.14A, whereas the moment capacity decreased significantly with increase in cover due to a reduced lever arm between the reinforcing bars. These results indicate that in some cases, providing extra cover may not be an adequate solution for enhancing the performance of a cross-section under high temperatures.

Variation of Capacity with Temperature

Fig. 7.15 shows the effect of elevated temperature on the moment capacity of different cross-sections. As an overall observation, the capacity of all the sections reduced extensively with increase in temperature. It is interesting to note the behavior of the rectangular and hollow rectangular sections having almost the same capacity at room temperature. However, the hollow rectangle loses almost all its capacity at high temperatures in a range of 800—1000°C. This is due to the fact that at such high temperatures the concrete near the cross-section boundary loses all its strength. Since there is no core to contribute towards the capacity of the section there is a marked reduction in the load-carrying capacity of the section. The rebars also fall in a high temperature zone thus losing all their strength and not contributing to the capacity of the section. It may also be noted here that all the cross-sections loose only 20—30% of their capacity up to 600°C, after which there is a drastic reduction in the load-

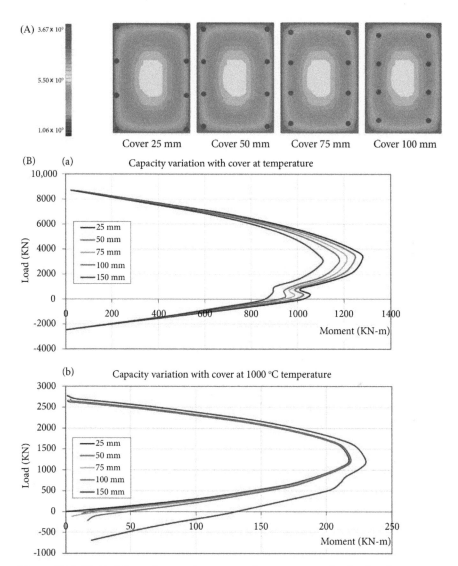

Figure 7.14 (A) Variation of concrete cover and temperature distribution for a rectangular RC section. (B) Effect of cover on the capacities of rectangular sections at room temperature (a) and elevated (b) temperature.

carrying capacity of the sections. Fig. 7.15 shows some cases of temperature distribution across the analyzed cross-sections.

Similar capacity interaction surfaces generated or computed using the proposed procedure can be directly used to determine the load or moment carrying capacity of the cross-section subjected to high

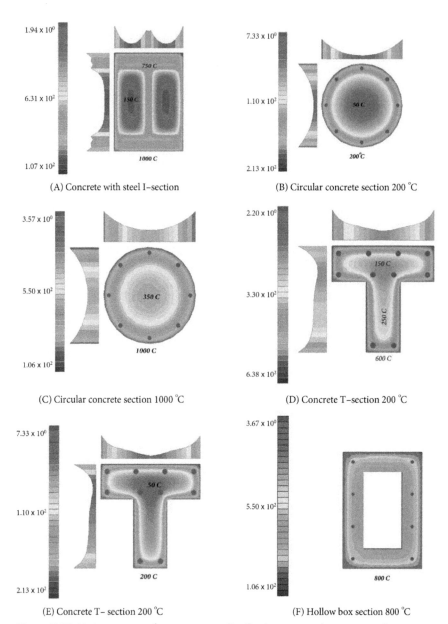

(A) Concrete with steel I–section

(B) Circular concrete section 200 °C

(C) Circular concrete section 1000 °C

(D) Concrete T–section 200 °C

(E) Concrete T– section 200 °C

(F) Hollow box section 800 °C

Figure 7.15 Various cases of temperature distribution across the cross-sections.

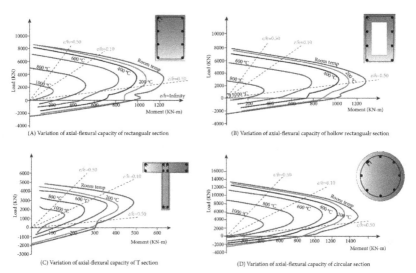

Figure 7.16 The variation of axial-flexural capacity with increasing temperature, for various cross-sections.

temperatures. This means that both beams and columns can be investigated for their ability to carry loads during an exposure to a certain, known intensity of fire. Several capacity interaction surfaces can be generated (e.g. Fig. 7.16), each corresponding to a particular exposure temperature or conditions. Moreover, Temperature versus Capacity relationships can also be developed that may be used directly to determine capacity for a particular loading condition Fig. 7.16.

For all four cross-sections, a more generalized and useful relationship between the capacity ratio and temperature is presented in Fig. 7.17. The capacity ratio is determined as the ratio of the capacity of a section at given temperature to that of the section at room temperature for a particular e/h ratio. Graphs were plotted for four e/h ratios for each cross-section. It can be seen that cross-sections perform better for low e/h ratios (high axial load, lesser bending) at elevated temperatures. It can also be seen that the curves typically consist of two distinct slopes, especially for low e/h ratios. There is a marked increase in slope at around 600°C. It means that the sections loose the first 20–30% of their load-carrying capacity over a temperature range of 20–600°C while the remaining capacity is lost when the temperature is increased to 1000°C. The capacity ratio can include both the change in capacity due to high temperature as well as change in loads due to temperature increase. Another

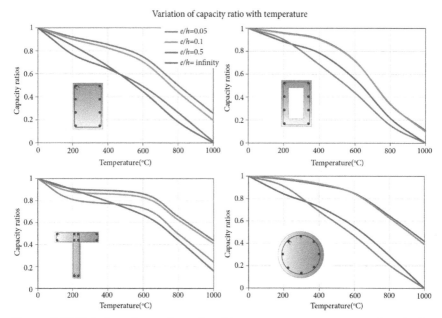

Figure 7.17 Variation of capacity ratio with temperature at constant e/h ratio for different cross-sections.

application of the capacity interaction surface generated by the proposed model is to determine the amount of cover needed or the thickness of insulation needed to provide a certain capacity ratio. Similar variations in other cross-section parameters, such as conductivity, density, types of aggregates, can be plotted against temperature to determine the appropriate values for a given capacity ratio at a particular temperature.

SOLVED EXAMPLES

Problem 1: A circular RC column section is shown in Fig. Ex. 7.1. Moment–curvature and P–M Interaction curves are shown in Fig. Ex. 7.2. Due to the change in the scope of building, both the ultimate moment and axial load capacity of the columns are required to be increased by a factor of 1.5. Plain concrete jacketing is considered as most economic option. Determine the thickness of the concrete required to enhance column capacity by factor of 1.5. Concrete compressive strength is 40MPa and yield strength of reinforcement bars is 400 MPa.

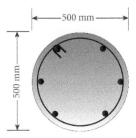

Figure Ex. 7.1 A circular RC column section.

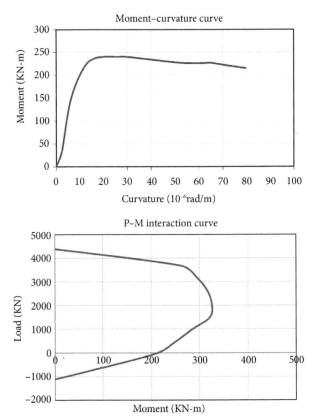

Figure Ex. 7.2 Moment–curvature and P–M interaction curves of given cross-section.

Solution

CSI Section Builder is used to model the cross-section. The whole procedure of modeling of cross-section, analysis, and results are illustrated below:

Material Properties

The following material properties are used to model the cross-sections shown in Fig. Ex. 7.3.

$f_c' = 40$ MPa

$f_y = 400$ MPa

$E_c = 30,000$ MPa

$E_s = 200,000$ MPa

$M - \phi$ and P$-$M Interaction Curves

After the modeling, the $M - \phi$ and P$-$M interaction curves are generated for the cross-sections and are as shown in Fig. Ex. 7.4. These curves give axial$-$flexural capacity and ductility of the cross-sections.

Results

From Fig. Ex. 7.4, it can be concluded the minimum thickness of 75 mm of plain concrete jacketing is required to achieve the capacity of the section by a factor of 1.5. Hence 75$-$mm plain concrete jacketing is recommended to improve this section. An important observation is that this enhancement in capacity is also coupled with reduction in ductility ratio, as indicated by a decrease in ultimate curvature. For high seismic activity areas, this reduction in ductility may be of serious concern.

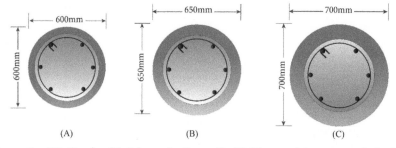

Figure Ex. 7.3 Circular RC Column Section with (A) 50-mm plain concrete jacketing, (B) with 75-mm plain concrete jacketing, and (C) with 100-mm plain concrete jacketing.

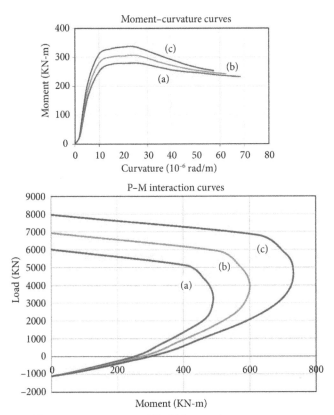

Figure Ex. 7.4 Moment–curvature and P–M interaction curves.

Problem 2: For the same circular RC column section as shown in Fig. Ex. 7.1, if RC jacketing is used to enhance capacity of the section, evaluate the minimum thickness of RC jacketing that is required to enhance capacity by factor of 1.5.

Solution
Same procedure is followed as illustrated in Example 7.1. CSI Section Builder is used to model the cross-section.

Material Properties

The following material properties are used to model the cross-sections shown in Fig. Ex. 7.5.

$f_c' = 40 \text{ MPa}$

$f_y = 400 \text{ MPa}$

$E_c = 30,000 \text{ MPa}$

$E_s = 200,000 \text{ MPa}$

$M - \phi \text{ and } P-M \text{ Interaction Curves}$

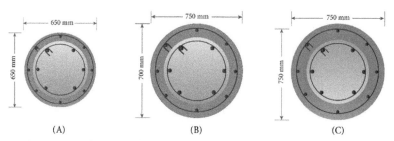

Figure Ex. 7.5 Circular RC Column Section (A) with 75-mm RC jacketing, (B) with 100-mm RC jacketing, and (C) with 100-mm RC jacketing.

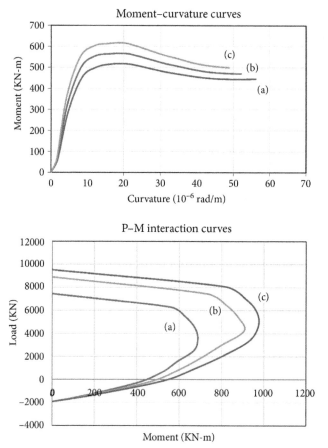

Figure Ex. 7.6 Moment–curvature and P–M interaction curves.

After the modeling, the $M - \phi$ and $P - M$ interaction curves are generated for the above cross-sections and are as shown in Fig. Ex. 7.6. These curves give axial–flexural capacity and ductility of the cross-sections.

Results

From Fig. Ex. 7.6, it can be concluded the minimum thickness of 75 mm of RC jacketing is required to achieve the capacity of the section by a factor of 1.5. In this case, we can also get the maximum ductility among all the possible retrofit sections.

Problem 3: A circular column cross-section of diameters 500 mm needs to be retrofitted. Three possible retrofitted cross-sections are shown in Fig. Ex. 7.7. Determine which of the following will give maximum axial–flexural and moment capacity by generating $M - \phi$ and PM Interaction diagrams.

Solution

Material Properties

The following material properties are used to model the cross-sections shown in Fig. Ex. 7.7.

For cross-section (a):
$f_c' = 25$ MPa
$f_y = 400$ MPa

For cross-section (b):
$f_c' = 25$ MPa (for original column)
$f_c' = 40$ MPa (for retrofitted sections)
$f_y = 400$ MPa

For cross-section (c):
$f_c' = 25$ MPa (for original column)
$f_y = 400$ MPa
$t = 5$ mm

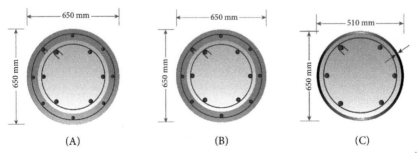

Figure Ex. 7.7 (A) Reinforced concrete jacketing with new concrete having same f_c' as old, (B) reinforced concrete jacketing with new concrete having two times the f_c' as old, and (C) column section confined by a steel tube.

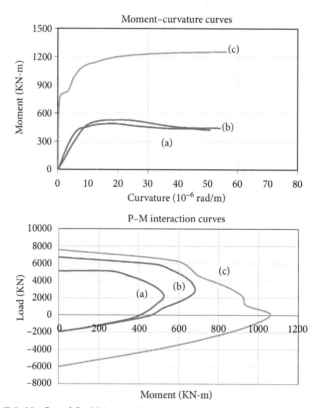

Figure Ex. 7.8 M−C and P−M interaction curves.

M − φ and P−M Interaction Curves

After the modeling, the $M - \phi$ and P−M interaction curves are generated which are as shown in Fig. Ex. 7.8. These curves give axial−flexural capacity and ductility of the cross-sections.

Results

From Fig. Ex. 7.8, it can be concluded that cross-section (c) has the maximum moment and axial−flexural capacity among all the three cross-sections. It is also worth mentioning that it also have high ductility among the given cross-section.

Problem 4: A steel I-section is shown in Fig. Ex. 7.9. Moment−curvature and PM interactions are shown in Fig. Ex. 7.10. Rectangular RC jacketing is being considered as a retrofitting option to make it a composite section. Determine the dimensions of rectangular jacketing required to

Figure Ex. 7.9 A steel I-section.

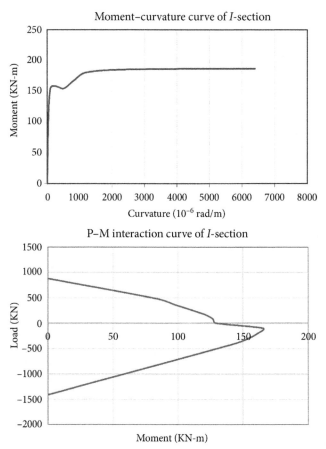

Figure Ex. 7.10 Moment–curvature and PM interaction diagrams.

enhance the axial—flexural and moment capacity by a factor of 3.0. Concrete compressive strength is 40 MPa and yield strength of reinforcement bars is 250 MPa.

Solution

Section Properties

Three dimensions are selected for the retrofitting of given I-Section. Fig. Ex. 7.11 shows the retrofitted cross-sections and properties are mentioned below:

Cross-Section (a):

$b = 325$ mm

$h = 450$ mm

$f_c' = 40$ MPa

$f_y = 250$ MPa

Cross-Section (b):

$b = 375$ mm

$h = 500$ mm

$f_c' = 40$ MPa

$f_y = 250$ MPa

Cross-Section (c):

$b = 425$ mm

$h = 550$ mm

$f_c' = 40$ MPa

$f_y = 250$ MPa

M − ϕ and P−M Interaction Curves

After the modeling, the $M - \phi$ and $P-M$ interaction curves are generated using CSI Section builder which are as shown in Fig. Ex. 7.12.

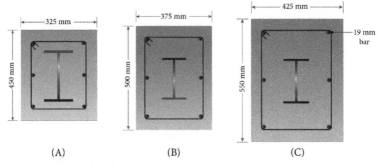

Figure Ex. 7.11 Retrofitted cross-sections.

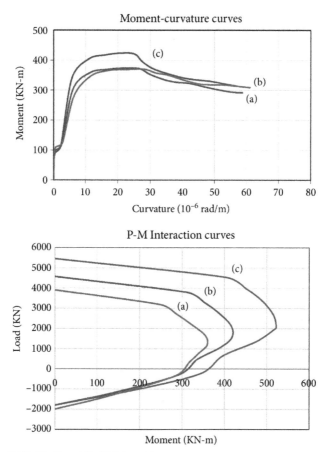

Figure Ex. 7.12 M—C and P—M interaction curves.

These curves give axial—flexural capacity and ductility of the cross-sections.

Results

From Fig. Ex. 7.12, it can be concluded that retrofitting using RC rectangular block of 325mm × 450mm is adequate enough to enhance the capacity of the given I-section with a factor of 3. However, the ductility of the retrofitted cross-section is reduced significantly.

Problem 5: RC concrete square column is shown in Fig. Ex. 7.13. Three possible retrofitted options are also shown in Fig. Ex. 7.14. Generate $M - \phi$ and P—M interaction curves for all the cross-section and

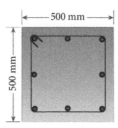

Figure Ex. 7.13 Square RC cross-section.

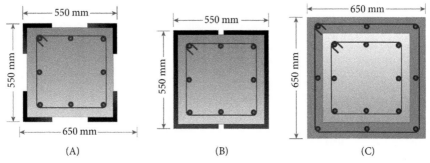

(A)   (B)   (C)

Figure Ex. 7.14 (A) Square RC cross-section retrofitted using Angle Sections, (B) square RC cross-section retrofitted using Channel Sections, and (c) square RC cross-section retrofitted using RC Section.

determine which of the following gives maximum axial—flexural and moment capacities.

Solution

Section Properties

  Angle Section:

  $b = 100$ mm

  $h = 100$ mm

  $t = 25$ mm

  $f_c' = 40$ MPa

  $f_y = 400$ MPa

  Channel Section:

  $b = 250$ mm

  $h = 600$ mm

  $t_w = 25$ mm

  $t_f = 40$mm

  $f_c' = 40$ MPa

  $f_y = 400$ MPa

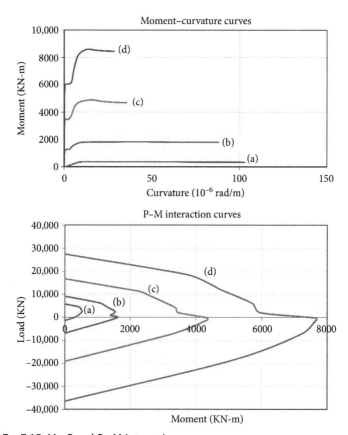

Figure Ex. 7.15 M—C and P—M interaction curves.

RC Cross-Section:

$b = 650$ mm

$h = 650$ mm

$f_c' = 40$ MPa

$f_y = 250$ MPa.

M − φ and P−M Interaction Curves

After the modeling, the $M - \phi$ and P—M interaction curves are generated using CSI Section builder which are as shown in Fig. Ex. 7.15. These curves gives axial—flexural capacity and ductility of the cross-sections.

Results

From Fig. Ex. 7.15, it can be concluded that RC cross-section when completely encased with RC box shows maximum moment and axial–flexural load capacities compared with other retrofitted options. However, significant reduction in section ductility is observed.

Problem 6: A hollow rectangular-reinforced concrete section is shown in Fig. Ex. 7.16. Concrete compressive strength is 25 MPa and yield strength of reinforcement bars is 250MPa. Determine the effect on capacity and ductility of the given cross-section under the following cases:

1. Hollow part is filled with plain concrete with concrete having compressive strength of $2f_c'$.
2. Hollow part is filled with additional reinforced concrete with having compressive strength of $2f_c'$.
3. Hollow section is filled with an I-section and plain concrete with same f_c'.

Solution

Section Properties

The retrofitted cross-sections are shown in Fig. Ex. 7.17 and the section properties are given below:

Cross-Section (a):
$f_c' = 50$ MPa
$f_y = 400$ MPa
Cross-Section (b):
$f_c' = 50$ MPa
$f_y = 400$ MPa
Cross-Section (c):
$b = 175$ mm (I−Section)
$h = 300$ mm (I−Section)

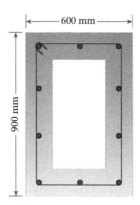

Figure Ex. 7.16 Hollow rectangular cross-sections.

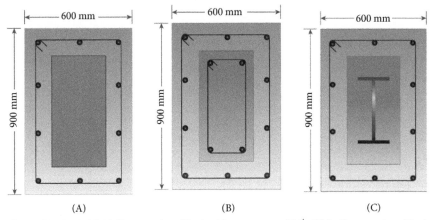

(A)   (B)   (C)

Figure Ex. 7.17 (A) Hollow section filled with concrete of $2f_c'$, (B) hollow section filled with RC of $2f_c'$, and (C) hollow section filled with I-Section and concrete.

$t_w = 7.5$ mm (I−Section)
$t_f = 12$ mm (I−Section)
$f_c' = 25$ MPa
$f_y = 400$ MPa
$M - \phi$ *and* $P-M$ *Interaction Curves*

After the modeling, the $M - \phi$ and $P-M$ interaction curves are generated using CSI Section builder which are as shown in Fig. Ex. 7.18. These curves give axial−flexural capacity and ductility of the cross-sections.

Results

From Fig. Ex. 7.18, it can be concluded that hollow section when filled with steel I-section and concrete of same f_c' results in maximum axial−flexural and moment capacities. However, ductility of the cross-section is reduced to almost half as compared with other retrofitted options.

UNSOLVED EXAMPLES

Problem 7: A rectangular RC section is shown in Fig. Ex. 7.19. Concrete compressive strength is 40 MPa and yield strength of reinforcement bars is 415 MPa. Determine the dimension of enlarged section if

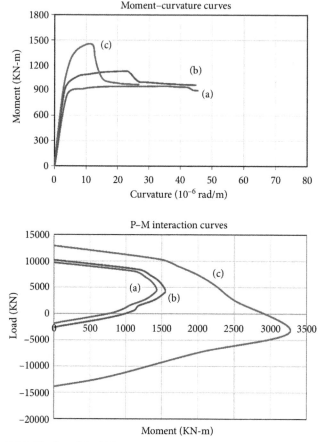

Figure Ex. 7.18 M−C and P−M interaction curves.

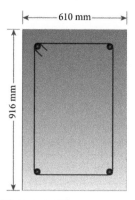

Figure Ex. 7.19 Rectangular RC cross-section.

the capacity is required to be enhanced by a factor of 1.5, keeping the same material strengths. Assume cover of 50 mm.

Problem 8: The column cross-section and its retrofitted cross-sections are shown in Fig. Ex. 7.20. Concrete compressive strength of original section is 25MPa and yield strength of reinforcement bars is 345 MPa. Generate moment−curvature and P−M Interaction curves for the following cross-section and determine which cross-section gives maximum axial−flexural and moment under the following section properties. Assume cover of 50 mm.

1. For cross-section (b): $f_c' = 50$ MPa and $f_y = 345$ MPa
2. For cross-section (c): $f_c' = 25$ MPa and $f_y = 415$ MPa.

Problem 9: An I-Section is shown in Fig. Ex. 7.21. Model the cross-sections and generate moment capacity and P−M Interaction curves if the given I-section is retrofitted with plain concrete with $f_c' = 50$ MPa, and with RC concrete of $f_c' = 25$ MPa and $f_y = 250$ MPa. Assume cover

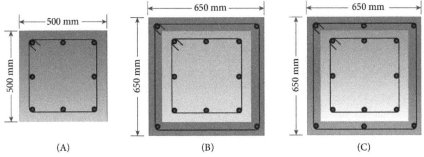

Figure Ex. 7.20 (A) Original cross-section, (B) retrofitted section 1, and (C) retrofitted section 2.

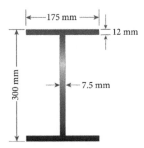

Figure Ex. 7.21 (A) I-Section, (B) I-Section retrofitted Concrete pipe and confined concrete inside the pipe, and (c) I-Section retrofitted RC pipe and confined concrete.

of 50 mm. Also determine effect of axial–flexural capacity, moment capacity, and ductility of the cross-sections.

REFERENCES

Anwar, N., Aung, T.H., Najam, F.A. (2016). Smart systems for structural response control—An overview. In 5th ASEP convention on concrete engineering practice and technology—a.concept'16, Manila, Philippines.

Kong, F. K., Evans, R. H., Cohen, E., & Roll, F. (1983). *Handbook of structural concrete*. London: Pitman Books Limited.

Lie, T. T., & Celikkol, B. (1991). Method to calculate the fire resistance of circular reinforced concrete columns. *ACI Materials Journal*, *88*(1), 84–91.

Sugano S. (1996). State-of-the-art in techniques for rehabilitation of buildings. Paper No. 2175. In Eleventh World Conference on Earthquake Engineering. ISBN: 0080428223. Acapulco, Mexico

Wu, B., Su, X.-p, Li, H., & Yuan, J. (2002). Effect of high temperature on residual mechanical properties of confined and unconfined high-strength concrete. *ACI Materials Journal*, *99*(4), 399–407.

FURTHER READING

ACI 216 (1995). *Building code requirements for structural concrete (ACE 318-95) and commentary (ACE 318R-95)*. American Concrete Institute.

American Society of Civil Engineers (ASCE) (2003). *Seismic evaluation of existing buildings*. Reston, VA: ASCE/SEI 31-03.

American Society of Civil Engineers (ASCE) (2007a). *Seismic rehabilitation of existing buildings*. Reston, VA: ASCE/SEI 41-06.

American Society of Civil Engineers (ASCE) (2007b). *Supplement No. 1, Seismic rehabilitation of existing buildings*. Reston, VA: ASCE/SEI 41-06.

ATC 40 (1996). Seismic evaluation and retrofit of concrete buildings. Applied Technology Council, report ATC-40, Redwood City.

CSI Section Builder (2001). *Released by Computers and Structures Inc*. Berkeley, CA: CSI.

CSICOL 8.4.0 (2008). *Released by Computers and Structures Inc*. Berkeley, CA: CSI.

EN 1998-3 (2005). Eurocode 8: Design of structures for earthquake resistance—Part 3: Assessment and retrofitting of buildings.

Federal Emergency Management Agency (FEMA) (1998). *Handbook for the seismic evaluation of buildings: A pre-standard*. Washington, DC: FEMA 310.

Federal Emergency Management Agency (FEMA) (2000). *Prestandard and commentary for the seismic rehabilitation of buildings*. Washington, DC: FEMA 356.

GEAR (1997). *General Engineering Assistant and Reference. Released by Asian Center for Engineering Computations and Software, ACECOMS*. Thailand: Asian Institute of Technology.

Section Designer (2012). *ETABS 2013. Released by Computers and Structures Inc*. Berkeley, CA: CSI.

CHAPTER EIGHT

Software Development and Application for the Analysis of Cross-Sections

INTRODUCTION

In last 50 years, the developments in computing tools, programming languages, information technology, and other related fields have resulted in a significant improvement in structural engineering practice. The development started with the birth of FORTRAN in 1954, which introduced many new possibilities to engineering computations, including finite element analysis (FEA). In 1960s, the pioneering FEA software (NASTRAN for the aerospace industry and SAP IV for structural engineering) was developed which led a whole new paradigm in engineering profession and computations. Since then, structural engineering software has been an integral part of the profession, enabling the practicing engineers to analyze and design complex structures, from the commencement of the Sydney Opera House in 1959 to the completion of the Burj Khalifa in 2009 and further.

Various software companies were established in last few decades focusing on the structural engineering software development. Some of the pioneering companies include Computers & Structures, Inc., in 1975, Bentley Systems, Inc., in 1984, and RISA Technologies, LLC, in 1987, and other various commercial software packages developed and introduced by these companies have now become an integral part of the structural engineering day-to-day design processes. Recently, tremendous development is happening in various fields related to computing software and hardware and its applications. Often these developments are done independently with a particular focus, including the development of new Human Computer Interfaces (HCI), the new display technologies and devices, the Brain Computer Interface (BCI), in Robotics, etc. The renewed interest and developments in Artificial intelligence, virtual reality, and augmented reality are changing the paradigms of interaction between

531

humans and computers and opening new possibilities. Machine learning, deep thinking, and big data are being utilized in almost all spheres of information technology. Several new integrated devices such as the smartphones, the tablet, and slate computers have become more popular. A similar development is happening in computing paradigm and software with the special focus on visualization. The wide spread use of internet, intranet, and more recently cloud computing is changing the way computers are being used, and information is being shared and stored.

Almost all research disciplines and fields are being benefitted by this exploding and sometimes miraculous developments in the IT industry. For example, medical sciences is one of the main area in which many of the computer technology and hardware are been integrated and converging to tackle a wide range application in diagnosis, biometrics, genetics, and biomedical engineering. The problems in structural engineering are also numerous and complex, and like medical sciences, also related to safety and livability of human beings.

This chapter focuses on computer-aided structural engineering analysis in general and design of cross-sections in particular and presents various recent advancements in the broad range of computing technologies and paradigms and their current as well as future potential for applications for solving the structural design-related problems. It also presents a general framework for the development of structural engineering applications and provides a systematic methodology to implement that framework for developing applications for cross-sectional analysis. Using CSiCol, a commercial package for the analysis of column cross-sections as an example, a brief introduction of various analysis capabilities, implementation issues for the determination of capacity of general sections, various programming techniques, and future perspectives will also be discussed.

A GENERAL FRAMEWORK FOR THE DEVELOPMENT OF STRUCTURAL ENGINEERING APPLICATIONS

This section presents an architectural framework for the development of integrated structural engineering applications using component-based software approach with special emphasis to the information processing context. This framework views the integrated design process in terms of various standard information processing packages which themselves are assembled from information processing components. The information processing approach is used to highlight the relationship of structural engineering and design process and the developments in information technologies. The details of this general framework will be discussed in subsequent subsections.

Component-Based Software Development

The mechanical and electronics industries have long embraced the idea of the component-based manufacturing of their products. For example, several components used in cars are identical across different models or even different brands. A similar concept is used in the manufacturing of computers, where the standard components are assembled and integrated into different computer models, brand names, and configurations.

The component-based manufacturing concept has been extended to the development of software applications in recent years. Component-based software development (CBSD) is a logical extension of object-oriented concepts and the corresponding object technologies, which have pervaded in the software industry since the early 1990s. The CBSD has also made it possible for parts of the same software to be developed by different organizations, and to assemble new applications from existing, specialized, as well as generalized components.

CBSD in Structural Engineering

The structural engineering discipline, and hence the scope of computer applications and software development, deals with a very broad range of activities and applications. The applications can be classified by the type of structure it handles and the extents of the overall design process. The applications can be as small and specific as a program for the design of reinforced concrete cross-sections, or as large and general as a fully integrated analysis, design, detailing, and costing program for structures. Current structural engineering software not only addresses the computational aspects but also focuses on interactive and graphical user interfaces, specialized pre- and postprocessing, graphic visualization, database management, and integration and interaction with other software systems.

It is becoming increasingly difficult for the researchers and developers of software for structural engineering to keep abreast of all the latest developments, both within their own field and in the general structural engineering industry. It therefore makes sense for individual researchers or teams of developers to focus on their own particular area of interest and expertise, and develop programs and software within their specialty. These software or programs, when developed using well-defined frameworks and patterns, can become software components that can then be connected in a variety of ways to develop and deploy complete systems or solutions. In a typical scenario, researchers in the finite element technology could focus on developing a dedicated structural analysis component. Experts in computer graphics could develop components for graphic display and manipulation of structural models, as well as visualization of response. Similarly, specific components could be developed for the design of concrete, steel, or composite structural members. These components could be used time and again in a variety of ways and be physically integrated into applications or used over the internet/network as web services or application servers.

Structural Design Information Space

Traditionally, the structural design process has been described and viewed in terms of distinct design steps, such as conception, modeling, analysis, design, detailing, drafting and costing. The structural design process is part of the overall planning and design process that leads to construction, use, and maintenance of construction projects. Specifically, within the discipline of structural engineering, the design information is divided into several subspaces; these include conceptual design information, modeling and analysis information, design information, detailing and drafting information, material quantities, and costing information. The whole is defined as the structural design information space. Fig. 8.1 shows that a significant information overlap (and communication) exists between the modeling and analysis, analysis and design, design and detailing, and detailing and drafting processes. It also illustrates the information flow paths within these information spaces.

The relationship and dependency of some of the information models are shown in Fig. 8.2. In fact, from an information processing perspective, the overall objective of the structural process is to produce enough information for the physical construction of the project, based on other information sources.

Information Processing and Information Packages

Object-oriented concepts have replaced most of the programming paradigms and data representations. Objects can be viewed as entities that

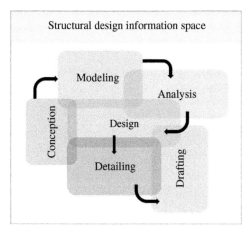

Figure 8.1 Structural design information space.

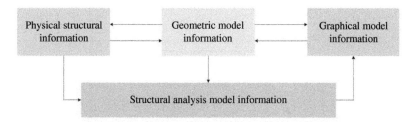

Figure 8.2 Relationship and dependency of some information models.

contain and process information. At higher level of abstraction, complete components and packages can be viewed as information containers and processors. Data processing has progressively given way to information processing, which is now giving way to knowledge processing, and ultimately to the processing of concepts and ideas. These new developments in the information technologies can be readily applied to the development of an information model for the structural design process.

Fig. 8.3 presents the information flow model for structural design process. The concept of an information bus is adopted for this model, where each information processing block (or design process step) in this case can be viewed as a "package," which accesses and updates the overall information space as needed. The information space and information flow described are used as the starting point for developing the architecture of a typical integrated structural engineering application. The information flow is therefore considered to occur on an information bus passing outside of the packages so that it becomes independent of the individual

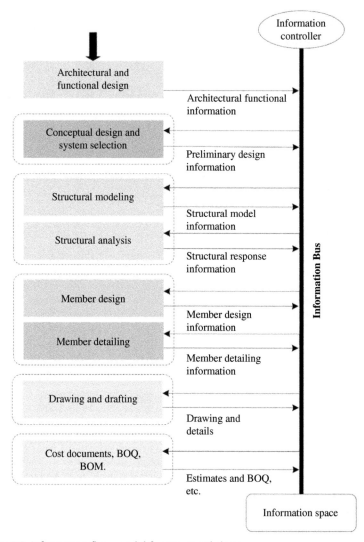

Figure 8.3 Information flow model for structural design process.

package processes or any predetermined information flow sequence. Each package then takes the information it needs from the information space through this bus and returns the processed information in the same way. In this manner, the interdependency of the packages can be reduced significantly, in which various processes can be started simultaneously without the need to follow a predefined flow.

Overall Application Architecture and Frameworks

The overall software architecture for an integrated structural design can be described in terms of three basic layers:

1. Application—made up of one or more packages. Each package handles one major information processing task in the overall design process.
2. Packages—made up of one or more integrated components. Each component provides specific services or handles dedicated tasks within the package.
3. Components—made up of one or more objects. Each object provides specific services or handles specific information processing tasks.

The frameworks required to govern the design and development of actual applications conforming to the above architecture are as follows:

1. A general framework for the whole software that defines the purpose of the software, the tasks to be assigned to each package, and the way in which they will be located and used.
2. A package framework for each package that defines the tasks to be assigned to each component and the way in which they will be located and used.
3. A component framework, which is at a lower level and governs the design of the components themselves, defines the behavior of the components as well their overall design in terms of objects, classes, services, messages, etc.

The standardization of frameworks and components is achieved through the use of design patterns. The design patterns are recurring, planned solutions that work across a range of problems. They allow the discussion of design issues at a higher level, enabling faster software design and a better understanding of the problem, and are used to present the frameworks and components.

Application Frameworks

The overall application framework pattern is shown in Fig. 8.4. The bounds of the information space will depend on the types of structures the software will handle and the level of integration it will provide. The information may be provided to the software through a fully interactive graphical user interface, command driven macros, simple text-based data files, or directly through data and information links with other software or systems.

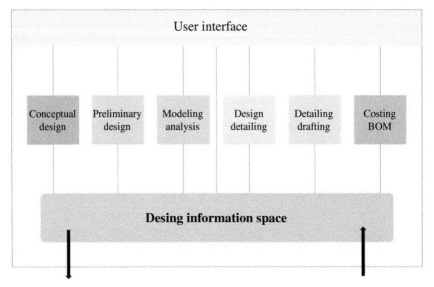

Figure 8.4 Conceptual structural design information package architecture.

The software should have the means to handle both sequential and random use case scenarios. A typical sequential use case follows the traditional steps in the structural design process, in which each package is activated, one by one, in a predetermined order. A random use case involves the activation of any package in random order. In this case, if the activated package needs some other information in order to proceed, it attempts to obtain this from the information bus. If the needed information is not available, either the user of the application or other responsible packages generate this information.

This process continues until all needed information is generated and processed. In this architecture, the validity and concurrency of the information must be ensured at all times by the application level functionality. The propagating effect of information change, therefore, needs to be implemented in the information model and its processing.

Package Frameworks

The specific design and functionality of the individual packages will vary significantly depending on the type of structure and level of integration of the packages (each package is basically an information processor). Upon activation, it takes the required information from the information space using the information bus, converts it to an appropriate context, processes this information using its components and produces new

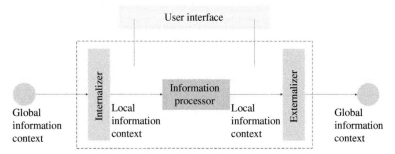

Figure 8.5 A general package framework.

information or updates the initial information, and finally converts it back to the global context and returns it to the main information space.

This framework pattern is shown in Fig. 8.5. In some ways, the general package framework is similar to the application framework, except that in this case the packages within the application are replaced by components within the package.

Implementation Issues

The actual implementation is generally geared towards, and is dependent upon, a particular hardware and software environments. The choice of the development and deployment environment will depend on several factors including:

1. Personal preferences, expertise, and training of the developers regarding various technologies and programming languages (*Java*, *VB*, *FORTRAN*, *C++*, *C#*, etc.)
2. Target users and their computing environment (IOS, Windows, Linux, Androids, etc.)
3. The deployment architecture of the final software system (stand alone, client server, distributed, internet based, cloud based, mobile, etc.)
4. Availability of existing software (and components) on the particular system and environment.

FRAMEWORK FOR THE DEVELOPMENT OF CROSS-SECTION ANALYSIS APPLICATIONS

The development of a software component for the design of member cross–sections using the above general framework is presented in this section. A component developed using the proposed patterns and model can be used in analysis, design, and detailing packages to handle reinforced concrete, partially prestressed concrete, steel—concrete composite, and steel sections. This component can provide the entire response parameters of the cross-section including the determination of geometric properties, elastic stresses, flexural capacity, moment—curvature, and ductility ratios. It can be further extended to handle the retrofitting and strengthening of cross-sections, the determination of fire-damage parameters, etc.

Almost every structural analysis and design software handles the design of member cross-sections, such as columns and beams. The functionality to design a typical cross-section for frame-type members can be encapsulated into a complete component that can be used both with the analysis package and the design package. When linked to the analysis package, it can provide the ability to create or define the cross-sections and to compute the properties needed for analysis. When linked with the design package, it can provide the functionality to design the cross-sections, check the stresses, compute the capacity ratios, and determine the performance and ductility parameters. The section design component can also be used in the detailing package to provide the cross-sections geometry, graphics and other detailing information obtained, or updated from the design process.

Functionality

The specific implementation of the component may be limited to the determination of a simple beam section properties, to a full performance-based analysis and design of general beam—column section used in nonlinear dynamic or pushover analysis. The proposed component is able to handle cross-sections of different materials and shapes, including reinforced concrete, prestressed concrete, hot rolled steel and composite sections, and is able to determine response ranging from computing geometric properties, to elastic stresses, section capacity, ductility, and performance indicators.

The component needs to be developed using standard design patterns so that their reuse is guaranteed. The pattern in Fig. 8.6 gives the functionality and interface of the SD Component.

The Computation Model for the Cross-Section Behavior

The section designer (SD) component is designed to provide a significant amount of varied output. Some of the outputs are related to particular

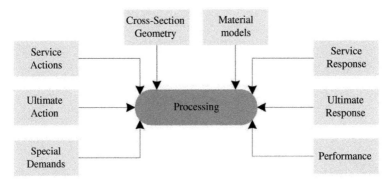

Figure 8.6 Pattern for section designer component.

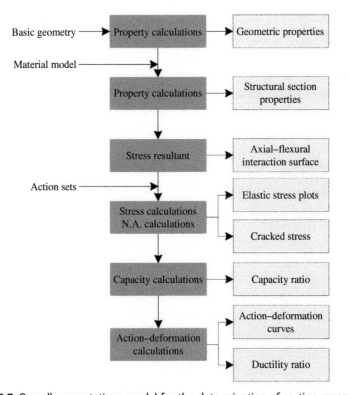

Figure 8.7 Overall computations model for the determination of section response.

input parameters, whereas others are related to the intrinsic property of the cross-section and can be obtained once the geometry and materials are defined. Fig. 8.7 shows a computation model in which the sequence of computation and the relationship of various response parameters are linked to the input parameters.

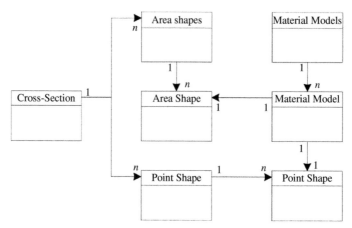

Figure 8.8 Overall object model for SD component.

It can be seen from Fig. 8.7 that the cross-section response is progressively built-up based on the available information either from previous steps or from additional input from the component client. This computation model ensures that the component can be used at various levels and for various purposes.

Object Model

The SD component consists of a collection of shape objects and a collection of material objects.

The cross-section geometry is handled by the shape object that represents both polygon shapes and rebars or prestressing strands. Each shape object is associated with one material object from a collection of various material objects encapsulating material properties and material behavior. The overall object model is depicted in Fig. 8.8. It can be seen that the shape objects have a one-to-many relationship with the section object whereas the material objects have one-to-one relationship with the shape objects. Therefore, the section is, in fact, composed of several shapes, each of a specific material. Defining the material model in separate objects rather than as properties of the shape objects allows for much greater flexibility in the handling of complex cross-sections made up of several shapes and materials.

Class Hierarchy

The SD component is expected to be used in graphical user interface environment. It is, therefore, imperative that the component contains and exhibits graphics capability as well.

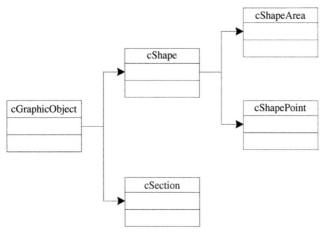

Figure 8.9 Hierarchy of the section class and shape class.

The shape class is further specialized into polygon shape class and point shape class to handle area and rebars, etc. A base material class is used to define the basic material properties and behavior, which is then specialized into concrete material and steel material to handle specific demands of these materials. For example, the concrete material class may be needed to handle creep and shrinkage in a specialized manner, not relevant to steel. Similarly, the steel material is further specialized into mild steel to handle structural steel, and rebar and into prestressing steel to handle prestressing strands and wires. The need for this specialization is to tackle the special behaviors such as relaxation and prestress losses, which is not relevant to the reinforcing or structural steel. The partial design of the section class, the shape class, and the material class is shown in Fig. 8.9 and Fig. 8.10, respectively.

Determination of Section Capacity

The capacity of a general cross-section made up of several shapes and materials as described above is determined using the basic concepts of the fiber section model and is presented in Chapter 3, Axial—Flexural Response of Cross-Sections. In this approach, the stress at each point (or fiber) in the cross-section is determined and then integrated to find the internal stress resultants. Although the concept is simple, its application requires handling of several aspects that are not as simple. For example, how to discretize the cross-section into appropriate fibers/mesh? How to determine stress at each point? And how to integrate the stresses accurately across the areas?

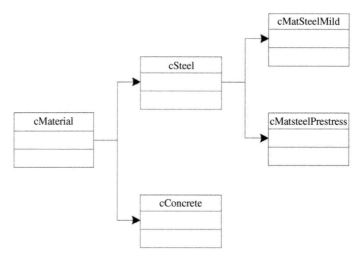

Figure 8.10 Material class.

Discretization of the cross-section converts a planer continuum to fibers or a mesh element of finite size and position. For a general cross-section, represented by a number of complex polygons and points, the points need not be discretized, and each is taken as an individual fiber.

Sample Implementation

The entire response parameters of cross-sections can be determined using this component, including geometric properties, elastic stresses, cracked stresses, moment−curvature curves, flexural capacity curves, and ductility ratios. The SD component receives the basic input parameters such as section dimension, amount of reinforcement, material properties, etc., from the client application. These properties are then used to determine the requested response of the section and the output is provided to the client.

In addition to this, various tools can be provided to facilitate the accurate graphical drawing and editing of section shapes such as rotate, flip, merge, and scale. Similarly, tool for addition and distribution of rebars and prestressing strands over the section are provided. The materials defined earlier may be assigned to any part of the section. Thus the component can allow for building of sections having nonuniform material properties for a single section.

Flexural Capacity Curves

Flexural capacities of sections can be reported in various different formats according to user preferences. These include capacity surface generation, P−M curves at various neutral axis angles, M−M curves at various axial load values and capacity ratios for defined loading conditions. These outputs can be generated and reported in both graphical and text formats and may be exported to other applications for postprocessing. Although the program should be able to generate capacity curves for any section and combination of different shapes and materials, it is important that this information is used with proper understanding and checks on the validity and applicability of such calculations.

Moment−Curvature Curves

The SD component generates the moment−curvature curves for a given value of axial load, direction and magnitude of moment, and a specified strain criterion. Various strain criteria such as maximum strain, failure of first rebar, failure of any or all parts of the section may be specified for the computation of the moment−curvature curves. The component may report the output in both text and graphical forms.

Stress Distribution Plots

The determination of combined normal stress for axial load, moment M_x, and moment M_y is based on elastic properties and the linear strain distribution assuming fully composite and connected behavior of various shapes and components in the section. Similarly, the shear stress can be computed using the general shear stress equation. The same equation is also used to calculate the shear area. All shapes in the section can be assumed to be fully connected and fully effective in resisting shear force. The shear stress distribution is usually computed along two orthogonal axes independently, assuming no interaction.

Available Commercial Section Analysis Applications

There are many commercial software available in market for the analysis and design of cross-sections. Some of the popular options are listed below:
1. CSiCOL (CSI, 2008) by Computers & Structures, Inc. (www.csia-merica.com)
2. CSI SectionBuilder (CSI, 2001) and Section Designer (CSI, 2012) by Computers & Structures, Inc. (www.csiamerica.com)

3. ACECOMS GEAR2003 by ACECOMS, AIT (acecoms.ait.asia)
4. RISASection (2015) by RISA Technologies, Inc. (www.risa.com)
5. IES ShapeBuilder by IES, Inc. (www.iesweb.com)
6. ASDIP Concrete by ASDIP Structural Software (www.asdipsoft.com)
7. CADRE Profiler by CADRE Analytic (www.cadreanalytic.com)
8. Dlubal SHAPE-MASSIVE by Dlubal Software, Inc. (www.dlubal.com)
9. spColumn by StructurePoint, LLC. (www.structurepoint.org).

Although most of these programs provide some convenient analysis options and capable of performing most section analysis calculations, most of the results presented in previous chapters of this book are generated using either CSI SectionBuilder or CSICOL. These two software allow the user to design complex cross-sections with multiple material properties. They are capable of calculating the basic and advanced geometric properties of a general section and also provides information on the section interaction diagrams and stress−strain plots. They also provide options for different design codes and unit standards. Another important application, RISASection can also analyze complex composite cross-sections and new shapes. It includes many shape editing tools as well as the ability to create reports with cross-section graphics. A brief introduction of cross-section modeling and analysis using CSICOL will be presented in the following section.

INTRODUCTION TO CSICOL

CSiCOL is a commercial software package for cross-section analysis and is developed by Computers and Structures (CSI), Inc. It is implemented based on the general framework presented earlier for software development and can be used for the analysis and design of reinforced concrete and composite cross-sections. CSiCOL provides a design wizard for guiding users through the column design process, from section inputs up to report generation. It provides the user the facility of using multiple design codes including ACI 318, BS 8110, and Eurocode 2 and in various unit systems. Design and analysis of column sections can also account for the slenderness effects and for multiple loading conditions.

CSiCOL is capable of generating P−M interaction surfaces, moment−curvature curves, section stresses, stress−strain diagrams, etc., for any general cross-section shape composed of any number of distinct materials. Automated design reports can be created following the analysis and design process, and may be customized according to the user preferences.

Modeling of Cross-Sections

A quick overview of various analysis capabilities, modeling procedure, and visualization of analysis results will be presented here.

There are a multitude of section shapes available including flanged, hollow, and standard steel sections (Fig. 8.11). It also allows the user to create new geometric shapes with various material properties.

CSiCOL provides various material behaviors for concrete and rebar (Fig. 8.12). The rebar sets available are the metric, inch, and ASTM standards, CSiCOL also allows for user-defined rebar values.

Many tools are available to edit column sections as shown in Fig. 8.13, including merging various shapes and adding fillets.

Fig. 8.14 shows the functions available in the shape editing tool, including a rebar calculator and the ability to distribute the reinforcement according to the requirements of the situation. Fig. 8.15 shows the rebar calculator with the editing options available for it.

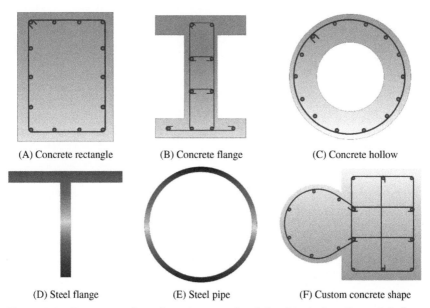

(A) Concrete rectangle   (B) Concrete flange   (C) Concrete hollow

(D) Steel flange   (E) Steel pipe   (F) Custom concrete shape

Figure 8.11 Concrete and steel sections can be defined including flanged, hollow, and custom shapes.

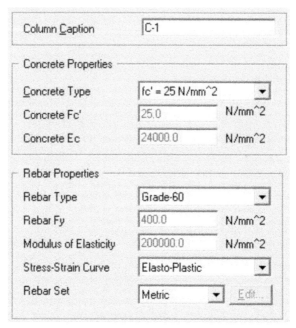

Figure 8.12 Column material properties.

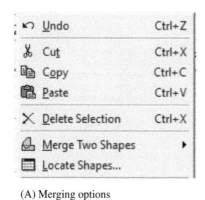

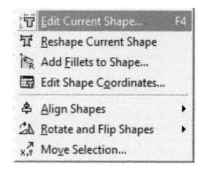

(A) Merging options   (B) Rotate, flip, move and add fillet options

Figure 8.13 Shape editing options.

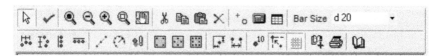

Figure 8.14 Rebar drawing options.

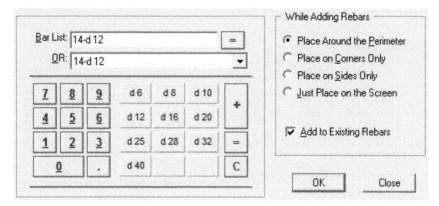

Figure 8.15 Rebar calculator.

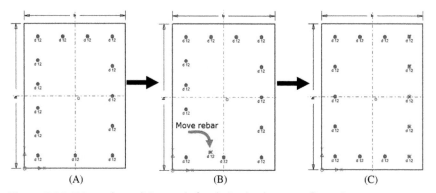

Figure 8.16 Using rebar editing tools for desired column configuration.

Fig. 8.16A−C shows the modeling of various configurations of longitudinal rebars in a rectangular cross-section. Rebars can be conveniently placed, moved, and distributed along the edge with precision. Using various editing tools (Fig. 8.14), users can edit and model any practical configuration of rebars around the column perimeter as shown in Fig. 8.16C.

Cross-Section Analysis

Depending upon the intended use of analysis results, CSiCOL can either check the capacity of current column sections or design a new section for the given loading criteria (Fig. 8.17).

Before starting the auto design process, various design options may need to be specified, including setting limitations on design changes such as keeping the column size constant throughout the design process or

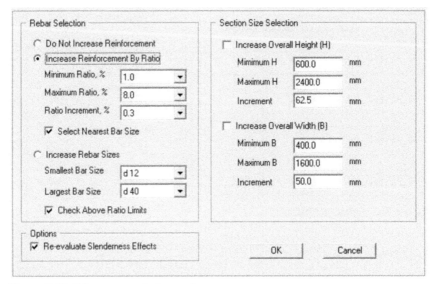

Figure 8.17 Column auto design options.

Mx-My Angle (Deg)	Load Vector	Capacity Vector	Capacity Ratio	N/A Angle (deg)	N/A Depth (mm)	Capacity Method	Remarks
34.3	N/A	N/A	0.87	303.2	269.6	4	OK

Figure 8.18 Capacity calculation results.

controlling the reinforcement ratios. Checking the "Re-evaluate Slenderness Effects" option allows to consider the slenderness effects previously defined for the section (Fig. 8.18).

When column auto design is run, CSiCOL iterates new sections according to design criteria specified.

Visualization of Analysis Results

Section properties is one of the basic outputs generated by CSICOL. These are divided into five categories (Fig. 8.19):
1. Overall dimensions (section bounds, centroid, etc.)
2. Basic properties
3. Principal properties
4. Additional properties
5. Global properties (inertia about global axes).

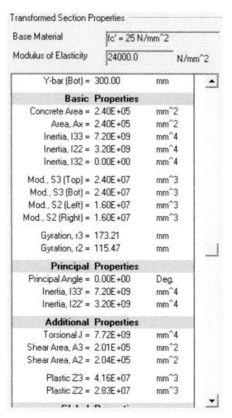

Figure 8.19 Geometric section properties.

The P—M interaction curve for example used in this section is shown in Fig. 8.20A. The applied loading combination with specified loading factors is also plotted on the capacity curve. In this case, the P—M point corresponding to applied loading is lying under the capacity curve, indicating that this section is safe. An M—M curve at an axial load of 800 kN is also shown in Fig. 8.20B. The moment—curvature behavior for this example section is shown in Fig. 8.20C. As discussed in Chapter 6, Ductility of Cross-Sections, various useful information can be obtained using this $M - \phi$ curve. Fig. 8.21 shows the full 3D axial—flexural capacity of cross-section in the form of P—M—M failure surface. All P—M and M—M curves are 2D slices of these failure surface at a constant third parameter.

The strain profile and stress diagrams for both concrete and rebars are shown in Fig. 8.22. It also shows the neutral axis location and reports the curvature of the column section for the given loading criteria.

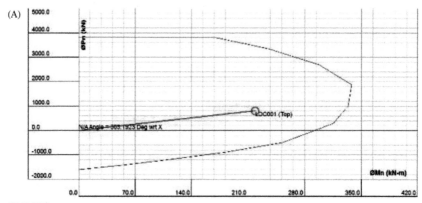

(A) P–M Curve

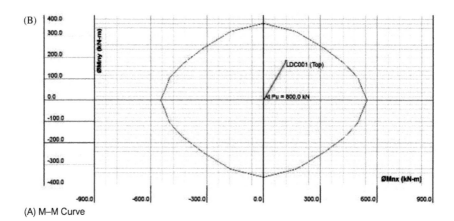

(A) M–M Curve

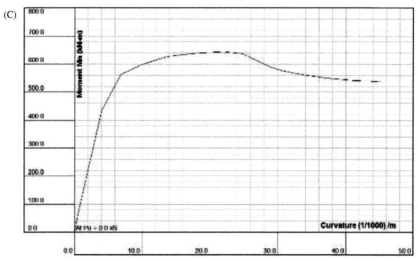

(A) Moment–Curvature Curve

Figure 8.20 Interaction diagrams (A and B) and moment–curvature (C).

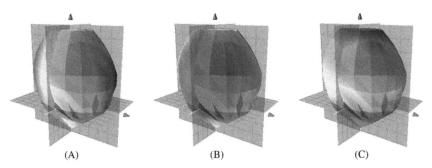

(A)   (B)   (C)

Figure 8.21 Interaction surfaces for (A) M_x, (B) M_y, and (C) P_u.

Figs. 8.23–8.25 show other visualizations of analysis results provided by CSiCOL. These include 2D and 3D views of ultimate section stresses at capacity, combined elastic stresses, equivalent loading point, and ultimate rebar stresses.

MOBILE DEVICES AND THE CLOUD FOR STRUCTURAL ENGINEERING APPLICATIONS

The 21st century was pioneered by the advent of mobile and cloud technology, which has brought about a new revolution in engineering applications. Today's smartphones have greater computational power than the entirety of NASA in 1969, when man set foot on the moon (Kaku, 2011).

Information and communications technology (ICT) has played a vital role in enhancing the efficiency and reducing the cost of information transfer, even before the advent of mobile and cloud computing. So far, mobile and cloud computing research in the civil engineering industry has been focusing on connecting the construction site office and on-site construction activities, as well as for monitoring the health of structures, with a fair degree of success. Mobile devices have now become essential tools in the construction industry, especially for BIM and construction management; two examples of such developments are described henceforth.

The first example is of Construction Opportunities for Mobile IT (COMIT), which is a research and development project that began in August 2003. The goal of the project is to bring together technology and construction companies to develop better ways of using mobile

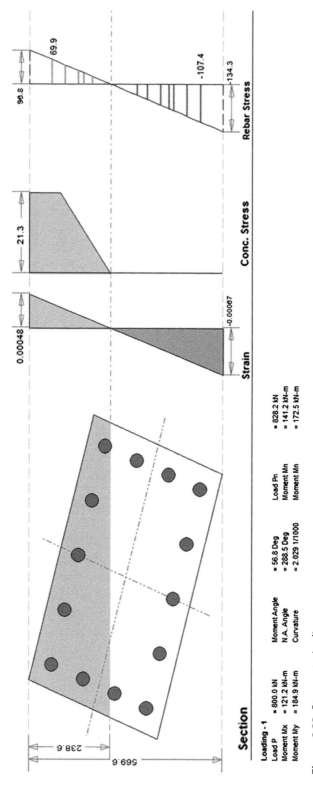

Figure 8.22 Stress—strain diagram.

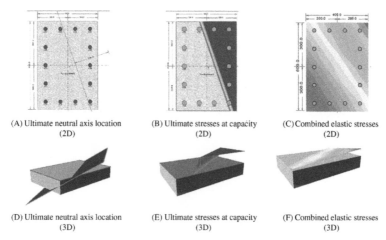

(A) Ultimate neutral axis location
(2D)

(B) Ultimate stresses at capacity
(2D)

(C) Combined elastic stresses
(2D)

(D) Ultimate neutral axis location
(3D)

(E) Ultimate stresses at capacity
(3D)

(F) Combined elastic stresses
(3D)

Figure 8.23 Cross-section stresses and neutral axis location.

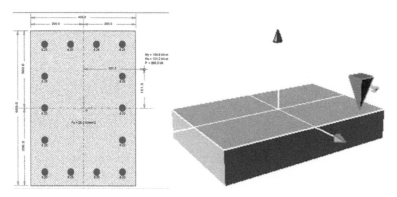

Figure 8.24 Section load point (2D and 3D views).

technology. The second example is Skanska AB, which is a global construction and development company based in Sweden, known for their extensive use of BIM for handling projects. Recently, Skanska implemented mobile computers for BIM into its work processes on-site and in the office from the conceptual stage to design through to the construction phase.

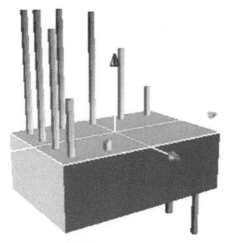

Figure 8.25 Ultimate rebar stresses (2D and 3D views).

Framework for Mobile and Cloud Usage in Structural Engineering Applications

A framework for the development and use of structural engineering applications on mobile platforms and the cloud is presented here. This framework takes into account mobile hardware performance limitations, current mobile application development paradigms, and information processing methods (information may be fed through the GUI, preset text file, custom data file, or shared file on the cloud, and also may be produced in a similar fashion). The overall framework architecture shown is Fig. 8.26 consists of layers structured according to their functionality. The composition of the inner layers develops its outer layers, additionally each outer layer provides support for its inner layer.

1. App platform is composed of one or more mobile apps and cloud services. The app platform manages the overall design process (e.g., develop and manage mobile apps and cloud services that complete the structural design process from conception to costing phase).

2. Apps and cloud services are composed of one or more components. The cloud services may execute one part or the overall process (e.g., computational services to handle the analysis phase of the overall process or a comprehensive structural engineering service to handle the whole process).

3. Components are composed of one or more objects. Components may execute a single aspect of a design process or a portion in that aspect, depending on the complexity of the mobile app (e.g., generating

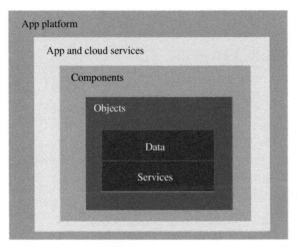

Figure 8.26 Overall framework architecture.

reports, displaying outputs, and transferring data are portions in the beam section design aspect in the design phase of the overall process).
4. Objects are composed of data from various sources and one or more services to process this data. These services are provided by the mobile device and the development environment. The objects layer stores data and provides functionalities for singular portions of the design process relevant to its component.

Mobile Framework

The overall framework is separated into a mobile framework and a cloud framework which follow the CBSD principles described in earlier sections. The implementation of the mobile framework shall be done using the components illustrated in Fig. 8.26.

Components Architecture

The components architecture represents all the basic components required to create comprehensive mobile applications as shown in Fig. 8.27.
1. The authentication component provides security and user logging features for mobile apps that make use of cloud services. This component can be an optional feature for mobile applications that do not require any cloud services (e.g., beam design app), however, it is considered as a basic component.
2. The graphical user interface (GUI) component provides visualizations of user information and renders geometric objects onto the screen. GUI component also handles the presentation and flow of the user interface.

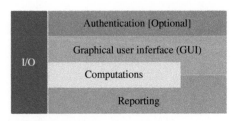

Figure 8.27 Components architecture.

3. The computations component handles all of the structural engineering calculations relevant for the design of structural members and provides support for unit conversions.
4. The reporting component is used to organize the results obtained for the user to view.
5. The input/output (I/O) component handles all processes related to creation, transfer, display, and storage of data, therefore for data-related methods, all other components collaborate with each other through the I/O component. Linkage to the cloud framework is provided inside the I/O component.

Cloud Framework

For the cloud framework, each component can be defined in terms of a service provided through and for the cloud. This framework is created and presented for the SaaS (Software-as-a-Service) model.

System Architecture

Fig. 8.28 illustrates the relationship and interactions of the cloud services with the mobile device and desktop workstation, which is defined as the STE SaaS model. Structural engineering applications are provided as a service by the cloud, having such services enables users to access their applications and data from any workstation, anywhere in the world (i.e., not having to install applications on new workstations, every time). Additionally, there is no need to upgrade applications, since the cloud service will handle this automatically, ensuring the quality of structural engineering applications.

The SaaS cloud contains the applications, content storage systems, and analysis engines required to conduct structural engineering activities. The authentication services are also held on the cloud for verifying user access and enabling specific privileges. The mobile device holds all the cloud

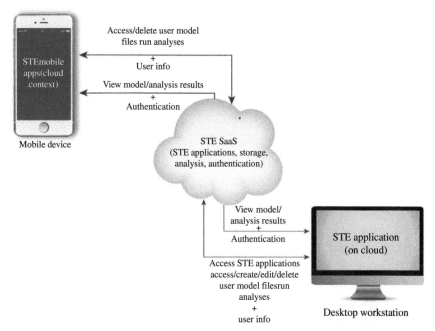

Figure 8.28 STE SaaS model system architecture.

context mobile apps, which can access the structural models held on the cloud. Cloud context mobile applications can enable any phase of the structural design process, except for the modeling phase. Furthermore, they can be used to view models and analysis results.

Components Architecture

STE SaaS components architecture contains all the services necessary to communicate with the mobile devices and desktop workstations in order to perform the structural design process.

1. Authentication service provides security and credential validation features to enable verified mobile devices and desktop workstations to use its other cloud services.
2. STE applications services hold a list of structural engineering softwares, which can be accessed by the user. This service manages and updates the softwares to ensure a good user experience.
3. Content management service performs functions required to store and manage user data, and enables a collaborative user environment to view and manipulate single or multiple structural model files.

4. Computations service manages all of the analysis engines and calculations relevant to structural engineering, and provides services to analyze structural models.

5. Web service enables the exchange of data between machines or applications over the internet, and provides support for interoperability between various mobile and desktop platforms. The computations service only communicates internally with the content management service (refer Figs. 8.29 and 8.30) and therefore does not require the web service.

The mobile and cloud frameworks are linked to each other through the web service component (on the cloud framework side) and the I/O component (on the mobile framework side) for information communications.

Fig. 8.30 shows the collaborations between the mobile and desktop clients with the cloud services, and between the services within the cloud (note the independence of the computations service from the rest of the framework).

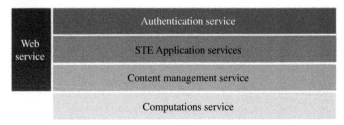

Figure 8.29 STE SaaS components architecture.

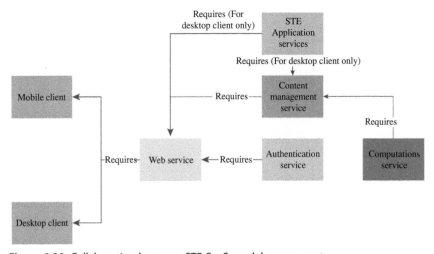

Figure 8.30 Collaboration between STE SaaS model components.

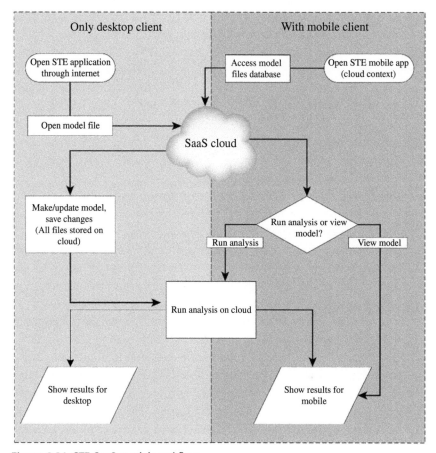

Figure 8.31 STE SaaS model workflow.

Fig. 8.31 describes the mobile–cloud interactions which allow the users to access the model files stored on the cloud through the mobile app, view and run analyses of structural models, and display results onto the mobile client.

CURRENT RESEARCH TRENDS AND FUTURE POTENTIAL

In recent years, extensive research and studies are being conducted in the fields of construction management, BIM, and structural health monitoring. Various new paradigms and potentials related to the use of mobile and cloud computing applications in structural engineering are also catching up recently. Many areas of computational mechanics and structural analysis can be benefitted from these exponential developments in information technology and computer science. However, a thorough understanding of various requirements and needs from construction industry is necessary prior to planning and development of mobile and cloud applications for structural engineering. The key questions to answer or the basic considerations while evaluating the favorability for a particular mobile and cloud application may include the following:

1. Which types of structural engineering mobile apps which need to be developed according to user preferences?
2. How the willingness or inclination of construction industry and practicing engineers towards using the mobile devices (for conducting structural engineering activities) can be accessed?
3. How to measure the level of awareness in the construction industry concerning mobile and cloud computing concepts and functions?
4. How to determine the current state of mobile and cloud computing in the construction industry from a practicing engineer's perspective?
5. Have the key parameters and their relative impact (for integrating mobile and cloud computing in structural engineering) identified?

The presented ideas and framework for the development and use of computer applications, mobile platforms, and cloud computing are just a first step towards a more versatile integration of structural engineering problems with solutions from software industry. This may open a whole new paradigm of various possible advancements in construction technology in coming years. Similarly, by making an optimum use of rapid developments in the fields of proficient computing tools, fast data processing units, and efficient numerical solvers, various current and future studies can come up with numerous new ways to ensure efficient and cost-effective design of structures.

REFERENCES

ACECOMS GEAR (2003). *General engineering assistant and reference. Released by Asian Center for Engineering Computations and Software, ACECOMS*. Thailand: Asian Institute of Technology.
CSI (2001). *Section Builder. Released by Computers and Structures Inc.*. Berkeley, CA: CSI.
CSI (2008). *CSICOL 8.4.0. Released by Computers and Structures Inc.*. Berkeley, CA: CSI.
CSI (2012). *Section Designer ETABS 2013. Released by Computers and Structures Inc.*. Berkeley, CA: CSI.

Kaku, M. (2011). *Physics of the future: How science will shape human destiny and our daily lives by the year 2100.* New York, USA: Doubleday Publishing.

RISA Technologies, LLC. (2015). RISA Technologies, LLC. Retrieved from, < http://risa.com/company.html > .

FURTHER READING

Adeli, H., & Yu, G. (1995). An integrated computing environment for solution of complex engineering problems using the object-oriented programming paradigm and a blackboard architecture. *Computers & Structures, 54*(2), 255–265.

Anwar, N. (1994). *Integrated analysis, design, detailing and costing of concrete buildings. 3rd regional conference on computer applications in civil engineering.* Kuala Lumpur, Malaysia: Penerbit Universiti Teknologi Malaysia.

Anwar, N., Kanok-Nukulchai, W., & Batanov, D. N. (2005). Component-based, information oriented structural engineering applications. *Journal of Computing in Civil Engieering, 19*, 45–57.

Chen, Y., & Kamara, J. M. (2011). A framework for using mobile computing for information management on construction sites. *Automation in Construction, ,* 776–788.

Chen, Z., & Chen, J. (2014). Mobile imaging and computing for intelligent structural damage inspection. *Advances in Civil Engineering, .*

COMIT—Construction Opportunities for Mobile IT. (2015). COMIT—Construction Opportunities for Mobile IT. Retrieved from COMIT Case Studies, < http://www.comit.org.uk > .

Harrison, R., Flood, D., & Duce, D. (2013). Usability of mobile applications: Literature review and rationale for a new usability model. *Journal of Interaction Science, 1,* 1.

Jones, P.H. (2006). Ove Arup: Masterbuilder of the Twentieth Century. New Haven: Yale University Press. Retrieved from, < https://en.wikipedia.org/wiki/Sydney_Opera_House#Design_and_construction > .

Kanok-Nukulchai, W. (1986). On a microcomputer integrated system for structural engineering practices. *Computers and Structures, 23*(1), 33–37.

Karahoca, A. (Ed.), (2012). *Advances and applications in mobile computing Rijeka* Rijeka: InTech Publishers.

Law, K.H., & Peng, J. (2000). Framework for collaborative structural analysis software development. Advanced Technology in Structural Engineering.

McKenna, F. T. (1997). *Object-oriented finite element programming: Frameworks for analysis, algorithms and parallel computing. Ph.D. Thesis.* Berkeley, CA: Department of Civil Engineering, University of California. Retrieved from: < http://opensees.berkeley.edu/OpenSees/doc/fmkdiss.pdf > .

Oskar, A. (2014). *Building blocks: Utilizing component-based software engineering in developing cross-platform mobile applications.* Stockholm, Sweden: KTH, School of Computer Science and Communication (CSC).

Pan, B., Xiao, K., & Luo, L. (2010). Component-based mobile web application of cross-platform. 10th international conference on computer and information technology (CIT-2010) (pp. 2072-2077). Bradford, West Yorkshire: IEEE.

Peng, J., McKenna, F.T., Fenves, G.L., & Law, K.H. (2000). An open collaborative model for development of finite element program. *Computing in Civil and Building Engineering,* pp. 1309–1316.

Romo, I. (2014). *The use of BIM and mobile computers in Skanska. International conference on information systems.* Auckland, New Zealand: International Conference on Information Systems.

Rucki, M. D., & Miller, G. R. (1996). An algorithmic framework for flexible finite element-based structural modeling. *Computer Methods in Applied Mechanics and Engineering*, 363–384.

Schellenberg, A., Mahin, S.A., & Fenves, G.L. (2007). A software framework for hybrid simulation of large structural systems. *Structures Congress.*

Shakshuki, E. M., Perchat, J., Desertot, M., & Lecomte, S. (2013). *Component based framework to create mobile cross-platform applications. The 3rd international conference on sustainable energy information technology (SEIT-2013)* (pp. 1004–1011). Halifax, Nova Scotia, Canada: Procedia Computer Science.

Stamatakis, W. (2000). *Microsoft visual basic design patterns.* Redmond, Washington: Microsoft Press.

Wilson, E. L., Bathe, K.-J., & Peterson, F. E. (1973). *NISEE e-Library: The earthquake engineering online archive. Retrieved from Pacific Earthquake Engineering Research (PEER).* Berkeley: Center, University of California. http://nisee.berkeley.edu/elibrary/Text/31001131.

Yu, L., & Kumar, A. V. (2001). An object-oriented modular framework for implementing the finite element method. *Computers and Structures*, *79*, 919–928.

Zhang, J. Y., Wilkiewicz, J., & Nahapetian, A. (Eds.), (2011). *Mobile computing, applications, and services* Los Angeles: MobiCASE.

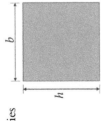

Properties			
A	bh	$b_f h_f + h_w b_w$	$2b_f h_f + h_w b_w$
$\bar{y}_t$	$\dfrac{h}{2}$	$\dfrac{b_f \dfrac{h_f^2}{2} + b_w h_w \left[\dfrac{h_w}{2} + h_f\right]}{\text{Area}}$	$\dfrac{h}{2}$
$\bar{x}_R$	$\dfrac{b}{2}$	$\dfrac{b_f}{2}$	$\dfrac{h_f b_f^2 + b_w h_w \left(b_f - \dfrac{b_w}{2}\right)}{\text{Area}}$
$I33$	$\dfrac{bh^3}{12}$	$\dfrac{b_f h_f^3}{12} + b_f h_f \left(\bar{y}_t - \dfrac{h_f}{2}\right)^2 + \dfrac{b_w h_w^3}{12} + b_w h_w \left[\bar{y}_t - \dfrac{h_w}{2}\right]^2$	$2\left[\dfrac{b_f h_f^3}{12} + b_f h_f \left(\bar{y}_t - \dfrac{h_f}{2}\right)^2\right] + \dfrac{b_w h_w^3}{12}$

I22	$\dfrac{hb^3}{12}$	$\dfrac{h_f b_f^3}{12} + \dfrac{h_w b_w^3}{12}$	$2\left[\dfrac{h_f b_f^3}{12} + h_f b_f\left(\bar{x}_R - \dfrac{b_f}{2}\right)^2\right]$ $+ \dfrac{h_w b_w^3}{12} + h_w b_w\left(\bar{x}_R - b_f + \dfrac{b_w}{2}\right)$
I32	0	0	0
J	$\dfrac{bh^3}{16}\left[\dfrac{16}{3} - 3.36\dfrac{h}{b}\left(1 - \dfrac{h^4}{12b^4}\right)\right]$	$K_1 + K_2 + \alpha D^4$ Refer to Appendix B for explanation	$K_1 + K_2 + \alpha D^4$ Refer to Appendix B for explanation
SA3	$\dfrac{5}{6}hb$	$\dfrac{5}{6}h_f b_f$	$2h_f b_f$
SA2	$\dfrac{5}{6}hb$	$b_w h$	$b_w h$
Z3	$\dfrac{bh^2}{4}$	$0.5\,\text{Area}\,(y_1 + y_2)$	$0.5\,\text{Area}\,(y_1 + y_2)$
Z2	$\dfrac{hb^2}{4}$	$0.5\,\text{Area}\,(x_1 + x_2)$	$0.5\,\text{Area}\,(x_1 + x_2)$

Properties

	H-section	L-section	Hollow box
A	$2b_\mathrm{f}h_\mathrm{f} + h_\mathrm{w}b_\mathrm{w}$	$b_\mathrm{f}h_\mathrm{f} + h_\mathrm{w}b_\mathrm{w}$	$2b_\mathrm{f}h_\mathrm{f} + 2h_\mathrm{w}b_\mathrm{w}$
$\overline{y}_\mathrm{t}$	$\dfrac{h}{2}$	$\dfrac{b_\mathrm{f}h_\mathrm{f}\left(h - \dfrac{h_\mathrm{f}}{2}\right) + b_\mathrm{w}\dfrac{h_\mathrm{w}^{2}}{2}}{\text{Area}}$	$\dfrac{h}{2}$
$\overline{x}_\mathrm{R}$	$\dfrac{b_\mathrm{f}}{2}$	$\dfrac{h_\mathrm{f}\dfrac{b_\mathrm{f}^{2}}{2} + b_\mathrm{w}h_\mathrm{w}\left(b_\mathrm{f} - \dfrac{b_\mathrm{w}}{2}\right)}{\text{Area}}$	$\dfrac{b_\mathrm{f}}{2}$
$I33$	$2\left[\dfrac{b_\mathrm{f}h_\mathrm{f}^{3}}{12} + b_\mathrm{f}h_\mathrm{f}\left(\overline{y}_\mathrm{t} - \dfrac{h_\mathrm{f}}{2}\right)^{2}\right] + \dfrac{b_\mathrm{w}h_\mathrm{w}^{3}}{12}$	$\dfrac{b_\mathrm{f}h_\mathrm{f}^{3}}{12} + b_\mathrm{f}h_\mathrm{f}\left(h - \overline{y}_\mathrm{t} - \dfrac{h_\mathrm{f}}{2}\right)^{2}$ $+ \dfrac{b_\mathrm{w}h_\mathrm{w}^{3}}{12} + b_\mathrm{w}h_\mathrm{w}\left[\overline{y}_\mathrm{t} - \dfrac{h_\mathrm{w}}{2}\right]^{2}$	$\dfrac{b_\mathrm{f}h^{3}}{12} - \dfrac{(b_\mathrm{f} - 2b_\mathrm{w})h_\mathrm{w}^{3}}{12}$
$I22$	$2\left(\dfrac{h_\mathrm{f}b_\mathrm{f}^{3}}{12}\right) + \dfrac{h_\mathrm{w}b_\mathrm{w}^{3}}{12}$	$\dfrac{h_\mathrm{f}b_\mathrm{f}^{3}}{12} + h_\mathrm{f}b_\mathrm{f}\left(\overline{x}_\mathrm{R} - \dfrac{b_\mathrm{f}}{2}\right)^{2}$ $+ \dfrac{h_\mathrm{w}b_\mathrm{w}^{3}}{12} + h_\mathrm{w}b_\mathrm{w}\left(b_\mathrm{f} - \dfrac{b_\mathrm{w}}{2} - \overline{x}_\mathrm{R}\right)$	$\dfrac{hb_\mathrm{f}^{3}}{12} - \dfrac{h_\mathrm{w}(b_\mathrm{f} - 2b_\mathrm{w})^{3}}{12}$

I32	0	—	0
J	$2K_1 + K_2 + 2\alpha D^4$ Refer to Appendix B for explanation	$K_1 + K_2 + \alpha D^4$ Refer to Appendix B for explanation	$\dfrac{2b_w h_f (b_f - b_w)^2 (h - h_f)^2}{b_f b_w + h h_f - b_w^2 - h_f^2}$
SA3	$\dfrac{5}{3} h_f b_f$	$h_f b_f$	$2b_w h$
SA2	$b_w h$	$b_w h$	$2b_f h_f$
Z3	$\dfrac{bh^2}{4}$	0.5 Area $(y_1 + y_2)$	0.5 Area $(y_1 + y_2)$
Z2	$\dfrac{hb^2}{4}$	0.5 Area $(x_1 + x_2)$	0.5 Area $(x_1 + x_2)$

Properties

A	$\dfrac{\pi}{4}d^2$	$\dfrac{\pi}{4}(d_e{}^2 - d_i{}^2)$
$\bar{y}_t$	$\dfrac{d}{2}$	$\dfrac{d_e}{2}$
$\bar{x}_R$	$\dfrac{d}{2}$	$\dfrac{d_e}{2}$
I33	$\dfrac{\pi d^4}{64}$	$\dfrac{\pi}{64}(d_e{}^4 - d_i{}^4)$
I22	$\dfrac{\pi d^4}{64}$	$\dfrac{\pi}{64}(d_e{}^4 - d_i{}^4)$
I32	0	0
J	$\dfrac{\pi d^4}{32}$	$\dfrac{\pi}{32}(d_e{}^4 - d_i{}^4)$
SA3	$0.9\pi\dfrac{d^2}{4}$	$\pi r(d_e - d_i)$
SA2	$0.9\pi\dfrac{d^2}{4}$	$\pi r(d_e - d_i)$
Z3	$0.5\ \text{Area}\ (y_1 + y_2)$	$0.5\ \text{Area}\ (y_1 + y_2)$
Z2	$0.5\ \text{Area}\ (x_1 + x_2)$	$0.5\ \text{Area}\ (x_1 + x_2)$

APPENDIX B: TORSIONAL CONSTANT FACTORS

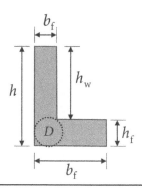

$$K = K_1 + K_2 + \alpha D^4$$

$$K_1 = h b_w^{\ 3} \left[\frac{1}{3} - 0.21 \frac{b_w}{h} \left(1 - \frac{b_w^{\ 4}}{12 h^4} \right) \right]$$

$$K_2 = (b_f - b_w) h_f^{\ 3} \left[\frac{1}{3} - 0.105 \frac{h_f}{(b_f - b_w)} \left(1 - \frac{h_f^{\ 4}}{192 (b_f - b_w)^4} \right) \right]$$

$$\alpha = 0.07 \frac{h_f}{b_w}$$

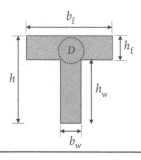

$$K = K_1 + K_2 + \alpha D^4$$

$$K_1 = b_f h_f^{\ 3} \left[\frac{1}{3} - 0.21 \frac{h_f}{b_f} \left(1 - \frac{h_f^{\ 4}}{12 b_f^{\ 4}} \right) \right]$$

$$K_2 = b_w^{\ 3} h_w \left[\frac{1}{3} - 0.105 \frac{b_w}{h_w} \left(1 - \frac{b_w^{\ 4}}{192 h_w^{\ 4}} \right) \right]$$

$$\alpha = 0.07\frac{t}{t_1}$$

where $t = b_f$ if $b_f < b_w$ and $t = b_w$ if $b_w < b_f$

and $t_1 = b_f$ if $b_f > b_w$ and $t_1 = b_w$ if $b_w > b_f$

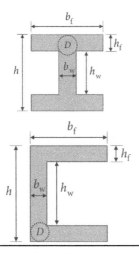

$$K = 2K_1 + K_2 + 2\alpha D^4$$

$$K_1 = b_f h_f^3 \left[\frac{1}{3} - 0.21\frac{h_f}{b_f}\left(1 - \frac{h_f^4}{12b_f^4}\right)\right]$$

$$K_2 = \frac{b_w^3 h_w}{3}$$

$$\alpha = 0.15\frac{t}{t_1}$$

where $t = b_f$ if $b_f < b_w$ and $t = b_w$ if $b_w < b_f$

and $t_1 = b_f$ if $b_f > b_w$ and $t_1 = b_w$ if $b_w > b_f$

INDEX

Note: Page numbers followed by "*b*", "*f*", and "*t*" refer to boxes, figures, and tables, respectively.

A

ACI 318 column design procedure, 354—355, 355*f*

ACI 318-14, 265
  flexural design procedure, 207—209, 208*f*
  procedure to determine skin reinforcement, 293*f*
  shear design procedure for rectangular, T- and L-shaped beams, 282*f*
  torsion design procedure for rectangular, T- and L-shaped beams, 288*f*

ACI Whitney Rectangle, 434

Action—deformation curves, 25, 393—400
  axial load—deflection curve, 399
  F—D relationship, 394—396
  moment—curvature curve, 396—399
  moment—rotation curve, 399
  torque—twist curve, 399—400

Actions, 24, 137—140

Adequate confinement, provision of, 488—489

Allowable Strength Design (ASD) procedure, 138—139

Allowable stress, 138—139, 204

American Concrete Institute (ACI), 10

American Society of Civil Engineers (ASCE), 10

Angle of twist and allowable torque, 119—122

Anisotropic materials, 55, 55*t*

Applied actions, deformations for, 32—33

Applied Technology Council (ATC), 10

Arbitrary force, 26

Area, cross-section, 74—75
  definition, 74
  mathematical computation, 74—75
  significance, 74

Average shear stresses in cracked RC beams, 267—268, 267*f*

Axial load, 32, 36
  effect on ductility, 428—430

Axial load—deflection curve, 399

Axial load—moment interaction curves, 434, 434*f*

Axial stiffness, 34—35, 40—41

Axial—flexual response of cross-sections, 137
  axial—flexural stress resultants, 153—157
    cross-section materials, unification of, 155
    cross-section shapes and configurations, unification of, 155—156
    design approaches and design codes, unification of, 156—157
    diversity of the problem and the need for unified approach, 153—155
    line-type structural members, unification of, 156
  biaxial axial—flexural stress resultants, computing, 175—189
    determination of stress from strain, 183—185
    final stress resultants, computation of, 185—188
    general stress resultant equations, use of, 188—189
    strain profile, determining, 178—183
  biaxial—flexural capacity, 200—207
    applied eccentricity vector, 202
    applied moment vector, 202—203
    cross-section capacity ratio, computing, 204—207
    definition of biaxial bending, 200—201
    resultant moment vector, 203

Axial—flexual response of cross-sections
(*Continued*)
  capacity interaction surface, 189—200
    capacity reduction factors, 197
    effect of material strengths and section
      depth on P—M interaction of RC
      sections, 197—200
    generation of the interaction surface,
      190
    visualization and interpretation,
      191—197
  code-based design for flexure, 207—211
  combined axial stress, 140—145
    combined stress ratio for axial stress,
      144—145
    usefulness and applicability of
      combined stress equation,
      144
  cross-section response, 137
    actions, stresses, stress resultants, and
      capacity, 137—140
  general stress resultant equations,
    157—169
    basic assumptions, their necessity, and
      validity, 157—158
    basic stress resultant equations,
      159—166
    generalized cross-section and
      materials, 167—169
    integrating design codes, 166—167
  interaction of stresses due to axial load
    and moment, 145—147, 146*f*
  principal stresses and Mohr's circle,
    147—153
    basic concept, 147—151
    significance of principal stresses in RC
      beam design, 152—153
  stress resultant equations, extended
    formulation of, 169—175
    strain distribution, determination of,
      174—175
    stress field, determination of,
      169—171
    stress profile, generation of, 172—173
Axial—flexural response, 36
Axial—flexural stress resultants,
  determination of, 432—434

B

Base isolation systems, 489—490
Bending moment, 32, 144—145
Bentley Systems, Inc., 531—532
Biaxial axial—flexural stress resultants,
  computing, 175—189
  determination of stress from strain,
    183—185
    discretization of cross-section and
      stress field, 183—185
    final stress resultants, computation of,
      185—188
    general stress resultant equations, use of,
      188—189
  strain profile, determining, 178—183
    neutral axis, concept of, 180—181
    neutral axis, depth of, 181—182
    practical strain distribution, 182—183
Biaxial bending, 200—201, 201*f*
Biaxial—flexural capacity, 200—207
  biaxial bending, definition of, 200—201
  cross-section capacity ratio, computing,
    204—207
  eccentricity vector, applied, 202
  moment vector, applied, 202—203
  resultant moment vector, 203
Bilinear elastoplastic model, 439—440
Bjerkeli et al. stress—strain model,
  420—421
Boundary conditions
  idealizations for modeling of actual
    restraining effects of, 346*f*
  nonsway condition, 344
  role in slenderness effects, 344—348
  sway condition, 344
  with varying degrees of end restraining
    effects for columns, 345*f*
Braced column condition, 345—346
Brain Computer Interface (BCI),
  531—532
Bridge piers, design of, 369—370
Brittle and ductile materials, 59—61
BS 8110 column design procedure, 356,
  357*f*
BS 8110 design procedure, 209, 210*f*
  for reinforced concrete beams,
    283*f*

Buckling and slenderness ratio, 342–344
Building blocks of structural mechanics, 24–35
  basic concepts and relationships, 24–25
  degree of freedom (DOF), 29–32, 30*f*
    DOFs, deformations, strains, and stresses, 30–31
    member cross-sections and, 29–30
    stress resultants and, 31–32, 31*f*
  linear, elastic stiffness relationships, 32–35
    deformations for applied actions, 32–33
    member stiffness and cross-sectional properties, 34–35
    restraining actions for assumed deformations, 34
  stiffness, concept of, 26–29
    nonlinearity of response and stiffness, 27–29
    structural equilibrium and role of stiffness, 26
Building codes, 8–11
  disaster resilience and environmental sustainability in, 11
  historical development, 10
  traditional, limitation of, 19–20

C

Cable structures, 3
Capacity interaction surface, 189–200
  capacity reduction factors, 197
  determination of, 507–514
  effect of material strengths and section depth on *P–M* interaction of RC sections, 197–200
  interaction surface, generation of, 190
    extended procedure, 190
    simplified procedure, 190
  visualization and interpretation, 191–197
    load–moment (PM) interaction curves, 195–197, 222*f*
    moment–moment interaction curve, 192–195, 222*f*
Capacity of cross-sections, 139–140
  variation with temperature, 510–514

Capacity ratio, 204–206, 206*f*, 207*f*
Capacity reduction factors, 197
Carbon fiber-reinforced plastic, 497–498
Carbon fiber-reinforced polymer, 499–501
Circular columns, 361
Circular shaft, shear stress of, 307–312
Class hierarchy, 542–543
Classification of cross-sections, 42–47
  based on geometry, 43–45, 44*f*
  based on material composition, 44*f*, 46, 47*f*
  based on method of construction, 46–47
  based on types of structural members, 42–43
  based on "compressed zone", 45–46
    compact sections, 45–46
    plastic sections, 46
    slender sections, 45
Cloud framework, 558–561
  components architecture, 559–561
  system architecture, 558–559
Code-based column design procedures, 354–357
Code-based design for flexure, 207–211
Code-based shear and torsion design of RC sections, 278–295
  design of RC beams for shear, 279–281
  design of RC beams for torsion, 281–287
  design of RC sections for combined shear–torsion, 287–288
  reinforcement requirement for combined axial–flexural and shear–torsion, 294–295
  special considerations for deep beams, 289–293
Column cross-sections, response and design of, 329
  column design process and procedures, 351–360
    Analysis and Design, 352, 352*f*
    code-based design procedures, 354–357
    structural designers, responsibility of, 357–360

Column cross-sections, response and
    design of (*Continued*)
  complexities involved in analysis and
    design of columns, 333–338
    loading and section, complexity space
      between, 337*f*
  RC columns, practical design
    considerations for, 360–369
    column stability and structural
      stability, 368–369
    designing for stability, 367–368
    designing for strength, 367
    effect of cross-section shape on
      column strength, 361–362
    effect of loading on column design,
      365–366
    selecting column cross-section shape,
      360–361
    selecting concrete strength for
      columns, 362–363
    selecting sizes and layout of
      longitudinal rebars, 363–364
    selecting sizes and layout of transverse
      rebars, 364–365
    selecting steel strength for columns,
      363
    strength versus stability of columns,
      366–368
  slenderness and stability issues in,
    338–351
    buckling and slenderness ratio,
      342–344
    consideration of slenderness,
      350–351
    effect of column slenderness on
      overall frame behavior, 349–350
    importance of slenderness effects,
      348–349
    role of boundary conditions,
      344–348
  solved examples, 372–386
  special cases and considerations,
    369–372
    designing shear walls as columns,
      370–371
    design of bridge piers, 369–370
    design of piles, 371–372

Columns, strengthening of, 501–502,
    502*f*
Columns, types of, 333–334, 333*f*
Columns as part of sway and nonsway
    frames, 344*f*
Combined axial–flexural and
    shear–torsion, reinforcement
    requirement for, 294–295
Combined shear stress, 254–255
Combined stress ratio for axial stress,
    144–145
Commercial section analysis applications,
    545–546
Common shapes, cross-sectional properties
    of, 565
Compatibility factored torsion moment,
    T-beam for, 322–324
Compatibility torsion, 269
Compatibility torsional moment,
    271*f*
Compatible strain distribution,
    179
Component-based software development
    (CBSD), 533
  in structural engineering, 533–534
Composite columns, 430–431
Compressed zone, classifications of cross-
    sections based on, 45–46
  compact sections, 45–46
  plastic sections, 46
  slender sections, 45
Compression reinforcement
  effect on ductility, 423–424
Compressive force, Poisson's effect for,
    411*f*
Computation model for cross-section
    behavior, 540–542
Computers & Structures, Inc., 531–532
Concrete column section, 501–502
Concrete confinement
  effect on ductility, 427, 428*f*
Concrete cover, 510
Concrete jacketing, 495–496
Concrete strength, 196–197
  selection for columns, 362–363
Concrete-filled steel tubes, 430–435
  various forms of, 433*f*

Confined concrete, stress–strain models
    for, 393–394, 412–423
    Bjerkeli et al. model, 420–421
    Kent and Park model, 416–417
    Li et al. stress–strain model, 422–423
    Mander's model, 415
    Scott et al. model, 417–419
    Yong et al. model, 419–420
Confinement coefficient, 415
Confinement of RC sections, 410–412
Conservation, 485
Construction, method of
    classifications of cross-sections based on,
        46–47
Construction Opportunities for Mobile IT
    (COMIT), 553–555
Crack spacing, determination
    of, 404f
Cracked properties, 92–95
    definition, 92
    mathematical computation, 93–95
    significance, 92–93
Cracked RC beams, average shear stresses
    in, 267–268
Creep, 57–59
Critical buckling load, 342
Cross-section analysis applications,
        framework for the development of,
        539–546
    available commercial section analysis
        applications, 545–546
    class hierarchy, 542–543
    computation model for the cross-section
        behavior, 540–542
    determination of section capacity,
        543–544
    flexural capacity curves, 545
    functionality, 540
    moment–curvature curves, 545
    object model, 542
    sample implementation, 544
    stress distribution plots, 545
Cross-section capacity ratio, computing,
        204–207
Cross-section coordinate axes, 71, 71f
Cross-sectional analysis, 35–36
    axial–flexural response, 36

ductility of cross-sections, 36
    shear and torsion response, 36
Cross-sectional area and bearing area,
        108–109
Cross-sectional design, 36–37
Cross-sectional properties, classification of,
        69–70
Cross-sectional shape, 35
    effect on ductility, 427
Cross-sections exposed to fire, analysis of,
        502–514
    capacity interaction surface,
        determination of, 507–514
    effect of concrete cover, 510
    section capacity, determination of,
        503–504
    structural cross-sections, effect of high
        temperature on, 502–503
    surface temperatures, determination of,
        504–505
    temperature distribution within cross-
        section, determination of, 506–507
    variation of capacity with temperature,
        510–514
CSiCOL, 546–553
    cross-section analysis, 549–550
    modeling of cross-sections, 547–549
    visualization of analysis results, 550–553

D
Deep beams
    failures associated with, 293f
    special considerations for, 289–293
    truss models for design of, 291–292,
        292f
Definition of cross-sections, 39–42
Deformations
    for applied actions, 32–33
    restraining actions for assumed
        deformations, 34
Deformation–strain relationship, 25
Degree of freedom (DOF), 29–32, 30f,
        68–69, 68f
    deformations, strains, and stresses, 30–31
    member cross-sections and, 29–30
    stress resultants and, 31–32, 31f

Demand-to-capacity (D/C) ratio, 20–21, 236–240
Derived cross-section properties, 84–91
    elastic section moduli, 87
    geometric, elastic, plastic, and shear centers, 84–87
    principal properties, 88–91
    radii of gyration, 88
Design process, significance of cross-sections in, 35
Designing the structures, 6–23
    analysis and design levels, 14–15
    building codes, 8–11
        disaster resilience and environmental sustainability in, 11
        historical development, 10
    design objectives and philosophy, 7–8
    from force-based to displacement-based design, 20–21
    performance-based design (PBD), 21–23, 21f, 22f, 23f
    traditional approaches to structural design, 15–19
        limit state design concept, 16–19
        ultimate strength design, 16
        working stress design, 15–16
    traditional building codes, shortcomings of, 19–20
    typical structural design process, 11–14
Diagonal tension, 265
Diameter of rebars, 424–426
Direct and shear stress on rotated element, 212–243
Directional behavior-based classification of materials, 54–55
Disaster resilience and environmental sustainability in building codes, 11
Displacement-based design, 21–23
Double confinement, 410–411
Dowel action, 266–267
Ductility of cross-sections, 36, 391
    action–deformation curves, 393–400
        axial load–deflection curve, 399
        F–D relationship, 394–396
        moment–curvature curve, 396–399
        moment–rotation curve, 399
        torque–twist curve, 399–400

concrete-filled steel tubes, 430–435
    ductility ratio, 392f
    factors affecting moment–curvature relationship and ductility of RC sections, 423–430
        axial load on ductility, 428–430
        compression reinforcement, 423–424
        confinement model for concrete, 427
        cross-sectional shape, 427
        longitudinal steel, yield strength, and diameter of rebars, 424–426
    moment–curvature curve, 401–409
        important outputs from, 402–406
        performance-based design of cross-sections, 406–409
        and stiffness, 401–402
    nonlinear analysis of structures, cross-sectional response in, 436–444
    reinforced concrete sections, ductility and confinement of, 410–423
        confinement of RC Sections, 410–412
        ductility and seismic performance, 410
        stress–strain models for confined concrete, 412–423
    solved examples, 444–475
    strain-based definition, 391
    strength-based design, limitations of, 391–392
    unsolved problems, 475–479

E
Earthquakes, resilience to, 11
EC 2 design procedure, 209, 212f
Eccentricity vector, applied, 202
Effective length factor, 343f, 344–345
Elastic center
    definitions, 84
    mathematical computation, 86–87
    significance, 86
Elastic response, basic, 253–262
    combined shear stress due to shear force and torsion, 258–259
    interaction of shear, torsion and bending stresses, 260–262

shear stresses, 253—258
  development of, in cross-section, 254f
  due to shear force, 255—256
  due to torsion, 256—258
Elastic section moduli, 87
  definition, 87
  mathematical computation, 87
  significance, 87
Elastic shear stress distribution in various
    cross-sections, 257f
Encasing, 495—496
Environmental sustainability in building
    codes, 11
Equilibrium factored torsion moment, T-
    beam for, 318—322
Equilibrium torsion, 268—269
Equilibrium torsional moments, examples
    of, 270f
Equivalent transformed properties, 91—92
  definition, 91
  mathematical computation, 91—92
  significance, 91
Esthetics, 490
ETABS, 370—371
Euler's buckling formula, 343—344

F

Factor of safety (FOS), 15—16, 139—140
Failure modes, identification of, 487—488
Federal Emergency Management Agency
    (FEMA), 10
Fiber model, 165—166
Fiber-reinforced cement (FRC), 498
Fiber-reinforced polymer (FRP)
  jacketing, 497
  laminates, 497—498
Finite element analysis (FEA), 26, 503, 531
Finite element solution for torsional
    constant, 83
First and second moment of areas, 112—114
Flexibility relationships, 32—33
Flexural capacity curves, 545
Flexural response, 251
Flexural stiffness, 346
Flexural strength, 498
Flexural stress magnification factors, 356—357

Flexure, code-based design for, 207—211
Force—deformation (F—D) relationship,
    394—396

G

Geometric and cross-sectional properties,
    difference between, 67—68
Geometric center, 85f
  definitions, 84
  mathematical computation, 86—87
  significance, 86
Geometrical hierarchy of structural
    members, 5f
Geometry, 48—52
  built-up and composite sections, 51—52
  classifications of cross-sections based on,
    43—45, 44f
  complex and arbitrary shapes, 52
  parametrically defined simple sections,
    49—50
  standard cross-sections, 49
Global axes, 70
Global retrofit, 491—492, 492f

H

Height-to-web thickness ratio, 99—100
  definition, 99—100
  mathematical computation, 100
  significance, 100
Hierarchy of structures, 1—6, 5f
  member cross-sections, 4—5
  physical structures, 1—4
    cable structures, 3
    mixed structures, 4
    skeletal structures, 3
    solid structures, 3—4
    spatial structures, 3
  structural materials, 6
  structural members, 4
High-strength concrete, 362—363
Hognestad's stress—strain model, 440—444
Hook's law, 26, 55—56
Horizontal shear stress in beam section,
    296—299
Human Computer Interfaces (HCI),
    531—532

I

Information and communications
    technology (ICT), 553
Integrated understanding, developing,
    41−42
Interaction surface, generation
    of, 190
    extended procedure, 190
    simplified procedure, 190
International Building Code (IBC), 10
International Code Council (ICC), 10
Isotropic materials, 54−55, 55*t*

J

Jacketing and encasement, 495

K

Kent and Park stress−strain model,
    416−417, 447−449

L

Length, specific
    calculation of, 58*b*
Li et al. stress−strain model, 422−423,
    459−462
Limit state design concept, 16−19, 18*t*
Linear elastic material, stress−strain
    behavior, 61
Linear inelastic material, stress−strain
    behavior, 61
Line-type members
    cross-sections of, 39, 40*f*, 42−43,
        54−55
    design of, 5, 5*f*
    unification of, 156
Line-type element, 4
Load Contour or Breslar Approximation,
    337−338
Load path, efficient and integrated
    provision of, 486
Load−deflection curve, 398*f*
    axial load−deflection curve, 399
Load−deformation curves, 396−398
Loading, 23, 432
    effect of, on column design, 365−366

Load−moment (PM) interaction curves,
    195−197, 222−235, 551
    effect of reinforcement ratio on,
        240−241
    effect of yield strength on, 241−243
Load−moment relationship, 340−341
Local member axes, 71
Local retrofit, 491−492, 492*f*
Local stiffness reductions, 488
Longitudinal rebars
    selecting sizes and layout of, 363−364
Longitudinal steel
    effect on ductility yield strength,
        424−426

M

Mander's stress−strain model, 415,
    451−459
Material composition, classifications of
    cross-sections based on, 44*f*, 46, 47*f*
Material failures, 488*t*
Materials, 52−67
    basic properties of, 54
    brittle and ductile materials, 59−61
    directional behavior-based classification
        of, 54−55
    material behavior, importance of,
        52−54
    material properties
        determining and defining, 56−58
    stress−strain behavior, classification
        based on, 61
    stress−strain curves, 62−67
        for steel, 63
        for unconfined concrete, 63−67
    structural, 6
Maximum bending stress, 122−123
Member cross-sections, 4−5
    and degree of freedom (DOF), 29−30
Member stiffness and cross-sectional
    properties, 34−35
Members, structural, 4
Merging
    using boundary intersection, 108
    using meshing, 107
Meshing, 104−105, 497−498
Mixed structures, 4

Mobile and cloud usage in structural engineering applications, framework for, 556—557
Mobile framework, 557—558
components architecture, 557—558
Modified Kent and Park confinement model, 449—451
Modified Mander Confined Stress—Strain Curve, 434
Modular ratio, 73—74
definition, 73
mathematical computation, 74
significance, 73
Mohr's circle, principal stresses and, 147—153, 150f
basic concept, 147—151
significance of principal stresses in RC beam design, 152—153
Mohr's stress circle, 264
Moment capacity, 159f, 206
variation with temperature for various cross-sections, 513f
Moment magnification according to ACI 318, 355—356, 355f
Moment vector, applied, 202—203
Moment—curvature curve, 396—399, 398f, 401—409, 436—437, 437f, 545
determination of, 466—475
important outputs from, 402—406
outputs from, 444—447
performance-based design of cross-sections, 406—409
and stiffness, 401—402
Moment—curvature relationship, factors affecting, 423—430
axial load on ductility, 428—430
compression reinforcement, 423—424
confinement model for concrete, 427
cross-sectional shape, 427
longitudinal steel, yield strength, and diameter of rebars, 424—426
Moment—moment interaction curve, 192—195
Moment—rotation curve, 399
Moments of area and moments of inertia, 75—76
definition, 75
mathematical computation, 75—76
significance, 75
Moments of inertia, 75—76, 114—118
definition, 75
effect of cross-sectional shape on, 125—127
mathematical computation, 75—76
significance, 75

N

National Earthquake Hazards Reduction Program (NEHRP), 10
Neutral axis (NA), 162, 181f, 182f, 200—201, 218—222
concept of, 180—181
depth of, 181—182
Nonlinear analysis of structures, cross-sectional response in, 436—444
Nonlinear elastic material, stress—strain behavior, 61, 62f
Nonlinear inelastic material, stress—strain behavior, 61, 62f
Nonlinear stress—strain model of concrete, 438b
Nonlinearity of response and stiffness, 27—29
Nonsway column condition, 345—346

O

Object model, 542
Orthotropic materials, 55, 55t
Overall structural design process, 12f, 41

P

P—Δ effects, 332, 337, 353
Performance-based design (PBD), 20—23, 21f, 22f, 156, 406—409, 437—444
performance levels in, 23f
professionals and their roles in, 409f
Physical structures, 1—4
cable structures, 3
mixed structures, 4
skeletal structures, 3
solid structures, 3—4
spatial structures, 3

Piles, design of, 371–372
Plastering, 494–495
Plastic center, 85*f*
  definitions, 85
  mathematical computation, 87
  significance, 86
Plastic section moduli, 83–84
  definition, 83
  mathematical computation, 84
  significance, 83
Point, polyline, and polygon method,
  100–104
Pointing, 494–495
Poisson's effect for compressive force, 411*f*
Polygon and polyline methods, accuracy
  of, 104
Polygon shapes, meshing of sections made
  up from, 105–108
  merging of shapes, 107–108
  meshing of single polygon, 106–107
Post-tensioning, 498
Practical strain distribution, 182–183
Primary purpose of any structure, 1
Principal axes, 114–118
  and principal stresses, 214–217
  and section deformation, 89*f*
Principal properties, 88–91
  definition, 88–89
  mathematical computation, 90–91
  significance, 89
Principal stresses, 264–267, 264*f*, 266*f*
  and maximum shear stresses, 299–303
  and Mohr's circle, 147–153
    basic concept, 147–151
    in RC beam design, 152–153
Properties, cross-section, 67–84
  classification of, 69–70
  cross-section area, 74–75
    definition, 74
    mathematical computation, 74–75
    significance, 74
  geometric and cross-sectional properties,
    difference between, 67–68
  modular ratio, 73–74
    definition, 73
    mathematical computation, 74
    significance, 73

  moments of area and moments of
    inertia, 75–76
    definition, 75
    mathematical computation, 75–76
    significance, 75
  plastic section moduli, 83–84
    definition, 83
    mathematical computation, 84
    significance, 83
  reference axis, 70–71
    cross-section coordinate axes, 71, 71*f*
    global axes, 70
    local member axes, 71
  section properties, computation of,
    72–73
  in section stiffness, 68–69
  shear areas, 76–78
    definition, 76
    mathematical computation, 77–78
    significance, 76–77
  torsional constant
    definition, 78–79
    finite element solution for, 83
    significance, 79–80
    for thin-walled open shapes, 79
  torsional equations, general
    mathematical computation, 81–83
  warping constant
    significance, 80–81
Pure shear, 264
Pushover analysis, 353, 398–399, 436, 437*f*

R
Radii of gyration, 88, 114–118
  definition, 88
  mathematical computation, 88
  significance, 88
Rebar calculator, 547, 549*f*
Rebuilding, 485
Rebuilding of London Act, 10
Rectangular beam for shear, design of,
  312–314
Reference axis, 70–71
  cross-section coordinate axes, 71, 71*f*
  global axes, 70
  local member axes, 71

Refurbishment, 485
Rehabilitation, 483, 485
Reinforced concrete (RC) beam design, principal stresses in significance of, 152–153
Reinforced concrete (RC) beams for shear, design of, 279–281
Reinforced concrete (RC) beams for torsion, design of, 281–287
Reinforced concrete (RC) columns, practical design considerations for, 360–369
column stability and structural stability, 368–369
designing for stability, 367–368
designing for strength, 367
effect of cross-section shape on column strength, 361–362
effect of loading on column design, 365–366
selecting column cross-section shape, 360–361
selecting concrete strength for columns, 362–363
selecting sizes and layout of longitudinal rebars, 363–364
selecting sizes and layout of transverse rebars, 364–365
selecting steel strength for columns, 363
strength versus stability of columns, 366–368
Reinforced concrete (RC) members, principal stresses and shear capacity of, 264–267
Reinforced concrete (RC) members, retrofitting of, 498–502
external posttensioning by steel, 500
external prestressing, 500
strengthening of beams and slabs in flexure, 498–499
strengthening of beams in shear, 499
strengthening of columns, 501–502
Reinforced concrete (RC) sections, ductility and confinement of, 410–423
confinement of RC sections, 395f, 410–412

ductility and seismic performance, 410
stress–strain models for confined concrete, 412–423
Bjerkeli et al. stress–strain model, 420–421
Kent and Park stress–strain model, 416–417
Li et al. stress–strain model, 422–423
Mander's stress–strain model, 415
Scott et al. stress–strain model, 417–419
Yong et al. stress–strain model, 419–420
Reinforced concrete (RC) sections, response of, 263–277
average shear stresses in cracked RC beams, 267–268
principal stresses and shear capacity of RC members, 264–267
response under torsion, 268–277
space truss analogy for reinforced concrete members for torsion, 273–277
torsional cracks in unreinforced concrete, 269–273
Reinforced concrete (RC) sections, specific properties of, 91–95
cracked properties, 92–95
definition, 92
mathematical computation, 93–95
significance, 92–93
equivalent transformed properties, 91–92
definition, 91
mathematical computation, 91–92
significance, 91
overview, 90–91
Reinforced concrete (RC) sections for combined shear–torsion, design of, 287–288
Reinforcement requirement, for combined axial–flexural and shear–torsion, 294–295
Reinforcing fibers, 497–498
Renovation, 485
Repair, defined, 485
Resilience to earthquakes, 11

Restoration, defined, 485
Restraining actions for assumed
	deformations, 34
Resultant moment vector, 203
Retrofitting of cross-sections, 483
	analysis of cross-sections exposed to fire,
		502—514
		capacity interaction surface,
			determination of, 507—514
		effect of concrete cover, 510
		section capacity, determination of,
			503—504
		structural cross-sections, effect of high
			temperature on, 502—503
		surface temperatures, determination
			of, 504—505
		temperature distribution within
			cross-section, determination of,
			506—507
		variation of capacity with
			temperature, 510—514
	concrete jacketing or encasing, 495—496
	definition, 485
	fiber-reinforced polymer (FRP)
		jacketing, 497
	FRP laminates, fiber-reinforced cement,
		and mesh, 497—498
	jacketing and encasement, 495
	overall retrofit process and strategies for
		structures, 485—494
		basic approaches, 491—492
		esthetic considerations, 490
		final selection of retrofit scheme,
			490—491
		identification of possible failure
			modes, 487—488
		increase in deformational capacity of
			sections and members, 488—489
		lateral thinking approach for,
			492—494
		preliminary retrofit design, 491
		provision of efficient and integrated
			load path, 486
		reducing force demands on sections
			and members, 489—490
		strengthening and stiffening of
			structures, 486—487

	technical strategies, 485—486
	reinforced concrete members,
		retrofitting of, 498—502
		external posttensioning by steel,
			500
		external prestressing, 500
		strengthening of beams and slabs in
			flexure, 498—499
		strengthening of beams in shear, 499
		strengthening of columns, 501—502
	solved examples, 514—527
	steel jacketing, 496—497
	surface treatment, plastering, and
		pointing, 494—495
	unsolved examples, 527—530
RISA Technologies, LLC, 531—532

S

SaaS (Software-as-a-Service) model,
	558—559, 559f, 560f, 561f
Safety, proportioning for, 17f
Sample implementation, 544
Scott et al. stress—strain model, 417—419
Second-order moment, 340
Section capacity, determination of,
	503—504, 543—544
Section designer (SD) component,
	540—541
Section properties, computation of,
	72—73
Section properties, numerical computations
	of, 100—108
	meshing of sections made up from
		polygon shapes, 105—108, 107f
	merging of shapes, 107—108
	meshing of a single polygon,
		106—107
	point, polyline, and polygon method,
		100—104
	using meshing to compute properties,
		104—105
Section stiffness, role of cross-section
	properties in, 68—69
Serviceability limit states, 18t
Shape and dimensions of column cross-
	section, selecting, 361—362

Shape factor, 76—77
Shapes
    complex and arbitrary shapes, 52
    merging of, 107—108
    thin-walled open shapes, 79
Shear and torsion, response and design for,
        36, 251
    basic elastic response, 253—262
    code-based shear and torsion design of
        RC sections, 278—295
        design of RC beams for shear,
            279—281
        design of RC beams for torsion,
            281—287
        design of RC sections for combined
            shear—torsion, 287—288
        reinforcement requirement for
            combined axial—flexural and
            shear—torsion, 294—295
        special considerations for deep beams,
            289—293
    reinforced concrete (RC) sections,
        response of, 263—277
        average shear stresses in cracked RC
            beams, 267—268
        principal stresses and shear capacity of
            RC members, 264—267
        response under torsion, 268—277
        space truss analogy for reinforced
            concrete members for torsion,
            273—277
        torsional cracks in unreinforced
            concrete, 269—273
    shear stresses, 253—258
        combined shear stress due to shear
            force and torsion, 258—259
        interaction of shear, torsion and
            bending stresses, 260—262
        shear stress due to shear force,
            255—256
        due to torsion, 256—258
    solved examples, 296—324
    unsolved problems, 324—327
Shear areas, 76—78
    definition, 76
    mathematical computation, 77—78
    significance, 76—77

Shear center, 85f, 123—125
    definitions, 85
    mathematical computation, 87
    significance, 86
Shear flow in rectangular tube, 274f
Shear force, 32, 36
    shear stress due to, 255—256
Shear reinforcement, 265
Shear stresses
    of circular shaft, 307—312
    distribution in various cross-sections,
        260f
    due to torsion, 303—306
Shear walls as columns, 370—371
Shear—moment interaction diagram,
        262f
Significance of cross-sections, 39
    in design process, 35
Skeletal structures, 3
Slenderness effects, 337, 338f, 339—342
    consideration of, 350—351
    effect of column slenderness on overall
        frame behavior, 349—350
    importance of, 348—349
    in nutshell, 339f
Slenderness ratio, 88, 342—344
Software development and application for
        analysis of cross-sections, 531
    cloud framework, 558—561
        components architecture, 559—561
        system architecture, 558—559
    cross-section analysis applications,
        framework for the development of,
        539—546
        available commercial section analysis
            applications, 545—546
        class hierarchy, 542—543
        computation model for the cross-
            section behavior, 540—542
        determination of section capacity,
            543—544
        flexural capacity curves, 545
        functionality, 540
        moment—curvature curves, 545
        object model, 542
        sample implementation, 544
        stress distribution plots, 545

Software development and application for
analysis of cross-sections (*Continued*)
CSiCOL, introduction to, 546−553
cross-section analysis, 549−550
modeling of cross-sections, 547−549
visualization of analysis results,
550−553
current research trends and future
potential, 561−562
mobile and cloud usage in structural
engineering applications,
framework for, 556−557
mobile framework, 557−558
components architecture, 557−558
structural engineering applications,
general framework for the
development of, 532−539
application frameworks, 537−538
component-based software
development (CBSD), 533−534
implementation issues, 539
information processing and
information packages, 534−536
overall application architecture and
frameworks, 537
package frameworks, 538−539
structural design information space,
534
Solid structures, 3−4
Space truss
analogy, 273−277
forces in various elements of, 276*f*
Spatial structures, 3
Special limit states, 18*t*
Specific length, 58*b*
Stability failures, 488*t*
Stability issues in columns
column stability and structural stability,
368−369
designing for stability, 367−368
strength versus stability of columns,
366−368
Steel, idealized stress−strain curves for, 63
Steel jacketing, 496−497
Steel sections, specific properties of,
95−100
height-to-web thickness ratio, 99−100

definition, 99−100
mathematical computation, 100
significance, 100
net area and effective net area, 96−98
overview, 95−96
width-to-thickness ratio, 98−99
definition, 98
mathematical computation, 99
significance, 98
Steel strength, 363
Stiffness, 26−29, 401−402
effective, 401−402
nonlinearity of response and, 27−29
structural equilibrium and role of
stiffness, 26
Stiffness matrix method, 26
Strain, determination of stress from,
183−185
discretization of cross-section and stress
field, 183−185
Strain distribution, determination of,
174−175
Strain profile, determining, 178−183
neutral axis, concept of, 180−181
neutral axis, depth of, 181−182
practical strain distribution, 182−183
Strain-dependent stress, 170
Strain-independent stresses, 190
Strains and stresses corresponding to each
degree of freedom, 253*f*
Strength evaluation, 484
Strength-based design, limitations of,
391−392
Strengthening, defined, 485
Strengthening of beams
in flexure, 498−499, 499*f*
in shear, 499, 500*f*
Strengthening of slabs in flexure, 498−499,
499*f*
Stress distribution plots, 545
Stress resultant equations, 157−169
basic assumptions, their necessity, and
validity, 157−158
basic equations, 159−166
extended formulation of, 169−175
strain distribution, determination of,
174−175

stress field, determination of, 169–171
 stress profile, generation of, 172–173
 generalized cross-section and materials, 167–169
 integrating design codes, 166–167
Stress resultant–action relationship, 25
Stress resultants, axial–flexural, 153–157
 cross-section materials, unification of, 155
 cross-section shapes and configurations, unification of, 155–156
 design approaches and design codes, unification of, 156–157
 diversity of the problem and the need for unified approach, 153–155
 line-type structural members, unification of, 156
Stress resultants and DOFs, 31–32
Stress-control methods, 139–140
Stresses, 137–139
 interaction of, due to axial load and moment, 145–147, 146f
Stress–strain behavior, classification based on, 61
Stress–strain curves, 62–67
 for steel, 63
 unconfined concrete, 63–67
Stress–strain diagram, 554f
Stress–strain models for confined concrete, 393–394, 412–423
 Bjerkeli et al. model, 420–421
 Kent and Park model, 416–417
 Li et al. stress–strain model, 422–423
 Mander's model, 415
 Scott et al. model, 417–419
 Yong et al. model, 419–420
Stress–strain relationship, 25, 170–171, 190
 determination of, 183–185
Strong columns–weak beams principle, 489
Structural cross-sections, 41
 effect of high temperature on, 502–503
 practical cases of fire exposure to various configurations of, 505f

Structural design, conceptual role of, 7f
Structural designers, 357–359
Structural engineering applications, general framework for the development of, 1, 532–539
 application frameworks, 537–538
 component-based software development (CBSD), 533
  in structural engineering, 533–534
 implementation issues, 539
 information processing and information packages, 534–536
 overall application architecture and frameworks, 537
 package frameworks, 538–539
 structural design information space, 534
Structural members, types of
 classifications of cross-sections based on, 42–43
Strut-and-tie model, 290–291
Surface temperatures, determination of, 504–505
Surface treatment, 494–495
Surface-type member, 4
Sway column condition, 345–346
Symmetrical rebars, 363–364

T
T-beam
 for compatibility factored torsion moment, 322–324
 for equilibrium factored torsion moment, 318–322
 for shear, 314–318
Temperature distribution within cross-section, determination of, 506–507
Thin-walled open shapes, torsional constant for, 79
TOR steel, 412–414
Torque, allowable
 determination of, 306–307
Torque–twist curve, 399–400, 400f
Torsion space truss analogy, with reinforcement, 275f
Torsional constant, 118–119
 definition, 78–79
 finite element solution for, 83

Torsional constant (*Continued*)
  significance, 79−80
  for thin-walled open shapes, 79
Torsional constant factors, 571
Torsional cracks in unreinforced concrete,
    269−273
Torsional equations, general, 81−83
Torsion−moment interaction diagram,
    263*f*
Torsion−shear interaction diagram,
    262*f*
Traditional approaches to structural design,
    15−19
  limit state design concept, 16−19
  ultimate strength design, 16
  working stress design, 15−16
Traditional building codes, shortcomings
    of, 19−20
Transformed section, properties of,
    110−112
Transverse rebars
  selecting sizes and layout of, 364−365
True capacity ratio, 205
Truss models, for design of deep beams,
    291−292, 292*f*
Typical structural design process,
    11−14

U
Ultimate limit states, 16−17, 18*t*
Ultimate strength design, 16, 156
Unbraced column condition, 345−346
Unconfined concrete, idealized
    stress−strain curves for, 63−67
Uniaxial bending, 200−201
Unreinforced concrete, torsional cracks in,
    269−273, 271*f*

W
Warping constant, 80−81
Warping stresses, 251−252
Width-to-thickness ratio, 98−99
  definition, 98
  mathematical computation, 99
  significance, 98
Working Strength Design, 138−140, 156
Working Stress Design, 15−16

Y
Yield strength, 424−426
Yong et al. stress−strain model, 419−420,
    462−466

Z
Zero-strain surface, 180−181

CPI Antony Rowe
Chippenham, UK
2018-01-22 18:30